NONLINEAR WAVE DYNAMICS

Selected Papers of the Symposium Held in Honor of **Philip L-F Liu's 60th Birthday**

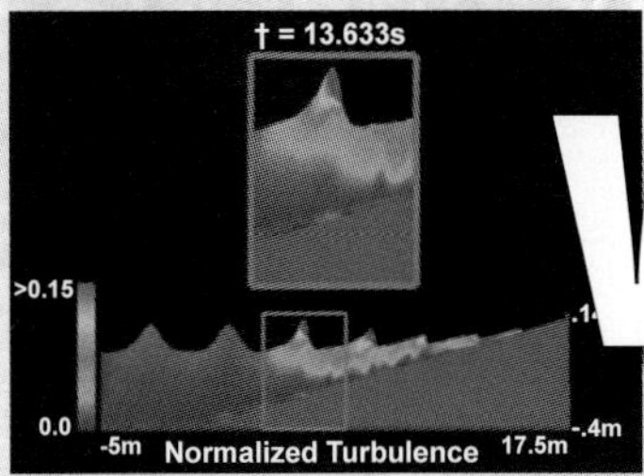

NONLINEAR WAVE DYNAMICS

edited by

Patrick Lynett

Texas A&M University, USA

World Scientific

NEW JERSEY • LONDON • SINGAPORE • BEIJING • SHANGHAI • HONG KONG • TAIPEI • CHENNAI

Published by

World Scientific Publishing Co. Pte. Ltd.
5 Toh Tuck Link, Singapore 596224
USA office: 27 Warren Street, Suite 401-402, Hackensack, NJ 07601
UK office: 57 Shelton Street, Covent Garden, London WC2H 9HE

British Library Cataloguing-in-Publication Data
A catalogue record for this book is available from the British Library.

NONLINEAR WAVE DYNAMICS
Selected Papers of the Symposium Held in Honor of Philip L.-F. Liu's 60th Birthday

ISBN-13 978-981-270-903-5
ISBN-10 981-270-903-7

Printed in Singapore by Mainland Press Pte Ltd

PREFACE

In September of 2006, research leaders in the field of coastal engineering, fluid mechanics, and wave theory meet at Cornell University to celebrate the 60th birthday of Prof. Philip L.-F. Liu. The symposium was attended by numerous research collaborators, and many of Prof. Liu's past students. This volume is dedicated to Phil Liu, and his many past and future achievements.

As a scholar and researcher, Prof. Liu has been at the forefront of a variety of topics in coastal engineering, including nonlinear wave theory, tsunamis generation, propagation and runup, wave breaking processes, sediment transport processes, and the interaction of waves with structures. His research approach integrates analytical, computational and experimental methodologies in elegant ways. The ASCE has awarded his research excellence with the Walter L. Huber Civil Engineering Research Prize in 1978, the John G. Maffatt-Frank E. Nichol Harbor and Coastal Engineering Award in 1997, and the International Coastal Engineering Award in 2004. Prof. Liu was also a recipient of the prestigious J.S. Guggenheim Fellowship in 1980.

As an educator, Prof. Liu has spent his entire career at Cornell University, where he has taught courses in the area of Fluid Mechanics, Computational Methods and Modeling, Coastal Engineering and Water Wave Theories. At Cornell, Prof. Liu has supervised more than 35 graduate (MS/PhD) students. Many of his former graduate students have gone on to leadership position in universities, research institutions, and consulting firms, both nationally and internationally. These former students are building the future of coastal engineering in the U.S. and abroad.

To promote excellence and advancement of coastal engineering, Prof. Liu has become the founding editor for a book series entitled "Advanced Series on Ocean Engineering" started in 1989 and a review series entitled "Advances in Coastal and Ocean Engineering" in 1993. Both series have provided authors a platform to disseminate current research information and knowledge and the published volumes have reached a wide range of audience, including undergraduates, graduate students, researchers and engineers.

Prof. Liu has made very significant contributions to the advancement of coastal engineering in the U.S. and throughout the world in the manner of research, teaching and professional leadership. This volume is a compilation of research papers, both review and new work, which have benefited in some way through the research Prof. Liu has completed. The reader will find that there are two main groups of papers presented here. The first group consists of papers written by some of Prof. Liu's long-term collaborators. This group starts off with a theoretical paper by Maarten Dingemans and continues with three coastal structure focused papers with authors including the notable researchers Hocine Oumeraci, Nobu Kobayashi, and Inigo Losada. Closing this first group are four papers on the subject of tsunamis and long waves. Prof. Liu's fellow organizers of the three International Workshop on Long-Wave Runup Models, Harry Yeh and Costas Synolakis, are among the authors.

The second group of papers is a series written by some of Prof. Liu's previous Ph.D. students. This section kicks off with a paper by Jerry Lennon, Prof. Liu's first Ph.D. student, and concludes with a paper by Tso-Ren Wu, one of the more recent graduates to come out of the Cornell program. Inside this group, the reader will find a wide range of topics, from groundwater flow to swash zone dynamics and transport, reflecting the breadth of topics in which Prof. Liu has been involved.

The attendees of the symposium thank Prof. Liu for his untiring enthusiasm to better understand the fundamental physics which govern all of the coastal engineering profession, and his leadership and education of this community.

Attendees of the Symposium Held in Honor of Philip L.-F. Liu's 60th Birthday

Collaborators

Tony Dalrymple
Bob Dean
Maarten Dingemans
David Farmer
Nobu Kobayashi
Inigo Losada
C.C. Mei
Jano Orfila
Hocine Oumeraci
Costas Synolakis
Gonzalo Simarro
Michelle Teng
Harry Yeh

Past Students

Khaled Al-Banaa
Kuang-An Chang
Shih-Chun Hsiao
Tom Hsu
Jerry Lennon
Patrick Lynett
T.-K. Tsay
Sueng-Buhm Woo
Tso-Ren Wu

CONTENTS

SOME REFLECTIONS ON THE GENERALISED LAGRANGIAN MEAN

M.W. DINGEMANS

formerly Delft Hydraulics
Boomkensdiep 11
8303 KW Emmeloord, the Netherlands
E-mail: Maarten.Dingemans@Deltares.NL

We discuss the application of the Generalised Lagrangian Mean method, devised by Andrews and McIntyre (1978a,b). This is the only method with which the interaction between waves and currents can be described in a consistent and precise way. Because the method is used only sparingly, we want to give attention to this important method.

1. Introduction

It is a pleasure to write this paper in honour to Phil Liu's sixtieth birthday. Almost twenty years ago, Phil came to Delft Hydraulics for half a year sabbatical and during that time he had a table in my already crowded room and managed to stay calm. Since then I visited Phil a number of times and always felt very welcome. Two occasions stand out. In the summer of 1989 I visited Cornell for nine weeks, starting a large project which was finished seven years later. I was taken good care of by Phil and his family. The second visit I want to mention here was in the summer of 1996, when I visited family Liu together with my three children, helping to cope with the loss of their mother and my wife earlier that year.

I present these reflections on GLM to Phil for his birthday. I regret it that I only am able to present some reflections, but I really want Phil and others to look into this important, but difficult method. Congratulations, Phil.

In water wave problems in which both waves and currents play a role, we can identify the following problems:

- Splitting between waves and currents is problematic for the case of depth-varying currents.
- It is difficult to identify the mean motion in an otherwise oscillating field.
- Splitting through averaging over the waves is only possible for currents which are uniform over the depth.
- Phillips' (1966, 1977) averaging method is only strictly valid for uniform currents.

Because of these difficulties another method is wanted. The GLM-method provides an answer to these difficulties. The GLM-method is treated extensively and completely in Andrews and McIntyre (1978a). A very good introduction to the main ideas of the method has been given by McIntyre (1980). Following his ideas, Dingemans (1997, section 2.10.6) has also discussed some properties of GLM.

The main properties of GLM can be stated as follows:

(1) Simplest idea of Lagrangian averaging is Stokes' classical idea of taking the time-mean following a single parcel of fluid.

- If we wish to speak of the Lagrangian-mean velocity at a point in space, the classical concept cannot apply in any exact sense because the parcel wanders away.
- Thus any exact theory of the Lagrangian-mean **field** must abandon the simple concept of the mean following a single fluid parcel.
- The result will be a **hybrid** Eulerian-Lagrangian description of the motion.

(2) What the GLM theory does provide is a precise analytical structure

- which makes clear what is involved in generalising the classical notions of Lagrangian mean and Stokes drift and
- which makes available certain tools whose use might help in the search for a fundamentally improved understanding of the physics involved.

The GLM method gives thus a **hybrid** Eulerian-Lagrangian description of the fluid motion. The perturbed position is denoted by $\Xi(\boldsymbol{x},t) = \boldsymbol{x} + \boldsymbol{\xi}(\boldsymbol{x},t)$. The essential idea of GLM is to **average over the perturbed position**[a]:

$$\overline{\phi}^L(\boldsymbol{x},t) = \langle \phi\{\boldsymbol{x} + \boldsymbol{\xi}(\boldsymbol{x},t), t\}\rangle \,.$$

The difference between the generalised Lagrangian and the Eulerian mean is the Stokes correction;

$$\langle \phi(\boldsymbol{x},t)\rangle^S = \langle \phi(\boldsymbol{x},t)\rangle^L - \langle \phi(\boldsymbol{x},t)\rangle \quad \text{or, shortly,} \quad \overline{\phi}^S = \overline{\phi}^L - \overline{\phi}\,.$$

When the quantity ϕ is the velocity $\boldsymbol{u}$, the Stokes correction $\overline{\boldsymbol{u}}^S$ is also referred to as the *Stokes drift*.

To define $\boldsymbol{\xi}$ for infinitesimal waves, it is customary to resort to the simplest possible approximation: $\partial_t \boldsymbol{\xi}(\boldsymbol{x},t) = \boldsymbol{u}'(\boldsymbol{x},t)$ where we have defined $\boldsymbol{u}'(\boldsymbol{x},t) \equiv \boldsymbol{u}(\boldsymbol{x},t) - \langle \boldsymbol{u}(\boldsymbol{x},t)\rangle$ and where $\langle \boldsymbol{\xi}(\boldsymbol{x},t)\rangle = 0$. The last relation formally expresses the idea that $\boldsymbol{\xi}$ should be a *disturbance-associated* quantity. For having the possibility that $\overline{\boldsymbol{\xi}(\boldsymbol{x},t)}$ be zero, the level of $\boldsymbol{x}$ should be adapted, see Andrews and McIntyre (1978a, Eqs. (2.6)-(2.8), p. 615).

[a]The quantity $\phi(\boldsymbol{x},t)$ is to be averaged and $\langle \phi(\boldsymbol{x},t)\rangle = \overline{\phi(\boldsymbol{x},t)}$ is the Eulerian mean at $\boldsymbol{x}$ in any of the usual senses (time, space, ensemble, etc.). The corresponding Lagrangian mean is now given by $\overline{\phi}^L(\boldsymbol{x},t)$.

The Lagrangian disturbance velocity $\boldsymbol{u}^\ell$ is defined in a similar way, $\boldsymbol{u}^\ell(\boldsymbol{x},t) = \boldsymbol{u}(\boldsymbol{x},t) - \left\langle \boldsymbol{u}^L(\boldsymbol{x},t)\right\rangle$. The disturbance ξ is now defined using the total derivative: $\left(\partial_t + \overline{\boldsymbol{u}}^L \cdot \text{grad}\right)\xi \equiv \overline{D}^L \xi = \boldsymbol{u}^\ell(\boldsymbol{x},t)$, where $\overline{D}^L$ is shorthand for the total-derivative operator. For shortness we also write the velocity at the perturbed position $\boldsymbol{u}(\boldsymbol{x}+\xi(\boldsymbol{x},t),t)$ as $\boldsymbol{u}^\xi$. We also have (see Andrews and McIntyre, 1978a, p. 615) the property $\overline{D}^L\left(\phi^\xi\right) = \left(\frac{D\phi}{Dt}\right)^\xi$ where $D = \partial_t + \boldsymbol{u}\cdot\text{grad}$ and $\overline{D}^L\Xi = \boldsymbol{u}^\xi$.

As an example, we will first apply the GLM method to derive the Stokes drift of water waves.

2. Stokes drift

The Stokes correction is defined as

$$\phi^S = \overline{\phi}^L - \overline{\phi}\,. \tag{1}$$

To obtain an expression for the Stokes correction in Eulerian variables we have to express it at the fixed point $\boldsymbol{x}$ instead of at the perturbed location $\boldsymbol{x}+\boldsymbol{\xi}$. Therefore, we have to Taylor $\boldsymbol{u}^S$ with respect to ξ; we always suppose $|\boldsymbol{\xi}|$ a small quantity of $\mathcal{O}(a)$, where a is a measure of the water-wave amplitude. The Taylor expansion of ϕ^ξ is:

$$\phi^\xi \equiv \phi(\boldsymbol{x}+\boldsymbol{\xi}(\boldsymbol{x},t),t) = \phi(\boldsymbol{x},t) + \xi_j\frac{\partial\phi}{\partial x_j} + \frac{1}{2}\xi_j\xi_k\frac{\partial^2\phi}{\partial x_j\partial x_k} + \mathcal{O}\left(a^3\right). \tag{2}$$

We have to take the mean:

$$\overline{\phi}^L = \langle\phi(\boldsymbol{x},t)\rangle + \left\langle \xi_j\frac{\partial\phi}{\partial x_j}\right\rangle + \frac{1}{2}\left\langle \xi_j\xi_k\frac{\partial^2\phi}{\partial x_j\partial x_k}\right\rangle + \mathcal{O}\left(a^3\right). \tag{3}$$

We write $\phi = \overline{\phi} + \phi'$ and thus $\langle\phi'\rangle = 0$. Therefore,

$$\overline{\phi}^S = \overline{\phi}^L - \overline{\phi} = \left\langle \xi_j\frac{\partial\phi'}{\partial x_j}\right\rangle + \frac{1}{2}\left\langle\xi_j\xi_k\right\rangle\left\langle\frac{\partial^2\overline{\phi}}{\partial x_j\partial x_k}\right\rangle + \mathcal{O}\left(a^3\right), \tag{4}$$

as is given by (Andrews and McIntyre, 1978a, Eq. (2.27) or Dingemans, 1997, Eq. (2.615)). We then have:

$$\overline{u}_m^S(\boldsymbol{x},t) = \overline{u}_m^L - \overline{u}_m = \left\langle \xi_j\frac{\partial u_m'}{\partial x_j}\right\rangle + \frac{1}{2}\overline{(\xi_i\xi_j)}\frac{\partial^2\overline{u}_m}{\partial x_i\partial x_j} + \mathcal{O}\left(a^3\right), \tag{5}$$

for $m = 1,2,3$. To estimate the magnitude of the Stokes drift, we introduce a number of simplifications. Suppose that we consider the case of a relatively narrow wave flume, with the axis in the $x_1 = x$ direction. Because we also consider a flume with a horizontal bottom, the variation of the velocity in the lateral $x_2 = y$

direction can be neglected as the current and waves are aligned in the x_1-direction. We thus suppose that the Stokes drift in this case can be approximated by:

$$\overline{u}_m^S \cong \left\langle \xi_1 \frac{\partial u_m'}{\partial x_1} \right\rangle + \left\langle \xi_3 \frac{\partial u_m'}{\partial x_3} \right\rangle + \frac{1}{2} \left\langle \xi_1^2 \right\rangle \frac{\partial^2 \overline{u}_m}{\partial x_1^2} + \frac{1}{2} \left\langle \xi_3^2 \right\rangle \frac{\partial^2 \overline{u}_m}{\partial x_3^2}$$
$$+ \langle \xi_1 \xi_3 \rangle \frac{\partial^2 \overline{u}_m}{\partial x_1 \partial x_3} + \mathcal{O}\left(a^3\right) . \tag{6}$$

To obtain an expression for $\overline{u}_m^S$, we have to find expressions for u_m', $\overline{u}_m$ and ξ_1, ξ_3.

2.1. *The particle paths*

Notice that $\boldsymbol{\xi}$ can be considered to be the path of the parcels, which is defined by

$$\xi_j = \int^t dt' \, u_j'(x_1 + \xi_1, x_2 + \xi_2, x_3 + \xi_3, t') \cong \int^t dt' \, u_j'(x_1, x_2, x_3, t') . \tag{7}$$

From linear theory we have (e.g., Dingemans, 1997, Eqs. (2.14))

$$\boldsymbol{u}'(\boldsymbol{x}, z, t) = \frac{g}{\omega} a \boldsymbol{k} \frac{\cosh k(h+z)}{\cosh kh} \cos\chi \, , \; w'(\boldsymbol{x}, z, t) = \frac{g}{\omega} ak \frac{\sinh k(h+z)}{\cosh kh} \sin\chi \tag{8}$$

with $\chi = \boldsymbol{k} \cdot \boldsymbol{x} - \omega t$ and where now $\boldsymbol{x} = (x_1, x_2)^T$ and a is the amplitude of the free-surface elevation. Without variation in the x_2-direction, we have $k_1 = k = |\boldsymbol{k}|$ and $\chi = kx - \omega t$. Estimates for the horizontal particle path ξ_1 and the vertical one, ξ_3, then follow from

$$\xi_1 = -\frac{gak_1}{\omega^2} \frac{\cosh k\,(h+z)}{\cosh kh} \sin\chi = -a \frac{\cosh k\,(h+z)}{\sinh kh} \sin\chi \tag{9a}$$

and

$$\xi_3 = \frac{gak}{\omega^2} \frac{\sinh k(h+z)}{\cosh kh} \cos\chi = a \frac{\sinh k(h+z)}{\sinh kh} \cos\chi \, . \tag{9b}$$

We see that $\langle \xi_1 \xi_3 \rangle = 0$ because $\cos\chi$ and $\sin\chi$ are in quadrature. For the current, we may neglect derivatives in the x_1 direction, while that is not permitted for the waves. The expression for the Stokes drift thus is given by:

$$\overline{u}_m^S \cong \left\langle \xi_1 \frac{\partial u_m'}{\partial x_1} \right\rangle + \left\langle \xi_3 \frac{\partial u_1'}{\partial x_3} \right\rangle + \frac{1}{2} \left\langle \xi_3^2 \right\rangle \frac{\partial^2 \overline{u}_m}{\partial x_3^2} + O\left(a^3\right) . \tag{10}$$

2.2. *An expression for the Stokes drift*

We first only derive an expression for $\overline{u}_1^S$. As is clear from (10), we have a contribution due to the waves and one due to the current. These two contributions will be investigated in the following subsections.

2.2.1. *The contribution to* $\overline{u}_1^S$ *due to the waves*

We now derive an expression for $\overline{u}_1^S$. It follows from (8) and (9) that

$$\left\langle \xi_1 \frac{\partial u_1'}{\partial x_1} \right\rangle = \frac{1}{2} a^2 k\omega \frac{\cosh^2 k(h+z)}{\sinh^2 kh} \quad \text{and} \quad \left\langle \xi_3 \frac{\partial u_1'}{\partial x_3} \right\rangle = \frac{1}{2} a^2 k\omega \frac{\sinh^2 k(h+z)}{\sinh^2 kh} , \tag{11}$$

where the factors 1/2 are due to $\left\langle \cos^2 \chi \right\rangle$ and $\left\langle \sin^2 \chi \right\rangle$. Similarly we obtain:

$$\left\langle \xi_3^2 \right\rangle = \frac{1}{2} a^2 \sinh^2 k(h+z) \sinh^2 kh . \tag{12}$$

The contribution to the Stokes drift due to the waves then is

$$\begin{aligned} \left(\overline{u}_1^S \right)_w &\cong \frac{1}{2} a^2 k\omega \frac{\cosh^2 k(h+z)}{\sinh^2 kh} + \frac{1}{2} a^2 k\omega \frac{\sinh^2 k(h+z)}{\sinh^2 kh} \\ &= \frac{1}{2} a^2 k\omega \frac{\cosh\left(2k\left(h+z\right)\right)}{\sinh^2 kh} . \end{aligned} \tag{13}$$

2.2.2. *The mean current*

For the mean current we assume that the eddy-viscosity concept may be used, resulting in (e.g., see Dingemans, 1997, pp. 307-308):

$$\rho \nu_T \frac{\partial \overline{u}}{\partial z} = -\boldsymbol{\tau}_c^b \frac{z}{h} \quad , \quad -h \le z \le 0 , \tag{14}$$

where $\boldsymbol{\tau}_c^b$ is the bottom shear stress belonging to the part of the current. We choose a function for ν_T which is quadratic over the depth:

$$\nu_T = -\kappa \frac{(z+h)\, z}{h} \overline{u}^* \quad , \quad -h + \delta_w \le z \le 0 , \tag{15}$$

with $\kappa \cong 0.40$ being Von Karman's constant, δ_w the thickness of the boundary layer due to the wave motion and $\overline{u}^*$ the friction velocity for the mean current, also defined by $\tau_c^b = \rho\, (\overline{u}^*)^2$. Integration of (14) to z then gives

$$\overline{u} = \frac{\overline{u}^*}{\kappa} \log\left(z+h\right) + c_1 \tag{16}$$

with log being the natural logarithm and c_1 an integration constant. This constant has to be determined experimentally (see Fredsøe and Deigaard, 1992, p. 21). For the case of a rough wall (the situation for which the thickness of the viscous sublayer is smaller than the Nikuradse roughness k_N), Nikuradse's method leads to a constant

$$c_1 = 8.5 - \frac{1}{\kappa} \log\left(k_N\right) , \tag{17}$$

and thus,

$$\overline{u} = \frac{\overline{u}^*}{\kappa} \log\left(\frac{z+h}{k_N/30}\right) = \frac{\overline{u}^*}{\kappa} \log\left(\frac{z+h}{z_0}\right) \quad \text{with} \quad z_0 = \frac{k_N}{30} , \tag{18}$$

for $-h + \delta_w \leq z \leq 0$.

We note that the logarithmic velocity profile is only valid outside the bottom-boundary layer, whose thickness may be estimated by (e.g., see Dingemans, 1997, Eq. (3.168c)):

$$\delta = 0.072 \left(A^3 k_N\right)^{1/4} , \tag{19}$$

where A is the length of the semi-axis of the bottom excursion and k_N is the Nikuradse length scale. The bottom velocity u^b (valid just outside the bottom boundary layer) follows from (8) and $\chi = kx - \omega t$ as

$$u^b = \frac{agk}{\omega \cosh kh} \cos\chi = \frac{a\omega}{\sinh kh} \cos\chi = A\omega\cos\chi \quad \text{so that} \quad A = \frac{a}{\sinh kh} . \tag{20}$$

2.2.3. *The contribution to $\overline{u}_1^S$ due to the current*

We now have:

$$\left|\frac{\partial u_1'}{\partial z}\right| = \frac{gak^2}{\omega} \frac{\sinh k(h+z)}{\cosh kh} \quad \text{with} \quad \max_{-h\leq z\leq 0} \left|\frac{\partial u_1'}{\partial z}\right| = \omega a k . \tag{21a}$$

$$\left|\frac{\partial \overline{u}_1}{\partial z}\right| = \frac{\overline{u}_*}{\kappa(h+z)} \quad \text{and} \quad \left|\frac{\partial^2 \overline{u}_1}{\partial z^2}\right| = \frac{\overline{u}_*}{\kappa\,(h+z)^2} . \tag{21b}$$

The contribution to the Stokes drift due to the current then is

$$\left(\overline{u}_1^S\right)_c = \left\langle \frac{1}{2}\xi_3^2 \frac{\partial^2 \overline{u}_1}{\partial z^2} \right\rangle = \frac{1}{4} \frac{a^2 \overline{u}^*}{\kappa\,(h+z)^2} \frac{\sinh^2 k(h+z)}{\sinh^2 kh} . \tag{22}$$

We can now estimate the terms in the right-hand member of Eq. (6). Using the expressions (9b) and (21), we obtain the following expression for the Stokes drift:

$$\begin{aligned}
\overline{u}_1^S &= \left(\overline{u}_1^S\right)_w + \left(\overline{u}_1^S\right)_c \cong \left\langle \xi_1 \frac{\partial u_1'}{\partial x} \right\rangle + \left\langle \xi_3 \frac{\partial u_1'}{\partial z} \right\rangle + \left\langle \frac{1}{2}\xi_3^2 \frac{\partial^2 \overline{u}_1}{\partial z^2} \right\rangle \\
&= \frac{1}{2} a^2 k\omega \frac{\cosh\left[2k(h+z)\right]}{\sinh^2 kh} + \frac{1}{4} \frac{a^2 \overline{u}^*}{\kappa\,(h+z)^2} \frac{\sinh^2 k(h+z)}{\sinh^2 kh} .
\end{aligned} \tag{23}$$

Close to the bed we have $z + h \cong 0$. In the limit for $z \to -h$, a Taylor expansion in $(h+z)$ of (23) yields

$$\begin{aligned}
\overline{u}_1^S \cong{}& \frac{1}{2} \frac{a^2 k\omega}{\sinh^2 kh} \left\{1 + k^2 (h+z)^2 + \cdots\right\} + \frac{1}{2} \frac{a^2 k\omega}{\sinh^2 kh} \left\{k^2 (h+z)^2 + \cdots\right\} \\
&+ \frac{1}{4\kappa} \frac{(ak)^2 \overline{u}^*}{\sinh^2 kh} \left\{1 + \frac{1}{3} k^2 (h+z)^2 + \cdots\right\} ,
\end{aligned} \tag{24a}$$

and thus,

$$\lim_{h+z\to 0} \overline{u}_1^S = \frac{1}{2}\frac{(ak)^2}{\sinh^2 kh}\left[\frac{\omega}{k} + \frac{\overline{u}^*}{2\kappa}\right]. \tag{24b}$$

The friction velocity $\overline{u}^*$ can be estimated from

$$\tau_c^b = \rho\left(\overline{u}^*\right)^2 \qquad \Rightarrow \qquad \overline{u}^* = \sqrt{\frac{\tau_c^b}{\rho}}, \tag{25a}$$

where the bottom shear stress for the current is estimated by (see Soulsby *et al.*, 1993):

$$\tau_c^b = \rho C_D \overline{u}^2 \quad \text{with} \quad C_D = \left(\frac{\kappa}{\log\left(\frac{h}{z_0}\right) - 1}\right)^2. \tag{25b}$$

It should be recognised that the bottom shear stress due to the current, τ_c^b, is modified by the presence of the waves. In Soulsby *et al.* (1993) parametrisations are given for a number of models for shear stress due to waves and currents acting simultaneously. Optimised parametrisations have been presented by Soulsby (1995), see also Soulsby (1997, pp. 68-70). The maximum shear stress is usually different from $\left|\boldsymbol{\tau}_c^b + \boldsymbol{\tau}_w^b\right|$, as is also found from the two examples further on. However, for an estimate of the Stokes drift due to the presence of a current, the estimate (25) will suffice.

Uittenbogaard and Klopman (2001) used the Stokes drift where also the component due to the current was included. Later, at Delft Hydraulics, when also some wind was added to the model, the results exploded. This was due to the second derivative to z, which has too much effect close to the surface. At the bottom its effect is negligible.

2.2.4. *Numerical examples*

We consider two numerical examples, with parameters as given in Table 1. We consider a so-called sea situation. We suppose a d_{50} distribution of sand of measure 250 μm and compute the ripple height of the sand bed by a method of Nielsen (1979), see Dingemans (1997, p. 304). The Nikuradse roughness parameter k_N is taken equal to the ripple height and z_0 subsequently follows by $z_0 = k_N/30$. This ripple height is usually larger than the grain size diameter d_{50}. The usual Nikuradse number is taken to be $2.5d_{50}$, which is the one for a smooth bottom.

For the laboratory measurement is taken the situation as measured by Klopman (1994). In this case the bottom roughness parameter z_0 was estimated as 0.04 mm (see Klopman, 1994, p. 31).

The bottom shear stress due to the waves, τ_w^b, is estimated by the usual quadratic friction law

$$\tau_w^b = \frac{1}{2}\rho f_w\left(u^b\right)^2, \tag{26}$$

Table 1. Parameters for the numerical examples

example	a [m]	T [s]	h [m]	$\overline{\overline{u}}$ [m/s]	z_0 [m]	k [m^{-1}]	$\overline{u}^*$ [cm/s]	u^b [m/s]	A [m]
sea (1)	2.00	7.00	10	0.50	7.93×10^{-5}	.1050	1.86	1.43	1.59
lab (2)	0.06	1.44	0.5	0.16	4×10^{-5}	2.350	.759	.179	.041

Table 2. Further parameters for the numerical examples

example	τ_c^b [N / m^2]	τ_w^b [N / m^2]	τ_{tot}^b [N / m^2]	$\langle \tau^b \rangle$ [N / m^2]	f_w -	C_D -
sea (1)	0.3559	8.236	10.007	1.036	0.0080	0.00139
lab (2)	0.0576	.6041	0.0355	.6041	0.0378	0.00225

where the friction parameter f_w is estimated from a formula due to Soulsby (1995) (see also Dingemans, 1997, Eq. (3.174)):

$$f_w = 1.39 \left(\frac{A}{z_0} \right)^{-0.52} . \tag{27}$$

The total and mean bottom shear stresses are estimated from the parametrisations of Soulsby *et al.* (1993) while using the parametrisations for Fredsøe's model. The mean current in these models is the current averaged over the vertical. Because we already had a mean current averaged over horizontal space, the further average over vertical space is denoted by double bar.

We plot the contribution from the waves and the mean current separately, together with the sum of these contributions, see Figures 1 and 2.

It is clear from these examples that the contribution of $\overline{u}$ to the Stokes drift is very small and may well be neglected in comparison to the contribution due to the waves. For the sea condition, with $\overline{\overline{u}} = 0.50$ m/s, the contribution to the

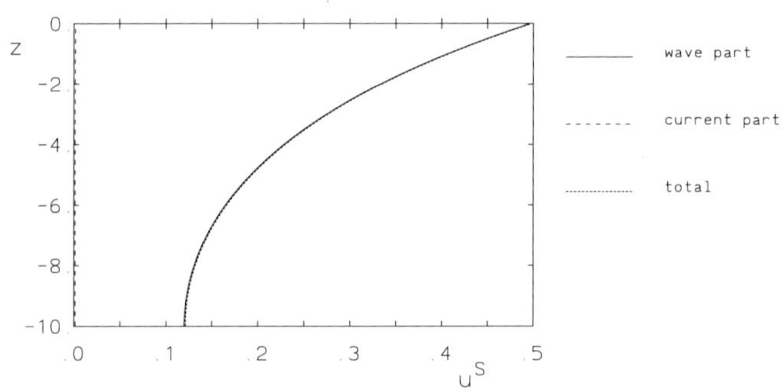

Figure 1. $\overline{u}^S$ in m / s as function of z, sea situation.

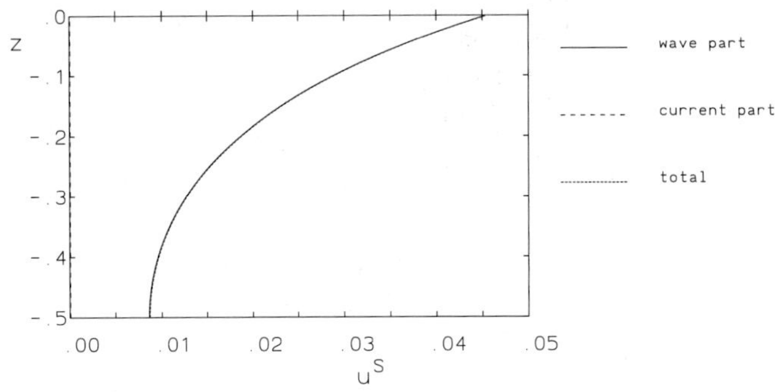

Figure 2. $\overline{u}^S$ in m / s as function of z, the laboratory situation.

Stokes drift from the current is 0.046 cm/s at the free surface and 0.033 cm/s at the bottom. It can be argued that a current of 0.50 m/s is not very large in coastal areas, but in the North Sea currents are typically not much larger than 1 m/s. Redoing the analysis with $\overline{\overline{u}} = 2$ m/s results in a friction velocity $\overline{u}^* = 7.45$ cm/s, roughly 4 times larger than was obtained for $\overline{\overline{u}} = 0.50$ m/s, which was to be expected. The contribution of the current to the Stokes drift is now 0.19 cm/s at $z = 0$ and 0.13 cm/s at $z = -h$. Compared with a maximum contribution of 50 cm/s contribution from the waves, the current contribution to the Stokes drift remains insignificant.

3. Application of GLM to water wave problems

Many problems in which water waves play a role occur in the coastal zone. Then the problems are the influence of water waves on the transport of sand or other materials in the coastal zone. A usual approach is to compute the wave behaviour in a spatial domain with some wave propagation model and determine the wave driving forces. These forces depend on the radiation stresses, but also an important contribution is due to wave breaking. As was shown by Dingemans *et al.* (1987), a direct application of the differentiation of the radiation stresses as resulting from the wave field can lead to large errors. An approach based on the determination of the dissipation gave a more robust approach. The driving forces are subsequently used as input for a current model, which is then used as input for sand-transport formulations. The answers of the current models are quite sensitive for the input of the driving forces. The whole approach is a first-order approach. Extensions to higher order are quite difficult. Notice that in this approach we consider *the effect of the waves on the current*, while in usual wave-current interaction problems only the effect of the fixed current on the waves is considered, yielding wave-refraction results.

The strength of the GLM-approach is that a mean current equation is obtained in which the effect of the water waves has been incorporated in a consistent exact way.

A simple introduction how to obtain the equations in GLM-form for water waves has been given in Dingemans (1997, §2.10.6), and is of course also to be found in Andrews and McIntyre (1978a). The resulting equations are given below.

For an incompressible fluid the Eulerian continuity equation $D\rho/Dt = 0$ becomes

$$\overline{D}^L \left(\overline{\rho}^L\right) = 0 . \tag{28a}$$

As remarked by Andrews and McIntyre (1978a, Appendix), a density ρ^J can be associated with the mean velocity $\overline{\boldsymbol{u}}^L$ such that a conservational form of the mass conservation equation is obtained:

$$\frac{\partial \rho^J}{\partial t} + \text{div}\left(\rho^J \overline{\boldsymbol{u}}^L\right) = 0 \quad \text{with} \quad \rho^J = J\overline{\rho}^L , \tag{28b}$$

and J is the Jacobian

$$J = \det\left(\frac{\partial \Xi_j}{\partial x_j}\right) = \det\left(\delta_{ij} + \frac{\partial \xi_i}{\partial x_j}\right) . \tag{28c}$$

It is shown by Andrews and McIntyre (1978a) that ρ^J so defined is a mean quantity. It can also be shown that the Lagrangian mean velocity field $\overline{\boldsymbol{u}}^L$ is not divergence-free as is the case for the Eulerian unaveraged velocity field:

$$\text{div}\left(\overline{\boldsymbol{u}}^L\right) = -\overline{D}^L \left(\log J\right) , \tag{28d}$$

where log is the natural logarithm.

A derivation of the averaged momentum equation in GLM-form in the way as has been given by Andrews and McIntyre (1978a, §3) is given by Dingemans (1997, Eq. (2.596)) as:

$$\overline{D}^L\left(\overline{u}_i^L\right) + \frac{1}{\rho_0}\frac{\partial \overline{p}^L}{\partial x_i} - \frac{\overline{\rho}^L}{\rho_0} g_i = \overline{D}^L\left(\overline{P}_i^L\right) + \frac{\partial}{\partial x_i}\left(\frac{1}{2}\left(\overline{u_m^\ell u_m^\ell}\right)\right) + \overline{P}_k^L \frac{\partial \overline{u}_k^L}{\partial x_i} , \tag{29a}$$

where the pseudo momentum $\overline{\mathbf{P}}^L$ is given by[b]

$$\overline{P}_i^L = -\left\langle \frac{\partial \xi_j}{\partial x_i} u_j^\xi \right\rangle = \left\langle \frac{\partial \xi_j}{\partial x_i} u_j^\ell \right\rangle . \tag{29b}$$

This now is the averaged momentum equation wholly written in terms of GLM quantities. It is noted that in the right-hand side of (29a) the terms are gathered which are quadratic in the fluctuations. For slowly-varying wave motion one can relate the pseudo-momentum to the wave action. Because

$$\frac{\partial}{\partial x_i}\left\langle \frac{1}{2} u_m^\ell u_m^\ell \right\rangle = \frac{\partial}{\partial x_i}\left\langle \frac{1}{2} u_m^\xi u_m^\xi \right\rangle - \frac{\partial}{\partial x_i}\left\langle \frac{1}{2} \overline{u}_m^L \overline{u}_m^L \right\rangle , \tag{30}$$

[b]Notice that the last equality immediately results from the definition $\boldsymbol{u}^\ell(\boldsymbol{x},t) = \boldsymbol{u}^\xi - \langle \boldsymbol{u}(\boldsymbol{x},t)\rangle^L$ and $\boldsymbol{u}^\xi = \boldsymbol{u}\{\boldsymbol{x} + \boldsymbol{\xi}(\boldsymbol{x},t), t\}$, because $\left\langle \overline{u}_j^\xi \, \xi_{j,i} \right\rangle = 0$.

it is possible to rewrite the mean-current equation (29a) in the following form:

$$\overline{D}^L \left(\overline{u}_i^L - \overline{\mathsf{P}}_i^L\right) + \frac{\partial \overline{u}_j^L}{\partial x_i} \left(\overline{u}_j^L - \overline{\mathsf{P}}_j^L\right) + \frac{\partial \overline{\pi}^L}{\partial x_i} + \frac{\overline{\rho}^L}{\rho_0} g_i = 0 \,, \tag{31a}$$

with the generalised pressure $\overline{\pi}^L$ defined by

$$\overline{\pi}^L = \frac{1}{\rho_0}\,\overline{p}^L - \frac{1}{2}\left\langle u_m^\xi u_m^\xi \right\rangle , \tag{31b}$$

which form has been derived directly by Andrews and McIntyre (1978a, Eq. (3.8)) and has also been used in this form by Leibovich (1980, Eq. (1))[c].

For the derivation of the wave action equation we have to introduce an ensemble average in the following way. We now suppose that $\overline{()}$ is an ensemble average and that each field $\phi(\boldsymbol{x}, t; \alpha)$ depends differentiably upon the ensemble label α so that $\partial\overline{\phi}/\partial\alpha = \langle \partial\phi/\partial\alpha \rangle = 0$ (see Andrews and McIntyre, 1978b). The wave action equation is obtained by scalar multiplication of the (shifted) Eulerian momentum equation $\rho_0\,(D\boldsymbol{u}/Dt)^\xi + (\mathrm{grad}\,p)^\xi = \rho^\xi \boldsymbol{g}$ with $\partial\boldsymbol{\xi}/\partial\alpha$. Then the wave action conservation equation is obtained in the form

$$\overline{D}^L A + \frac{1}{\rho^J}\,\mathrm{div}\,\boldsymbol{B} = 0 \quad \text{with} \quad A = \left\langle \boldsymbol{u}^\ell \cdot \frac{\partial \boldsymbol{\xi}}{\partial \alpha} \right\rangle \quad \text{and} \quad B_j = \left\langle K_{ij} p^\xi \frac{\partial \xi_j}{\partial \alpha} \right\rangle , \tag{32}$$

where K_{ij} is the cofactor[d] of the Jacobian J. Notice that we take here the plus sign in the definition for the wave action. This is due to the definition of the ensemble label. In fact, also in Andrews and McIntyre (1978b) two different signs have been used (their Eqs. (2.7a) and (2.19)) because in the two cases the meaning of α was different.

When also the Coriolis acceleration $\boldsymbol{\Omega}^{(c)} \wedge \overline{\boldsymbol{u}}^L$ is accounted for, then Eqs. (31) change into (see Andrews and McIntyre, 1978a and Leibovich, 1980):

$$\overline{D}^L \left(\overline{u}_i^L - \overline{\mathsf{P}}_i^L\right) + \frac{\partial \overline{u}_j^L}{\partial x_i} \left(\overline{u}_j^L - \overline{\mathsf{P}}_j^L\right) + 2\left(\boldsymbol{\Omega}^{(c)} \wedge \overline{\boldsymbol{u}}^L\right)_i + \frac{\partial \overline{\pi}^L}{\partial x_i} + \frac{\overline{\rho}^L}{\rho_0} g_i = 0 \tag{33a}$$

with

$$\overline{\pi}^L = \frac{1}{\rho_0}\,\overline{p}^L - \frac{1}{2}\left\langle \boldsymbol{u}^\xi \cdot \boldsymbol{u}^\xi \right\rangle + \left\langle \left(\boldsymbol{\Omega}^{(c)} \wedge \boldsymbol{\xi}\right) \cdot \boldsymbol{u}^\xi \right\rangle . \tag{33b}$$

---

[c]We write $\overline{\pi}^L$ instead of π as is written by Andrews and McIntyre (1978a) and Leibovich (1980) to have a consistent notation and to stress the fact that here π is a GLM quantity.

[d]From Pipes and Harvill (1970, p. 84) we have: The cofactor of an element of a determinant a_{ij} is the minor of that element with a sign attached to it determined by the numbers i and j which fix the position of a_{ij} in the determinant $|a|$. The sign is chosen by the equation $A_{ij} = (-1)^{i+j} M_{ij}$ where A_{ij} is the cofactor of the element a_{ij} and M_{ij} the minor of the element a_{ij}. The minor M_{ij} of the determinant $|a| = \det(a_{ij})$ is obtained by deleting the i-th row and the j-th column from the determinant $\det(a_{ij})$.

Here $\boldsymbol{\Omega}^{(c)}$ is the angular velocity of the rotating frame fixed to the earth. Notice that this definition is the usual one, e.g., see also Pedlosky (1979, §1.6). Because of the Coriolis force, also the definition of the pseudo momentum $\overline{\mathsf{P}}^L$ changes into:

$$\overline{\mathsf{P}}_i^L = -\left\langle \frac{\partial \xi_j}{\partial x_i} \left[u_j^\xi + \left(\boldsymbol{\Omega}^{(c)} \wedge \boldsymbol{\xi} \right)_j \right] \right\rangle . \tag{33c}$$

Equation (29) can be considered to be a *total current* equation, while (33) is a *mean momentum* equation. Ardhuin *et al.* (2007) call Eq. (29) the aGLM and Eq. (33) the GLM equation.

It has been argued in Dingemans (1997, §2.10.6) that it is advantageous to use the GLM-form Eq. (29) instead of the usual Eulerian forms. A disadvantage is that to get the results in the usual Eulerian form, it is necessary to Taylor the computational results to the order wanted. This is because one is not yet used to thinking in GLM-quantities. In this respect it is the same as in the sixties of the last century when engineers had to change from thinking in significant wave height to spectral quantities; another example is the use of Haar and Walsh transforms, which are especially suited for wave-slamming problems as well as heart-beat problems in which case one has to change from the usual notion of Fourier frequencies to the so-called sequencies (see, e.g., Beauchamp, 1975).

The important advantage of using the GLM-approach is that one does not need to make a new model as is the case for an Eulerian one when wave effects to a higher order need be included in the formulation. In the latter case the order of the differential equations may change which necessitates a wholly new numerical effort. In the GLM-case one only needs to Taylor the results of the (unchanged) computation to a higher order of approximation.

4. The CL equation

The CL equation is called after Craik and Leibovich, who, in 1976, published a paper on a model for Langmuir circulations. The basis for their model was that the current field consisted of irrotational water wave velocity $\boldsymbol{u}'$ and a much smaller current field $\boldsymbol{v}$, $\boldsymbol{q} = \varepsilon \boldsymbol{u}' + \varepsilon^2 \boldsymbol{v}$ with ε a measure for the wave slope. In fact their model was one in which the current was generated by the waves. The vorticity equation is

$$\boldsymbol{\omega}_t = \mathrm{curl}\,(\boldsymbol{q} \wedge \boldsymbol{\omega}) + \nu_e \nabla^2 \boldsymbol{\omega} \,, \tag{34a}$$

where $\nu_e = \alpha\varepsilon^2$ and α is independent of ε. With $\boldsymbol{v} = \boldsymbol{v}_0 + \varepsilon \boldsymbol{v}_1 + \cdots$, $\mathrm{curl}\,\boldsymbol{v} = \boldsymbol{\omega}_0 + \varepsilon\boldsymbol{\omega}_1 + \cdots$ and $\overline{\boldsymbol{\omega}}_0 = (\xi, \eta, \zeta)^T$, Craik and Leibovich chose coordinates such that $\boldsymbol{u}^S = \left(u^S(y,z), 0, 0 \right)^T$, $\overline{\boldsymbol{v}}_0 = (u, v, w)^T$ and $\overline{\boldsymbol{\omega}}_0 = \mathrm{curl}\,\overline{\boldsymbol{v}}_0 = (\xi, \eta, \zeta)^T$. Only solutions for $\overline{\omega}_0$ and $\overline{\boldsymbol{v}}_0$ which are independent of x are sought. Then the vorticity equation simplifies to

$$\alpha \nabla^2 \xi + \eta \partial_y u^S + \zeta \partial_z u^S = v \partial_y \xi + w \partial_z \xi \quad \text{where} \quad \nabla^2 \equiv \partial_y^2 + \partial_z^2 \,. \tag{34b}$$

By introducing a stream function ψ defined by $v = \partial_z \psi$ and $w = \partial_y \psi$, the final model used by Craik and Leibovich was

$$\alpha \nabla^4 \psi + \frac{\partial\left(u, u^S\right)}{\partial\left(y, z\right)} = \frac{\partial\left(\nabla^2 \psi, \psi\right)}{\partial\left(y, z\right)} \quad \text{and} \quad \alpha \nabla^2 u = \frac{\partial(u, \psi)}{\partial(y, z)} \,. \tag{34c}$$

Leibovich (1977) gave a vorticity equation

$$\frac{\partial \overline{\boldsymbol{\omega}}}{\partial t} + \left(\overline{\boldsymbol{v}} + \boldsymbol{u}^S\right) \cdot \operatorname{grad} \overline{\boldsymbol{\omega}} = \overline{\boldsymbol{\omega}} \cdot \operatorname{grad}\left(\overline{\boldsymbol{v}} + \boldsymbol{u}^S\right) - \frac{g}{\rho_r} \operatorname{curl}\left(\overline{\rho} \boldsymbol{k}\right) + \nu_T \operatorname{div} \operatorname{grad} \overline{\boldsymbol{\omega}} \,, \tag{35a}$$

with $\overline{\boldsymbol{\omega}} = \operatorname{curl} \overline{\boldsymbol{v}}$ and $\boldsymbol{k}$ a unit vector in the z-direction. Integration of (35a) yields (Leibovich's equation (7)):

$$\frac{\partial \overline{\boldsymbol{v}}}{\partial t} + \overline{\boldsymbol{v}} \cdot \operatorname{grad} \overline{\boldsymbol{v}} + \operatorname{grad} \pi = \boldsymbol{u}^S \wedge \overline{\boldsymbol{\omega}} - \frac{g}{\rho_r} \overline{\rho} \boldsymbol{k} + \nu_T \operatorname{div} \operatorname{grad} \overline{\boldsymbol{v}} \,, \tag{35b}$$

where the buoyancy is supposed to be not large compared to $\boldsymbol{\omega} \cdot \nabla \boldsymbol{v}$ and π was given in words as "including the mean kinetic energy of the wave motion in addition to the averaged pressure". This means that

$$\pi = \frac{1}{\rho} \overline{p} + \frac{1}{2} \left\langle \boldsymbol{u}' \cdot \boldsymbol{u}' \right\rangle \,. \tag{35c}$$

Leibovich (1980) showed that the GLM equation (29a) can be approximated to the Craik-Leibovich equation in Eulerian coordinates as:

$$\partial_t \overline{\boldsymbol{u}} + \overline{\boldsymbol{u}} \cdot \operatorname{grad} \overline{\boldsymbol{u}} + 2\boldsymbol{\Omega} \wedge \left(\overline{\boldsymbol{u}} + \overline{\boldsymbol{u}}^S\right) + \operatorname{grad}\left(\pi + \overline{\boldsymbol{u}} \cdot \overline{\boldsymbol{u}}^S\right) = \overline{\boldsymbol{u}}^S \wedge \operatorname{curl} \overline{\boldsymbol{u}} + \nu \operatorname{div} \operatorname{grad} \overline{\boldsymbol{u}} \,, \tag{36a}$$

$$\text{with} \quad \pi = \rho^{-1} \overline{p} - \frac{1}{2} \left\langle \boldsymbol{u}^\xi \cdot \boldsymbol{u}^\xi \right\rangle + \left\langle (\boldsymbol{\Omega} \wedge \boldsymbol{\xi}) \cdot \boldsymbol{u}^\xi \right\rangle \,, \tag{36b}$$

where the Lagrangian form $\boldsymbol{u}^\xi$ still has to be expanded. That the minus sign is clearly correct follows in fact directly from the GLM current equation (29). It has to be stressed that the sign of the contribution of the kinetic wave energy to the pressure is not important for the appearance of the instability, which is also clear from the fact that these instabilities (Langmuir circulations) were obtained from the vorticity equation and curl grad $\equiv 0$.

We now consider the Eq. (37) as the standard form for the CL equation.

$$\frac{\partial \overline{\boldsymbol{u}}}{\partial t} + \left(\overline{\boldsymbol{u}} \cdot \operatorname{grad}\right) \overline{\boldsymbol{u}} + \operatorname{grad} \overline{\pi} = \overline{\boldsymbol{u}}^S \wedge \operatorname{curl} \overline{\boldsymbol{u}} \quad \text{with} \quad \overline{\pi} = \frac{\overline{p}}{\rho} + gz - \left\langle \frac{1}{2} \boldsymbol{u}' \cdot \boldsymbol{u}' \right\rangle \,. \tag{37}$$

Dingemans *et al.* (1996) and Dingemans and Radder (2000) presented the CL equation as (37), but with $+\frac{1}{2}\left\langle \boldsymbol{u}' \cdot \boldsymbol{u}' \right\rangle$ in the pressure term π. We note that the vortex force may be approximated for a flume, with x directed along the flume, by

$$\overline{\boldsymbol{u}}^S \wedge \overline{\omega} \cong \left(0, \overline{u}^S \frac{\partial \overline{u}}{\partial y}, \overline{u}^S \frac{\partial \overline{u}}{\partial z}\right)^T \,. \tag{38}$$

After an idea of A.C. Radder, an explanation for the results of flume experiments of Klopman (1994)[e] of waves on a following and opposing current was sought by application of the CL equation. In those experiments the vertical current profile turned back for the following case and turned forward for the opposing case, compared to the case of no waves, see the left side of Fig. 3.

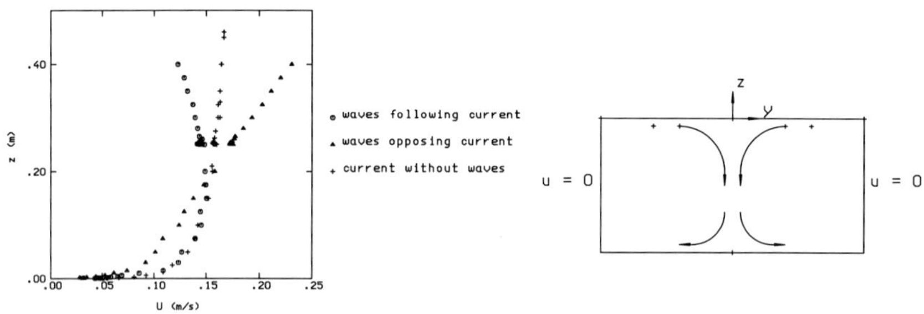

Figure 3. Mean current profiles after Klopman (1994) (*left*); cell formation in the flume (*right*).

An explanation of this behaviour can be given in the following way. We consider waves and current in a laboratory flume. In the lateral (y) direction, changes in the velocity field should occur. This has to be generated by the wall effects. Suppose that a circulation cell has developed. Suppose that the cell is such that in the middle of the flume and the flow is in downward direction. The situation is as sketched in Figure 3. In the upper part of the fluid, water with a small amount of momentum is transported towards the middle of the flume and consequently decelerates, while in the lower part of the fluid the reverse situation occurs. In the upper half of the fluid the downward velocity increases and in the lower half it decreases. In the middle of the flume $\partial v/\partial y = 0$ because of symmetry. From the continuity equation $u_x + w_z = 0$ in the middle of the flume then follows that $\partial u/\partial x < 0$ in the upper half and $\partial u/\partial x > 0$ in the lower half of the fluid. Consequently, the profile $u(z)$ turns back in the upper part of the fluid and increases in the lower part. The result is a more uniform horizontal velocity profile in the vertical.

The opposite case, of waves against the current can not be explained in this way. There is now nowhere a larger horizontal transport of momentum transported to the centre of the flume.

Dingemans *et al.* (1996) described a numerical experiment on basis of the CL equation (37) where viscosity terms were added. They were able to get a single circulation cell of Langmuir type for a normal wave flume. However, for the numerical experiment with a very wide flume (50 m wide) also a single cell developed instead of a number of them. A reasonable explanation is a very large influence

[e]Similar experiments were performed by Kemp and Simons (1982,1983), with similar results.

of the sidewards boundaries because of a too rough modelling. A second reason is due to the upwind numerical scheme used, which gives too much numerical damping for the instabilities to grow.

Because of the numerical result of cell formation, the experiments of Klopman (1994), in which only was measured in the middle of the flume, were repeated (Klopman, 1997), in which case also measurements were performed in a lateral plane. Indeed, the new measurements also showed cell formation.

Based on a 1D GLM formulation, Groeneweg and Klopman (1998) showed that a direct application of GLM gave numerical results which compared very well with the measurements, see their Fig. 3. Also for waves on an opposing current the results were good. Later results of a 2D GLM approach were shown by Groeneweg and Battjes (2003). These authors concluded that while the predictions in streamwise direction were very good, the predictions in lateral direction were only qualitatively, not quantitatively correct. They inferred that the longitudinal contribution to the wave-induced driving force is dominant over the lateral contributions.

5. Langmuir-type circulations through an instability mechanism

As discussed for example in Faller and Caponi (1978) and Csanady (1994), the important point of the contribution of Craik and Leibovich (1976) was to point out that the wave motion interacts with the current through the Stokes drift. As the Stokes drift is a residue of irrotational motion, it does not contribute itself to vorticity. Two versions of Craik-Leibovich models can be discerned: **CL1** and **CL2**.

CL1 is based upon the interaction of a cross-wave pattern with a shear flow driven by a constant wind stress. The variation of the Stokes drift in lateral direction (being maximum along the traces of the wave crest intersections) twists the vorticity of the wind-driven shear flow into the vorticity of the Langmuir circulation. A review of Langmuir circulation has been given by Leibovich (1983). As estimate of the cell spacing one would obtain $\lambda_c = \lambda_w/(2\tan\theta)$ where λ_w is the downwind spacing of the wave crests and θ is the half-angle of the intersection of the two wave trains.

CL2 The more or less uniform average Stokes drift of a complex wave pattern is believed to act through an instability mechanism. One has the following feed-back mechanism: an initial weak lateral variation of streamwise velocity u, acted upon by Stokes drift, generates vorticity, which in its turn enhances the u-variation. Apart from the initial small perturbation, the shear flow is supposed to be uniform in the (x, y)-plane.

5.1. *A third, nonlinear, instability mechanism*

A third, nonlinear instability mechanism was devised by A.C. Radder and has been published in Dingemans and Radder (2000). The basic steps are as follows.

The basis is the CL equation (37) plus a viscosity term[f];

$$\partial_t\overline{\boldsymbol{u}}+(\overline{\boldsymbol{u}}\cdot\mathrm{grad})\,\overline{\boldsymbol{u}}+\mathrm{grad}\,\overline{\pi}=\overline{\boldsymbol{u}}^S\wedge\overline{\boldsymbol{\omega}}+\rho^{-1}\,\mathrm{div}\,\overline{\boldsymbol{\sigma}}' \quad \text{with} \quad \overline{\pi}=\overline{p}/\rho+gz+\frac{1}{2}\left\langle\boldsymbol{u}'\cdot\boldsymbol{u}'\right\rangle\,, \tag{39}$$

where $\mathrm{div}\,\overline{\boldsymbol{\sigma}}' \equiv \partial\overline{\sigma}'_{ik}/\partial x_k$. Although no viscosity is taken into account in the usual CL-equation formulations, it is advantageous to do so. Viscosity in these equations is needed for obtaining shear in the mean-current equations, which, in its turn, is needed to generate the vortex-force term. We take the (eddy) viscosity coefficient to be isotropic because of the scales on which the flow occurs here. Applying the Boussinesq-hypothesis, the stresses $\overline{\sigma}'_{ik}$ are approximated as $\rho^{-1}\overline{\sigma}'_{ik} = \nu_T\left(\partial\overline{u}_i/\partial x_k + \partial\overline{u}_k/\partial x_i\right)$, while the eddy viscosity ν_T has still to be determined.

One of the explanations of Langmuir circulations rests upon the supposition that an instability mechanism in the CL equation is responsible for the generation of these vortex rolls. We suppose that the mean current $\overline{\boldsymbol{u}}$ is disturbed with a velocity $\hat{\boldsymbol{u}}$. These perturbations are supposed to be of periodic nature, i.e., $\hat{\boldsymbol{u}}$ obeys a WKBJ-type of behaviour which is natural, also in view of the resulting (periodic) vortex-roll motions. Then $\overline{\boldsymbol{u}} = \boldsymbol{U}+\hat{\boldsymbol{u}}$, where $\hat{\boldsymbol{u}}$ is the disturbance which is responsible for the formation of the vortex rolls and $\boldsymbol{U}$ is the velocity of the basic state. For the other quantities $\overline{\pi}$ and $\overline{\boldsymbol{\omega}}$ in the CL equation the same kind of perturbations are assumed to exist, viz. a basic state (denoted with capitals) and a perturbed state (denoted by hatted variables). All quantities, except the Stokes drift are perturbed. It has been shown (Dingemans, 1999) that the perturbation of the eddy viscosity has no effect on the present results.

We now insert the expressions $\overline{\boldsymbol{u}} = \boldsymbol{U} + \hat{\boldsymbol{u}}$, etc., in the CL equation and the continuity equation. The wave related quantities like $\boldsymbol{u}'$ and $\overline{\boldsymbol{u}}^S$ are not perturbed. Because of the periodicity of the perturbed quantities, averaging over a period and length large to the characteristic period and length of the perturbations, yields for the basic state:

$$\partial_t\boldsymbol{U} + (\boldsymbol{U}\cdot\mathrm{grad})\,\boldsymbol{U} + \langle(\hat{\boldsymbol{u}}\cdot\mathrm{grad})\,\hat{\boldsymbol{u}}\rangle + \mathrm{grad}\,\Pi = \boldsymbol{U}^S\wedge\boldsymbol{\Omega} + \rho^{-1}\,\mathrm{div}\,\boldsymbol{\Sigma}'\,. \tag{40}$$

Notice that $\langle(\hat{\boldsymbol{u}}\cdot\mathrm{grad})\,\hat{\boldsymbol{u}}\rangle$ are the similar to Reynolds stresses in turbulence studies. The evaluation of these stresses is the subject of this study. It will appear later that these stresses can be determined *analytically*, which does not necessitate a very fine numerical mesh; the basic current equation can be solved with the commonly used large mesh when no Reynolds stresses are present. The equation for the perturbation is obtained by subtraction of (40) from the full equation:

$$\begin{aligned}&\partial_t\hat{\boldsymbol{u}} + (\hat{\boldsymbol{u}}\cdot\mathrm{grad})\,\boldsymbol{U} + (\boldsymbol{U}\cdot\mathrm{grad})\,\hat{\boldsymbol{u}} + (\hat{\boldsymbol{u}}\cdot\mathrm{grad})\,\hat{\boldsymbol{u}} - \langle(\hat{\boldsymbol{u}}\cdot\mathrm{grad})\,\hat{\boldsymbol{u}}\rangle + \mathrm{grad}\,\hat{\pi}\\ &\quad = \boldsymbol{U}^S\wedge\hat{\omega} + \rho^{-1}\,\mathrm{div}\,\hat{\boldsymbol{\sigma}}'\,.\end{aligned} \tag{41}$$

[f]We write σ'_{ik} with the prime, denoting the part of the stress tensor without the pressure, see Dingemans (1997, p. 4).

The continuity equation splits in one for the basic state and one for the perturbed velocity:

$$\operatorname{div} \boldsymbol{U} = 0 \quad \text{and} \quad \operatorname{div} \hat{\boldsymbol{u}} = 0 \,. \tag{42}$$

In order to solve the system, a simplified approach is used. Firstly, we suppose that the basic current $\boldsymbol{U}$ is uniform in the horizontal directions, $\boldsymbol{U} = \boldsymbol{U}(z,t)$ and, moreover, no vertical component exists: $\boldsymbol{U} = (U(z,t), V(z,t), 0)^T \equiv \boldsymbol{U}^h$. This means that (nearly) horizontal nearly-uniform shear flows are considered, and therefore, also the bottom is (nearly) horizontal, i.e., $\nabla h(x,y) = 0$ where $\nabla = (\partial_x, \partial_y)^T$. Secondly, the wave-induced perturbation $\hat{\boldsymbol{u}}$ is supposed to be single-periodic in one specific direction θ. Thirdly, with θ the angle between the positive x-axis and the path of propagation s and n the lateral direction, we also suppose that $\partial \hat{\boldsymbol{u}}(x,y,z,t)/\partial n = 0$. Fourthly, the Stokes drift $\boldsymbol{U}^S$ is supposed to be only a function of depth, i.e., $\boldsymbol{U}^S = \boldsymbol{U}^S(z) = \left(U^S(z), V^S(z), 0\right)^T$. As the eddy viscosity is also a function of space and time through its dependence on the friction velocity, we now suppose that $\overline{\nu}_T = \overline{\nu}_T\left(\left|\boldsymbol{U}^*\left(\boldsymbol{X},T\right)\right|, z\right)$ with $\boldsymbol{X} = \delta \boldsymbol{x}$ and $T = \delta t$ and $\delta \ll 1$. The simplified momentum equations then become (details in Dingemans, 1999):

$$\partial_t \boldsymbol{U}^h + \left\langle \partial_{x_j} \left(\hat{u}_j \hat{\boldsymbol{u}}\right)\right\rangle^h + \nabla \Pi_0 = \partial_z \left(\overline{\nu}_T \partial_z \boldsymbol{U}^h\right) \quad \text{where} \quad \Pi_0 = \overline{P}/\rho + \frac{1}{2}\left\langle \tilde{\boldsymbol{u}} \cdot \tilde{\boldsymbol{u}}\right\rangle \,. \tag{43}$$

We note that in the present approximation the vortex force has only a vertical component and therefore plays no role in the horizontal mean momentum equations. For the perturbed velocity we get:

$$\partial_t \hat{\boldsymbol{u}}^h + \hat{w} \partial_z \boldsymbol{U}^h + \left(\boldsymbol{U}^h \cdot \nabla\right) \hat{\boldsymbol{u}}^h + \nabla \hat{\pi} = \left(\boldsymbol{U}^S \wedge \hat{\boldsymbol{\omega}}\right)^h + \rho^{-1}\left(\operatorname{div} \hat{\boldsymbol{\sigma}}'\right)^h \tag{44a}$$

where the viscosity term is a function on $\hat{\boldsymbol{u}}$ and $\overline{\nu}_T$. The vertical momentum equation becomes:

$$\begin{aligned} \partial_t \hat{w} + \left(\boldsymbol{U}^h \cdot \nabla\right) \hat{w} + \partial_z \hat{\pi} = \left(U^S \hat{\omega}_2 - V^S \hat{\omega}_1\right) + 2\partial_z \left(\overline{\nu}_T \left(\partial_z \hat{w}\right)\right) \\ + \overline{\nu}_T \partial_{x_j} \left(\partial_{x_j} \hat{w} + \partial_z \hat{u}_j\right) \quad j = 1,2 \,. \end{aligned} \tag{44b}$$

5.2. *Linear stability analysis*

A solution of Eqs. (44) is sought now. Following Cox (1997), an asymptotic solution is sought by applying a long-wave expansion. This expansion is based on the observation that Langmuir circulations have a much larger horizontal extent (in the direction perpendicular to the circulation) than the extent of the circulation cells. The boundary conditions for the perturbed velocities then are:

$$\partial_z \hat{u} = \partial_z \hat{v} = w = 0 \quad \text{at} \quad z = 0 \quad \text{and} \quad \partial_z \hat{u} = \partial_z \hat{v} = w = 0 \quad \text{at} \quad z = -h \,. \tag{45}$$

It is noted that the conditions (45) do not comply with the no-slip conditions, which should apply for viscous flow as is considered here. It seems reasonable to limit this stability analysis to the bulk of the fluid, just outside the bottom boundary layer. We now assume a slow growth rate σ and the expansions of $\hat{\boldsymbol{u}}$ is:

$$\hat{\boldsymbol{u}}(\boldsymbol{x}, z, t) = \boldsymbol{u}'(z)\,\mathrm{e}^{\vartheta} \quad \text{with} \quad \boldsymbol{u}'(z) = \hat{\boldsymbol{u}}_0(z) + \varepsilon\hat{\boldsymbol{u}}_1(z) + \varepsilon^2\hat{\boldsymbol{u}}_2(z) + \cdots \tag{46a}$$

where

$$\vartheta(\boldsymbol{x}, t) = i\varepsilon\tilde{\boldsymbol{k}} \cdot \boldsymbol{x} + \varepsilon\sigma t \quad \text{and} \quad \sigma = \sigma_1 + \varepsilon\sigma_2 + \cdots \tag{46b}$$

with $\tilde{\boldsymbol{k}} = \left(\tilde{k}_1, \tilde{k}_2\right)^T$ the scaled wave number vector: $\tilde{\boldsymbol{k}} = \boldsymbol{k}/\varepsilon$ and $\left|\tilde{\boldsymbol{k}}\right| = \mathcal{O}(1)$. For $\hat{\pi}$ we use a similar expansion. In this way one focuses attention to the most unstable wave numbers $\boldsymbol{k}$, which are $\mathcal{O}(\varepsilon)$.

These expansions are substituted in the linearised momentum equations for the perturbations. From the zeroth order it follows that the horizontal components of $\hat{\boldsymbol{u}}_0$ are constant and the vertical velocity component is zero, i.e., $\hat{\boldsymbol{u}}_0 = (u_0, v_0, 0)^T = \boldsymbol{u}_0$. In first-order the vertical velocity is still zero and from the bottom condition follows a condition between the unknown constants, $\tilde{k}_1 u_0 + \tilde{k}_2 v_0 = 0$. Averaging the first-order momentum equations over the depth, mean quantities are introduced denoted by a double overbar, $\overline{\overline{\boldsymbol{U}}} = \left(\overline{\overline{U}}, \overline{\overline{V}}, 0\right)^T = h^{-1}\int_{-h}^{0} \boldsymbol{U}(z)dz$ and similarly for $\overline{\overline{\boldsymbol{U}^S}}$. These equations are solved for π_0 and σ_1. As σ_1 is imaginary (as it should be because $u_0 \neq 0$ and $v_0 \neq 0$) we write $\sigma_1 = i\sigma_1^{(i)}$. We find $\pi_0 = \boldsymbol{u}_0 \cdot \overline{\overline{\boldsymbol{U}^S}}$ and $\sigma_1^{(i)} = \tilde{\boldsymbol{k}} \cdot \left(\overline{\overline{\boldsymbol{U}}} + \overline{\overline{\boldsymbol{U}^S}}\right)$.

In second order we differentiate the second-order continuity equation and substitute expressions for u_1 and v_1 taken from the unaveraged first-order equations. For w_2 we then obtain the differential equation:

$$\partial_z\left(\overline{\nu}_T \partial_z^2 w_2\right) = \left|\tilde{\boldsymbol{k}}\right|^2 \boldsymbol{u}_0 \cdot \left(\overline{\overline{\boldsymbol{U}^S}} - \boldsymbol{U}^S\right). \tag{47}$$

The right-hand side is thus zero when no shear is present (i.e., when $\boldsymbol{U}^S$ is constant over the depth). Recapitulating, we have the unknown constants u_0 and v_0 with relation $\tilde{k}_1 u_0 + \tilde{k}_2 v_0 = 0$ between them, and solutions for π_0 and σ_1. For the first non-zero vertical velocity component we have the differential equation (47). In next section we consider the energy equation for the perturbed velocities in order to close the system.

5.3. *The Landau-Stuart equation*

The energy equation for the perturbed velocities $\hat{\boldsymbol{u}}$ follows by scalar multiplication by $\hat{\boldsymbol{u}}$ of the momentum equation for the perturbed velocities. Introducing the mean kinetic energy by $\mathcal{K} = \iiint dxdydz\frac{1}{2}\hat{\boldsymbol{u}} \cdot \hat{\boldsymbol{u}} \equiv \int_{-h}^{0} dz \left\langle \frac{1}{2}\hat{\boldsymbol{u}} \cdot \hat{\boldsymbol{u}} \right\rangle$, the total change in kinetic energy may be written down.

An amplitude $\mathsf{A}_0 = \sqrt{u_0^2 + v_0^2}$ is introduced and we also write $\varepsilon^2 w_2(z) = \varepsilon^2 \tilde{k}^2 m_2(z) = k^2 m_2(z)$ where $\tilde{k}^2 = \tilde{k}_1^2 + \tilde{k}_2^2$. As zeroth-order solution of the expansion (46) we now have

$$\hat{\boldsymbol{u}}(\boldsymbol{x}, z, t) = \frac{1}{2}\left(\tilde{k}_2/\tilde{k}, -\tilde{k}_1/\tilde{k}, \varepsilon^2\tilde{k}^2 m_2(z)\right)^T \mathsf{A}_0\, \mathrm{e}^{\vartheta} + CC\,. \tag{48}$$

Following Stuart (1958), we now suppose the amplitude A_0 to be a function of time, $\mathsf{A}_0 = \mathsf{A}_0(t)$. Using the expansion (48) in the expression for the kinetic energy leads to the so-called Landau-Stuart equation:

$$\frac{d\mathsf{A}_0^2}{dt} = 2\alpha\mathsf{A}_0^2 - \ell\mathsf{A}_0^4 \text{ with exact solution } \mathsf{A}_0^2 = 1 \Big/ \left[\frac{\ell}{2\alpha} + \left(\frac{1}{\mathsf{A}_0^2} - \frac{\ell}{2\alpha}\right)\mathrm{e}^{-2\alpha t}\right], \tag{49}$$

where the coefficients α and ℓ consist of expressions in m_2, the Stokes drift, etc., see Dingemans (1999). We have $\ell > 0$, but the sign of α is not clear beforehand. An exact solution is found by rewriting Eq. (49) in one for A_0^{-2}, which equation turns out to be linear (Drazin and Reid, 1981). When $\alpha > 0$, the solution (49) approaches the equilibrium solution, $\mathsf{A}_0^2 \to \mathsf{A}_e^2 = 2\alpha/\ell$ for $t \to \infty$. When $\alpha < 0$, $\mathsf{A}_0 \to 0$ for $t \to \infty$.

With $q = \tilde{k}_1/\tilde{k}_2$, we determine the critical direction $q_c = \tan\varphi_c$ for which maximum growth of the perturbations occurs. When considering infinitesimal perturbations, the Landau-Stuart equation can be linearised to give $\max_q d\mathsf{A}_0^2/dt = 2\alpha\mathsf{A}_0^2$. Maximum growth is obtained for $d\alpha/dq = 0$ together with the condition that $d^2\alpha/dq^2 < 0$. Using the expressions for the coefficients of the Landau-Stuart equation, the value of q_c can be determined, see Dingemans (1999).

The Reynolds stresses $\langle\hat{w}\hat{u}\rangle = \varepsilon^2\langle w_2\hat{u}\rangle$ and $\langle\hat{w}\hat{v}\rangle$ can now be calculated. Returning to unscaled variables we see that for the case that $\alpha > 0$ we have $\langle\hat{w}\hat{u}\rangle = \frac{1}{2}k^2\mathsf{A}_0^2 m_2/\sqrt{1+q^2}$ and $\langle\hat{w}\hat{v}\rangle = \frac{1}{2}k^2\mathsf{A}_0^2 q m_2/\sqrt{1+q^2}$. When $\alpha < 0$ the Reynolds stresses are zero. For the amplitude A_0 we now use the equilibrium solution $\mathsf{A}_e = \sqrt{2\alpha/\ell}$. From the resulting expressions (see Dingemans, 1999) it appears that, to leading order, the Reynolds stresses do not depend on $k = \varepsilon\tilde{k}$, meaning that they are independent of the extent of the circulation cells (the size being proportional to $1/k$).

As a test for the theory, Dingemans and Radder (2000) considered an example from flume experiments by Klopman (1994). Measurements show the influence of waves on currents, see Figure 4. Taking the logarithmic velocity profile $U(z) = (\overline{u}^*/\kappa)\log\{(z+h)/z_0\}$ for $-h + z_0 \le z \le 0$, we have $\overline{u}^*/\kappa = 0.018$ with $z_0 = 0.04$ mm and $h = 0.5$ m. For this case an approximate calculation of the radiation stress, using long-wave approximations, yields a current contribution $u_w = -0.24h\log(2 + z/h)$. Determination of the mean current to be the same in the no-waves and waves case yields a constant $c = 0.0464$ m/s. In Figure 4 is given the $U(z)$ and the curve for $U(z) + u_w(z) + c$. It is clear that the effect of a backwards leaning velocity profile for following waves is included in the present theory.

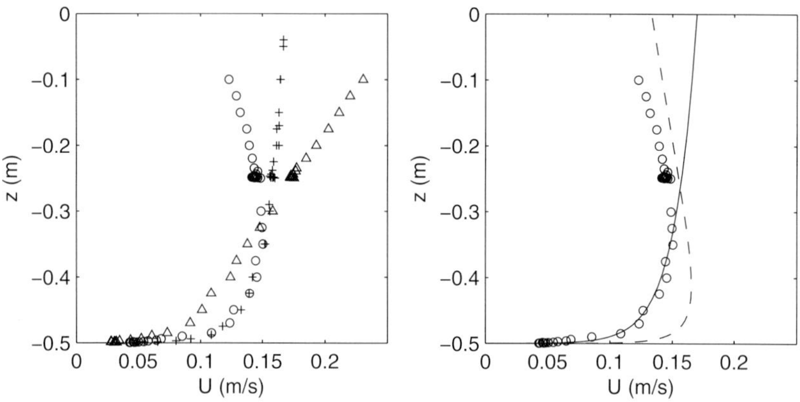

Figure 4. Left: Klopman's (1994) measurements; +: current without waves, ∘: waves following the current, △: waves opposing the current. Right: drawn line: logarithmic current profile, interrupted line: total velocity, circles: Klopman's measurements for following waves.

6. Derivation of extended CL equation

Here we want to show a derivation based on the GLM-equation with Coriolis, Eq. (33). We consider first the pseudo momentum. The pseudo momentum of (33c) consists of average of the sum of two terms. As a consequence of the first postulate of Andrews and McIntyre (1978a, p. 614), this is the same as the sum of averages over each contribution. We thus have

$$\overline{\mathsf{P}}_i^L = -\left\langle \frac{\partial \xi_j}{\partial x_i} u_j^\xi \right\rangle - \left\langle \frac{\partial \xi_j}{\partial x_i} \left(\boldsymbol{\Omega}^{(c)} \wedge \boldsymbol{\xi} \right)_j \right\rangle . \tag{50}$$

For the first part of the pseudo momentum we then have

$$\overline{\mathsf{P}}_i^{L\,(1)} = -\left\langle \frac{\partial \xi_j}{\partial x_i} u_j^\ell \right\rangle = \left\langle \xi_j \frac{\partial u_j^\ell}{\partial x_i} \right\rangle - \left\langle \frac{\partial}{\partial x_i} \left(u_j^\ell \xi_j \right) \right\rangle . \tag{51}$$

Consider first the first term at the right-hand side of (50). We have

$$\left\langle \xi_j \frac{\partial u_j^\ell}{\partial x_i} \right\rangle = \left\langle \xi_j \frac{\partial u_i^\ell}{\partial x_j} \right\rangle + \left\langle \left(\boldsymbol{\xi} \wedge \operatorname{curl} \boldsymbol{u}^\ell \right)_i \right\rangle = \left\langle (\boldsymbol{\xi} \cdot \operatorname{grad}) \, \boldsymbol{u}^\ell \right\rangle_i + \left\langle \boldsymbol{\xi} \wedge \boldsymbol{\omega}^\ell \right\rangle_i \tag{52}$$

where has been written $\boldsymbol{\omega}^\ell = \operatorname{curl} \boldsymbol{u}^\ell$. For the first part of the pseudo momentum we now have

$$\overline{\mathsf{P}}_i^{L\,(1)} = \left\langle \xi_j \frac{\partial u_i^\ell}{\partial x_j} \right\rangle + \left\langle \left(\boldsymbol{\xi} \wedge \boldsymbol{\omega}^\ell \right) \right\rangle_i - \left\langle \frac{\partial}{\partial x_i} \left(u_j^\ell \xi_j \right) \right\rangle . \tag{53}$$

To transform $\boldsymbol{u}^\ell$ into Eulerian quantities, we note that (e.g., Dingemans (1997, Eq. (2.618a))

$$\boldsymbol{u}^\ell = \boldsymbol{u}' + \xi_j \frac{\partial \overline{\boldsymbol{u}}}{\partial x_j} + \mathcal{O}\left(a^2 \right) . \tag{54}$$

Then we can write[g]

$$\left\langle \xi_j \frac{\partial u_i^\ell}{\partial x_j} \right\rangle = \left\langle \xi_j \frac{\partial u_i'}{\partial x_j} \right\rangle + \frac{\partial}{\partial x_j} \left(\overline{\xi_j \xi_k} \frac{\partial \overline{u}_i}{\partial x_k} \right) + \mathcal{O}\left(a^3\right) . \tag{55}$$

With $\overline{\boldsymbol{u}}^S$ given in (5) we can write

$$\begin{aligned} \overline{\mathsf{P}}_i^{L\,(1)} &= \overline{u}_i^S + \frac{1}{2} \overline{\xi_j \xi_k} \frac{\partial^2 \overline{u}_i}{\partial x_j \partial x_k} + \left\langle \xi_j \frac{\partial \xi_k}{\partial x_j} \frac{\partial \overline{u}_i}{\partial x_k} \right\rangle \\ &\quad + \left\langle \left(\boldsymbol{\xi} \wedge \boldsymbol{\omega}^\ell \right)_i \right\rangle - \left\langle \frac{\partial}{\partial x_i} \left(u_j^\ell \xi_j \right) \right\rangle + \mathcal{O}\left(a^3\right) . \end{aligned} \tag{56}$$

The fourth and fifth term of the right-hand side of (56) have to be estimated yet. We have[h]

$$\left\langle \frac{\partial}{\partial x_i} \left(u_j^\ell \xi_j \right) \right\rangle = \left\langle \frac{\partial}{\partial x_i} \left[\overline{D}^L \left(\frac{1}{2} \xi_j \xi_j \right) \right] \right\rangle . \tag{57}$$

The right-hand side now represents the spatial variation of the material (Lagrangian) derivative of the quantity $\frac{1}{2} \langle \boldsymbol{\xi} \cdot \boldsymbol{\xi} \rangle$. It seems reasonable to expect that this variation is a small quantity and may therefore be neglected for the present investigations. The neglect of the term $\left\langle \left(\boldsymbol{\xi} \wedge \boldsymbol{\omega}^\ell \right) \right\rangle_i$ is not so easy to justify. In principle the term is of $\mathcal{O}\left(a^2\right)$ because both $|\boldsymbol{\xi}|$ and $\left|\boldsymbol{u}^\ell\right|$ are of $\mathcal{O}(a)$. In the water it plays a role in situations with wave breaking. In the air the justification of its neglect is still harder to give and we therefore refrain from neglecting this term now. Using the representation (56), we then have for $\overline{\mathsf{P}}_i^{L\,(1)}$ the following approximation:

$$\overline{\mathsf{P}}_i^{L\,(1)} = \overline{u}_i^S + \frac{1}{2} \overline{\xi_j \xi_k} \frac{\partial^2 \overline{u}_i}{\partial x_j \partial x_k} + \left\langle \xi_j \frac{\partial \xi_m}{\partial x_j} \frac{\partial \overline{u}_i}{\partial x_m} \right\rangle + \left\langle \left(\boldsymbol{\xi} \wedge \boldsymbol{\omega}^\ell \right)_i \right\rangle + \mathcal{O}\left(a^3\right) . \tag{58}$$

We still have to rewrite $\boldsymbol{\omega}^\ell$. Because $\boldsymbol{\omega}^\ell = \operatorname{curl} \boldsymbol{u}^\ell$, the use of relation (54) permits us to write

$$\boldsymbol{\omega}^\ell \equiv \operatorname{curl} \boldsymbol{u}^\ell = \operatorname{curl} \boldsymbol{u}' + \operatorname{curl} \left(\xi_j \frac{\partial \overline{\boldsymbol{u}}}{\partial x_j} \right) + \mathcal{O}\left(a^2\right) \equiv \boldsymbol{\omega}' + \boldsymbol{\omega}'' + \mathcal{O}\left(a^2\right) . \tag{59}$$

As it is not yet clear on which grounds to neglect the Coriolis contribution to the pseudo momentum, the approximation of the total pseudo momentum then is

$$\begin{aligned} \overline{\mathsf{P}}_i^{L} &= \overline{u}_i^S + \frac{1}{2} \overline{\xi_j \xi_k} \frac{\partial^2 \overline{u}_i}{\partial x_j \partial x_k} + \left\langle \xi_j \frac{\partial \xi_m}{\partial x_j} \frac{\partial \overline{u}_i}{\partial x_m} \right\rangle + \left\langle \left(\boldsymbol{\xi} \wedge \boldsymbol{\omega}^\ell \right)_i \right\rangle \\ &\quad + \left\langle \frac{\partial \xi_j}{\partial x_i} \left(\boldsymbol{\Omega}^{(c)} \wedge \boldsymbol{\xi} \right)_j \right\rangle + \mathcal{O}\left(a^3\right) . \end{aligned} \tag{60}$$

[g]For details, see Dingemans (2001).

[h]This equality can be shown by noting that $\overline{D}^L \left(\frac{1}{2} \xi_j \xi_j \right) = \xi_j \overline{D}^L \xi_j = \xi_j u_j^\ell$ because of the definition of $\boldsymbol{u}^\ell$, e.g., see Dingemans (1997, Eq. (2.577d)).

Because we can consider $\partial\xi_j/\partial x_i$ to be of second order in a (strictly only $\operatorname{div}\boldsymbol{\xi} = \mathcal{O}\left(a^2\right)$, not its individual components) and recognising the extreme smallness of the Coriolis rotation $\left|\Omega^{(c)}\right|$, it seems to be permitted to ignore the last term in (60). For the moment we keep this term still in the considerations.

Then we also have:

$$\overline{u}_i^L - \overline{\mathsf{P}}_i^L = \overline{u}_i - \frac{1}{2}\overline{\xi_j\xi_k}\frac{\partial^2\overline{u}_i}{\partial x_j\partial x_k} - \left\langle \xi_j \frac{\partial\xi_m}{\partial x_j}\frac{\partial\overline{u}_i}{\partial x_m}\right\rangle - \left\langle \left(\boldsymbol{\xi}\wedge\boldsymbol{\omega}^\ell\right)_i\right\rangle$$
$$- \left\langle \frac{\partial\xi_j}{\partial x_i}\left(\boldsymbol{\Omega}^{(c)}\wedge\boldsymbol{\xi}\right)_j\right\rangle + \mathcal{O}\left(a^3\right) . \tag{61}$$

Expressions for $\left(\partial\overline{u}_j^L/\partial x_j\right)\left(\overline{u}_j^L - \overline{\mathsf{P}}_j^L\right)$ and $\overline{D}^L\left(\overline{u}_i^L - \overline{\mathsf{P}}_i^L\right)$ then follow easily. These expressions can be simplified further by noting that $\left|\overline{\boldsymbol{u}}^S\right| = \mathcal{O}\left(a^2\right)$, $\left|\boldsymbol{\omega}'\right| = \mathcal{O}(a)$ and $\left|\boldsymbol{\omega}''\right| = \mathcal{O}(a)$. We introduce the term $\overline{\mathcal{U}}_i^{(\omega)}$ denoting the vorticity contributions

$$\overline{\mathcal{U}}_i^{(\omega)} = \left\langle \left\{\boldsymbol{\xi}\wedge\left(\boldsymbol{\omega}'+\boldsymbol{\omega}''\right)\right\}_i\right\rangle + \left\langle \frac{\partial\xi_j}{\partial x_i}\left(\boldsymbol{\Omega}^{(c)}\wedge\boldsymbol{\xi}\right)_j\right\rangle , \tag{62a}$$

and the abbreviation $\mathcal{W}_i$ by[i]

$$\overline{\mathcal{W}}_i = \overline{u}_i - \frac{1}{2}\overline{\xi_j\xi_k}\frac{\partial^2\overline{u}_i}{\partial x_j\partial x_k} - \left\langle \xi_j \frac{\partial\xi_m}{\partial x_j}\frac{\partial\overline{u}_i}{\partial x_m}\right\rangle \equiv \overline{u}_i + \mathcal{V}_i . \tag{62b}$$

We also note that (see also Eq. (52))

$$\overline{u}_j^S\frac{\partial\overline{u}_i}{\partial x_j} = \overline{u}_j^S\frac{\partial\overline{u}_j}{\partial x_i} - \left(\overline{\boldsymbol{u}}^S\wedge\overline{\boldsymbol{\omega}}\right)_i . \tag{62c}$$

Because also

$$\overline{u}_j\frac{\partial\overline{\mathcal{W}}_i}{\partial x_j} = \overline{u}_j\frac{\partial\overline{\mathcal{W}}_j}{\partial x_i} - \left(\overline{\boldsymbol{u}}\wedge\operatorname{curl}\overline{\boldsymbol{\mathcal{W}}}\right)_i , \tag{62d}$$

we obtain for the two terms of the GLM equation (33a):

$$\overline{D}^L\left(\overline{u}_i^L - \overline{\mathsf{P}}_i^L\right) + \frac{\partial\overline{u}_j^L}{\partial x_i}\left(\overline{u}_j^L - \overline{\mathsf{P}}_j^L\right) = \frac{\partial}{\partial t}\left(\overline{\mathcal{W}}_i - \overline{\mathcal{U}}_i^{(\omega)}\right)$$
$$+ \frac{\partial}{\partial x_i}\left[\overline{u}_j\left(\overline{\mathcal{W}}_j + \overline{u}_j^S - \overline{\mathcal{U}}_j^{(\omega)}\right)\right] - \left(\overline{\boldsymbol{u}}^S\wedge\overline{\boldsymbol{\omega}}\right)_i - \left(\overline{\boldsymbol{u}}\wedge\operatorname{curl}\left(\overline{\boldsymbol{\mathcal{W}}} - \overline{\boldsymbol{\mathcal{U}}}^{(\omega)}\right)\right)_i . \tag{63}$$

[i]We notice that $\overline{\mathcal{W}}_i$ resembles the Lagrangian velocity $\overline{u}_i^L$, but is not the same. We have

$$\overline{\mathcal{W}}_i = \overline{u}_i^L - \left\langle \xi_j\frac{\partial u_i'}{\partial x_j}\right\rangle - \left\langle \xi_j\frac{\partial}{\partial x_j}\left(\xi_m\frac{\partial\overline{u}_i}{\partial x_m}\right)\right\rangle .$$

To rewrite the GLM mean-current equation in terms of Eulerian quantities we still have to rewrite the generalised pressure in Eulerian quantities.

The GLM expression for the generalised pressure $\overline{\pi}^L$ has been given in Eq. (33b). First we Taylor the term $\left\langle \boldsymbol{u}^\ell \cdot \boldsymbol{u}^\ell \right\rangle$ and find that

$$\begin{aligned}\left\langle \boldsymbol{u}^\ell \cdot \boldsymbol{u}^\ell \right\rangle &= \langle \boldsymbol{u} \cdot \boldsymbol{u} \rangle + 2 \left\langle u_i \xi_j \frac{\partial u_i}{\partial x_j} \right\rangle + \left\langle \xi_j \xi_m \frac{\partial u_i}{\partial x_j} \frac{\partial u_i}{\partial x_m} \right\rangle \\ &\quad + \left\langle \xi_j \xi_k u_i \frac{\partial^2 u_i}{\partial x_j \partial x_k} \right\rangle + \mathcal{O}\left(a^3\right) . \end{aligned} \tag{64}$$

We note that it is possible to write $\left(\boldsymbol{\Omega}^{(c)} \wedge \boldsymbol{\xi}\right) \cdot \boldsymbol{u}^\xi = \left(\boldsymbol{\xi} \wedge \boldsymbol{u}^\xi\right) \cdot \boldsymbol{\Omega}^{(c)}$. Because the Coriolis vector $\boldsymbol{\Omega}^{(c)}$ can be considered to be a constant vector on the spatial scales we consider, we then thus have:

$$\left\langle \left(\boldsymbol{\Omega}^{(c)} \wedge \boldsymbol{\xi}\right) \cdot \boldsymbol{u}^\xi \right\rangle = \left\langle \left(\boldsymbol{\xi} \wedge \boldsymbol{u}^\xi\right) \right\rangle \cdot \boldsymbol{\Omega}^{(c)} . \tag{65}$$

We also find $$\left\langle \boldsymbol{\xi} \wedge \boldsymbol{u}^\xi \right\rangle = \langle \boldsymbol{\xi} \wedge \boldsymbol{u} \rangle + \langle \boldsymbol{\xi} \wedge (\boldsymbol{\xi} \cdot \mathrm{grad})\, \boldsymbol{u} \rangle + \mathcal{O}\left(a^3\right) . \tag{66}$$

Some simplification is still possible by splitting the Eulerian velocity in its mean and a perturbation, $\boldsymbol{u} = \overline{\boldsymbol{u}} + \boldsymbol{u}'$ with $\boldsymbol{u}' = \mathcal{O}(a)$. The generalised pressure $\overline{\pi}^L$ then simplifies to:

$$\begin{aligned}\overline{\pi}^L &= \frac{1}{\rho}\left[\overline{p} + \left\langle (\boldsymbol{\xi} \cdot \mathrm{grad})\, p' \right\rangle + \frac{1}{2}\overline{\xi_j \xi_k} \frac{\partial^2 \overline{p}}{\partial x_j \partial x_k}\right] - \frac{1}{2}\left[\overline{\boldsymbol{u}} \cdot \overline{\boldsymbol{u}} + \left\langle \boldsymbol{u}' \cdot \boldsymbol{u}' \right\rangle\right] \\ &\quad - \left\langle \xi_j \frac{\partial}{\partial x_j}\left(\overline{u}_m u_m'\right) \right\rangle - \frac{1}{2}\left\langle \xi_j \xi_k \frac{\partial \overline{u}_m}{\partial x_j} \frac{\partial \overline{u}_m}{\partial x_k} \right\rangle - \frac{1}{2}\overline{u}_m \left\langle \xi_j \xi_k \frac{\partial^2 \overline{u}_m}{\partial x_j \partial x_k} \right\rangle \\ &\quad + \left\langle \boldsymbol{\xi} \wedge \boldsymbol{u}' \right\rangle \cdot \boldsymbol{\Omega}^{(c)} + \langle \boldsymbol{\xi} \wedge (\boldsymbol{\xi} \cdot \mathrm{grad})\, \overline{\boldsymbol{u}} \rangle \cdot \boldsymbol{\Omega}^{(c)} + \mathcal{O}\left(a^3\right) . \end{aligned} \tag{67}$$

We notice that in applications in the coastal zone, we usually measure $(x, y) \equiv (x_1, x_2)$ on wave length scale. Because the current varies over much larger scales, we can write $\overline{\boldsymbol{u}} = \overline{\boldsymbol{u}}(\beta x_1, \beta x_2, x_3, \beta t)$ with $\beta \ll 1$ being an ordering parameter which we suppose to be of the same order as a. For $\overline{p}$ we also have $\overline{p} = \overline{p}(\beta x_1, \beta x_2, x_3, \beta t)$. When then terms of $\mathcal{O}\left(a^3, \beta a^2\right)$ are neglected, relation (67) simplifies considerably. Writing $x_3 = z$ we obtain:

$$\begin{aligned}\overline{\pi}^L &= \frac{1}{\rho}\left[\overline{p} + \left\langle (\boldsymbol{\xi} \cdot \mathrm{grad})\, p' \right\rangle + \frac{1}{2}\overline{\xi_3^2} \frac{\partial^2 \overline{p}}{\partial z^2}\right] - \frac{1}{2}\left[\overline{\boldsymbol{u}} \cdot \overline{\boldsymbol{u}} + \left\langle \boldsymbol{u}' \cdot \boldsymbol{u}' \right\rangle\right] \\ &\quad - \left\langle \xi_3 \frac{\partial}{\partial z}\left(\overline{u}_m u_m'\right) \right\rangle - \frac{1}{2}\left\langle \xi_3^2 \left(\frac{\partial \overline{u}_m}{\partial z}\right)^2 \right\rangle - \frac{1}{2}\overline{u}_m \left\langle \xi_3^2 \frac{\partial^2 \overline{u}_m}{\partial z^2} \right\rangle \\ &\quad + \left\langle \boldsymbol{\xi} \wedge \boldsymbol{u}' \right\rangle \cdot \boldsymbol{\Omega}^{(c)} + \left\langle \boldsymbol{\xi} \wedge \xi_3 \frac{\partial \overline{\boldsymbol{u}}}{\partial z} \right\rangle \cdot \boldsymbol{\Omega}^{(c)} + \mathcal{O}\left(a^3, \beta a^2\right) . \end{aligned} \tag{68}$$

We then finally obtain the Eulerian form of the GLM equation (33) in the following form, in which we did not substitute $\overline{\pi}^L$ for simplicity:

$$\frac{\partial}{\partial t}\left(\overline{u}_i + \overline{\mathcal{V}}_i - \overline{\mathcal{U}}_i^{(\omega)}\right) + \frac{\partial}{\partial x_i}\left[\overline{u}_j\left(\overline{u}_j + \overline{u}_j^S + \overline{\mathcal{V}}_j - \overline{\mathcal{U}}_j^{(\omega)}\right)\right] + \frac{\partial \overline{\pi}^L}{\partial x_i} + g_i + \mathcal{S}_i = 0\ , \tag{69a}$$

where the approximation $\overline{\rho}^L \approx \rho_0$ has been used and the term $\mathcal{S}_i$ consists of vorticity contributions, and is given by:

$$\mathcal{S}_i = -\left(\overline{\boldsymbol{u}}^S \wedge \overline{\boldsymbol{\omega}}\right)_i - \left(\overline{\boldsymbol{u}} \wedge \operatorname{curl}\left(\overline{\boldsymbol{u}} + \overline{\boldsymbol{\mathcal{V}}} - \overline{\boldsymbol{\mathcal{U}}}^{(\omega)}\right)\right)_i + 2\left(\boldsymbol{\Omega}^{(c)} \wedge \left(\overline{\boldsymbol{u}} + \overline{\boldsymbol{u}}^S\right)\right)_i\ , \tag{69b}$$

where $\boldsymbol{\mathcal{V}}$ and $\boldsymbol{\mathcal{U}}^{(\omega)}$ are defined in (62b) and (62a).

Notice that with definitions (62a) and (62b) for $\overline{\boldsymbol{\mathcal{U}}}^{(\omega)}$ and $\overline{\boldsymbol{\mathcal{V}}}$ the mean pseudo momentum $\overline{\mathbf{P}}^L$ given in (60) can also be written as

$$\overline{\mathbf{P}}^L = \boldsymbol{u}^S - \overline{\boldsymbol{\mathcal{V}}} + \overline{\boldsymbol{\mathcal{U}}}^{(\omega)}\ , \tag{70}$$

showing that $-\overline{\boldsymbol{\mathcal{V}}} + \overline{\boldsymbol{\mathcal{U}}}^{(\omega)}$ indicates the difference between the pseudo momentum and the Stokes drift.

We now call Eq. (69) the *extended* Craik-Leibovich equation.

6.1. *Discussion of the extended CL equation*

Comparing Eq. (69) with the standard CL equation (37), we see that, apart from the Coriolis terms, also other differences appear. Neglecting the Coriolis terms, a number of differences with CL equation (37) remain. Most of these differences occur because we consider the velocity $\boldsymbol{u}$ to be of $\mathcal{O}(1)$, while in the original use of the Craik-Leibovich equation the currents were supposed to be much smaller than the the wave contributions, so, $|\boldsymbol{u}| = \mathcal{O}\left(a^2\right)$.

Consider at first the term $\overline{\boldsymbol{\mathcal{U}}}^{((\omega))}$, which now reads, in absence of Coriolis, as $\boldsymbol{\omega}' + \boldsymbol{\omega}'' = \langle \boldsymbol{\xi} \wedge \operatorname{curl}\left(\boldsymbol{u}'\right)\rangle + \langle \boldsymbol{\xi} \wedge \operatorname{curl}\left[(\boldsymbol{\xi} \cdot \operatorname{grad})\,\overline{\boldsymbol{u}}\right]\rangle$. The first term is the rotation of water waves, and this is usually taken small in some sense. It is, in fact, not easy to decide on the smallness of rotation of water waves. There exist two good definitions, due to Truesdell (1953, 1954), a kinematic one and a dynamic one. For a discussion see also Dingemans (1977, §5.11.12). Neglecting this contribution, we still have the rotation due to the shear of the current, $\boldsymbol{\omega}''$. There is no clear reason why this term could be neglected, and therefore we argue that this term should be kept in the CL equation.

Consider now the term $\boldsymbol{\mathcal{V}}$. From its definition (62b) and expression (70) for the pseudo momentum, it is clear that the contribution $\boldsymbol{\mathcal{V}}$ can not be simply ignored, as it is nominally of $\mathcal{O}\left(a^2\right)$. In many applications it can be supposed that the current is slowly varying in horizontal space. In that case one has $\overline{\boldsymbol{u}}(\beta x, \beta y, z, t)$ with $\beta \ll 1$, say $\beta = \mathcal{O}(a)$. Then the derivatives to the horizontal coordinates can be neglected as they are $\mathcal{O}\left(a^2\beta^2\right)$ for the first term of $\boldsymbol{\mathcal{V}}$ and $\mathcal{O}\left(a^2\beta\right)$ for the second term. The contributions to the vertical derivatives remain.

In the second term of the left-hand side of (69a) we have the contribution $\operatorname{grad}\left(\overline{\boldsymbol{u}} \cdot \overline{\boldsymbol{u}}^S\right)$. This contribution is also present in Leibovich's equation (36), but not in (37). This term $\operatorname{grad}\left(\boldsymbol{u} \cdot \boldsymbol{u}^S\right)$ is combined with the gradient of the first and second term of the second line in expression (67) for the generalised pressure. With the definition (5) for the Stokes drift then is obtained:

$$\frac{\partial}{\partial x_i}\left(\overline{u}_m\overline{u}_m^S - \left\langle \xi_j \frac{\partial}{\partial x_j}\left(\overline{u}_m u_m'\right)\right\rangle - \frac{1}{2}\overline{u}_m\left\langle \xi_j\xi_k\frac{\partial^2\overline{u}_m}{\partial x_j \partial x_k}\right\rangle\right)$$
$$= -\frac{\partial}{\partial x_i}\left(\left\langle \xi_j u_m'\right\rangle \frac{\partial \overline{u}_m}{\partial x_j}\right) . \tag{71}$$

In the second term of Eq. (69a) we have the term $\operatorname{grad}(\overline{\boldsymbol{u}} \cdot \overline{\boldsymbol{u}})$. A similar term is present in the first line of expression (67). Combined we have

$$\frac{1}{2}\operatorname{grad}(\overline{\boldsymbol{u}} \cdot \overline{\boldsymbol{u}}) = (\overline{\boldsymbol{u}} \cdot \operatorname{grad})\,\overline{\boldsymbol{u}} + \overline{\boldsymbol{u}} \wedge \operatorname{curl}\overline{\boldsymbol{u}} . \tag{72}$$

The term $\overline{\boldsymbol{u}} \wedge \operatorname{curl}\overline{\boldsymbol{u}}$ disappears against a similar term in the term $\boldsymbol{\mathcal{S}}$.

With these simplifications, we have the following momentum equation with the velocity variable $\overline{\boldsymbol{u}}$:

$$\frac{\partial\overline{\boldsymbol{u}}}{\partial t} + (\overline{\boldsymbol{u}} \cdot \operatorname{grad})\,\overline{\boldsymbol{u}} + \operatorname{grad}\left(\frac{1}{\rho}p - \frac{1}{2}\left\langle \boldsymbol{u}' \cdot \boldsymbol{u}'\right\rangle\right) + \boldsymbol{g} + \overline{\boldsymbol{u}}^S \wedge \operatorname{curl}\overline{\boldsymbol{u}} = \boldsymbol{\mathcal{T}} , \tag{73a}$$

with $\boldsymbol{\mathcal{T}}$ given by

$$\boldsymbol{\mathcal{T}} = \overline{\boldsymbol{u}} \wedge \operatorname{curl}\left(\overline{\boldsymbol{\mathcal{V}}} - \overline{\boldsymbol{\mathcal{U}}}^{(\omega)}\right) - \operatorname{grad}\left(\frac{1}{\rho}\left[\left\langle(\boldsymbol{\xi}\cdot\operatorname{grad})\,p'\right\rangle + \frac{1}{2}\overline{\xi_j\xi_k}\frac{\partial^2\overline{p}}{\partial x_j\partial x_k}\right]\right)$$
$$-\frac{\partial}{\partial t}\left(\overline{\boldsymbol{\mathcal{V}}} - \overline{\boldsymbol{\mathcal{U}}}^{((\omega))}\right) - \operatorname{grad}\left[\overline{\boldsymbol{u}}\left(\overline{\boldsymbol{\mathcal{V}}} - \overline{\boldsymbol{\mathcal{U}}}^{((\omega))}\right)\right] + \frac{\partial}{\partial x_i}\left(\left\langle \xi_j u_m'\right\rangle \frac{\partial\overline{u}_m}{\partial x_j}\right) , \tag{73b}$$

where we kept writing $\overline{\boldsymbol{\mathcal{U}}}^{((\omega))}$ instead of $\overline{\boldsymbol{\omega}}''$ for simplicity.

We now notice the following differences with the CL equation (37):

- In (67) a second-order correction to the pressure is found. How important this term is for applications is hard to say. This term is expected to be not important for Langmuir circulations, because the instability mechanism does not depend on the pressure. However, it is important to get a correct prediction of the mean current. The importance of this term has to be investigated further.
- Three terms of $\boldsymbol{\mathcal{T}}$ have contributions with $\overline{\boldsymbol{\mathcal{V}}} - \overline{\boldsymbol{\mathcal{U}}}^{((\omega))}$, which is the difference between the pseudo momentum and the Stokes drift. It is not possible to say if these terms are important, they appear as second-order terms from

an asymptotic evaluation. However, as was the case with the current contribution to the Stokes drift, which also was of second order, but proved to be insignificant from numerical examples. It should be investigated whether the difference $\overline{\mathbf{P}}^L - \overline{\boldsymbol{u}}^S$ is also insignificant. This will be investigated in a future project. For the situation that the current is slowly varying in horizontal space *and* the shear is very small so that the vertical derivatives are also very small, the difference may be neglected. However, we do not want to restrict ourselves to currents with small shear.

It should also be noted that we suppose the mean current field of $\mathcal{O}(1)$, while in the original CL-equation investigations the current was taken to be of $\mathcal{O}\left(a^2\right)$. In that situation we simply have $\overline{\mathbf{P}}^L - \overline{\boldsymbol{u}}^S = 0$.

- The last term of $\boldsymbol{\mathcal{T}}$ remains. The importance of this term has also to be evaluated numerically.

The GLM mean-current equation has been written in Eulerian variables up to second order of the wave amplitude. A number of extra terms compared to the originally CL equation were found, of which the importance is not immediately clear. When some second-derivative terms are indeed important, then this has a major influence on the numerical evaluation of the equation. Because of such reasons it is advantageous to keep computing in the GLM-defined equations.

To obtain the results in Eulerian quantities, Tayloring has been used exclusively. It was mentioned that the Stokes drift with second-oder vertical derivatives gave problems when also wind was accounted for. It might be useful to consider different approximation schemes to relate GLM quantities to their corresponding Eulerian quantities, such as Padé approximations.

7. General discussion

We gave a few reflections on the use and applicability of the GLM method. Much more could be said of it. At first the expression for Stokes drift was derived, and in our opinion this is a very simple derivation. The application to water waves was subsequently discussed. We especially focused on the influence of waves on currents, not the usual wave-current refraction which accounts for the current effect on the waves. It should be mentioned that the influence of waves on currents is especially useful for applications in the coastal zone. There the movement of sediment has to be determined by the combined effect of waves and currents. It was argued that it is especially useful to do current computation in the GLM frame. The CL equation and also the GLM formulation were found to be able to explain experiments with currents and waves. The main effect was due to the presence of the vortex force in the CL equation, which force is also present in the GLM current equation.

Subsequently, a third, nonlinear, instability mechanism in the Craik-Leibovich equation was described for secondary flow problems of Langmuir-type. By supposing that the mean current (in which wave-effects are incorporated) can be written as a basic current and a (small) perturbation, the latter give secondary flows of Langmuir-type when instability occurs. It is found that the Reynolds stresses due to the presence of these secondary flows can be determined *analytically*, when instability occurs. Then it is possible to compute the current field numerically with the same large numerical mesh as is the case when these Reynolds stresses are absent.

Lastly, an Eulerian description in which also Coriolis effect was accounted for, was derived up to third order in the wave amplitude. This result, Eq. (69), was used by Radder (2001) and Dingemans (2001) in the air above the waves to find an expression for the wind stress. It should be mentioned that the analysis for this case, follows closely the case of §5, now with the mean velocity variable given by $\boldsymbol{\mathcal{U}} = \boldsymbol{U} + \overline{\boldsymbol{\mathcal{V}}} - \overline{\boldsymbol{\mathcal{U}}}^{((\omega))}$. Later Radder (2002) extended the equation with effects of thermal stratification.

The extended CL equation (69) and the corresponding expression (67) for the generalised pressure are different from the usual expressions of the Craik-Leibovich equation. Our $\mathcal{O}(1)$ assumption for the mean velocity explains almost all differences.

The main purpose of this paper, apart from the homage to Philip Liu, is to focus attention to the very important method of the Generalised Lagrangian Mean of Andrews and McIntyre (1978a,b). It should also be mentioned that this methods permits one to define wave action also for non-slowly-varying wave fields, as is necessary with the present one due to Whitham and Lighthill.

Acknowledgments

Fifteen years ago I started with a small project, together with my colleague Rob Uittenbogaard to look into the possibilities of GLM formulations and we wrote a small note for Rijkswaterstaat, where Anne Radder was our contact. Later the present author and Anne Radder worked very closely together, where often Anne had some ideas and I had to check the validity of his ideas. I learned much of this cooperation. Also drafts of the present paper were discussed by both Rob and Anne. Thank you both, Anne and Rob.

References

1. Andrews, D.G. and McIntyre, M.E., 1978a. An exact theory of nonlinear waves on a Lagrangian mean flow. *J. Fluid Mechanics* **89(4)**, pp. 609-646.
2. Andrews, D.G. and McIntyre, M.E., 1978b. On wave action and its relatives. *J. Fluid Mechanics* **89(4)**, pp. 647-664.
3. Ardhuin, F., Rascle, N. and Belibassakis, K., 2007. Explicit wave-averaged primitive equations using a Generalized Lagrangian Mean. *Preprint* http://arXiv/physics/0702067v1, 42 pp., 8 Feb 2007. Version 3 appeared at 31 July 2007. The final paper appeared in *Ocean Modelling* **20**, pp. 35-60 (2008).
4. Beauchamp, K.G., 1975. *Walsh Functions and Their Applications.* Academic Press, 1975, 236 pp.
5. Cox, S.M., 1997. Onset of Langmuir circulation when shear flow and Stokes drift are not parallel. *Fluid Dynamics Research* **19**, pp. 149-167.

6. Craik, A.D.D. and Leibovich, S., 1976. A rational model for Langmuir circulations. *J. of Fluid Mech.* **73(3)**, pp. 401-426.
7. Csanady, G.T., 1994. Vortex pair model of Langmuir circulation. *J. of Marine Res.* **52(4)**, pp. 559-581.
8. Dingemans, M.W., 1997. *Water Wave Propagation over Uneven Bottoms.* World Scientific, Singapore, 967 pp.
9. Dingemans, M.W., 1999. 3D wave-current modelling; a model for secondary circulations. *WL|Delft Hydraulics Report* Z2612, 68 pp.
10. Dingemans, M.W., 2001. 3D wave-current modelling; Part 2: extension and application to air-sea interaction. *WL|Delft Hydraulics Report* X251, Aug. 2001, 87 pp.
11. Dingemans, M.W., van Kester, J.A.Th.M., Radder, A.C. and Uittenbogaard, R.E., 1996. The effect of the CL-vortex force in 3D wave-current interaction. *25th Int. Coastal Engineering Conference*, Orlando, Florida, pp. 4821-4832.
12. Dingemans, M.W. and Radder, A.C., 2000. The use of the CL equation as a model for secondary circulations. *Proc. 15th Int. Workshop on Water Waves and Floating Bodies*, Editors T. Miloh and G. Zilman, Cesarea, Israel, March 2000, pp. 36-39.
13. Dingemans, M.W., Radder, A.C. and de Vriend, H.J., 1987. Computation of the driving forces of wave-induced currents. *Coastal Engineering* **11**, pp. 539-563.
14. Drazin, P.G. and Reid, W.H., 1981. *Hydrodynamic Stability.* Cambridge University Press, 527 pp.
15. Faller, A.J. and Caponi, E.A., 1978. Laboratory studies of wind-driven Langmuir circulations. *J. Geophys. Res.* **83(C8)**, pp. 3617-3633.
16. Fredsøe, J. and Deigaard, R., 1992. *Mechanics of Coastal Sediment Transport.* World Scientific, Singapore, 369 pp.
17. Groeneweg, J. and Battjes, J.A., 2003. Three-dimensional wave effects on a steady current. *J. of Fluid Mech.* **478**, pp. 325-343.
18. Groeneweg, J. and Klopman, G., 1998. Changes of mean velocity profiles in the combined wave-current motion described in a GLM formulation. *J. of Fluid Mech.* **370**. pp. 271-296.
19. Kemp, P.H. and Simons, R.R., 1982. The interaction between waves and a turbulent current: waves propagating with the current. *J. Fluid Mech.* **116**, pp. 227-250.
20. Kemp, P.H. and Simons, R.R., 1983. The interaction between waves and a turbulent current: waves propagating against the current. *J. Fluid Mech.* **130**, pp. 73-89.
21. Klopman, G., 1994. Vertical structure of the flow due to waves and currents; Laser-Doppler flow measurements for waves following or opposing a current. *Delft Hydraulics* Report H840.30, Part II, Feb. 1994.
22. Klopman, G., 1997. Secondary circulation of the flow due to waves and current. *Delft Hydraulics*, Report Z2249.
23. Leibovich, S., 1977. Convective instability of stably stratified water in the ocean. *J. of Fluid Mech.* **82(3)**, 1977, pp. 561-581.

24. Leibovich, S., 1980. On wave-current interaction theories of Langmuir circulations. *J. of Fluid Mechanics* **99(4)**, pp. 715-724.
25. Leibovich, S., 1983. The form and dynamics of Langmuir circulations. *Ann. Rev. of Fluid Mech.* **15**, pp. 391-427.
26. McIntyre, M.E., 1980. Towards a Lagrangian-mean description of stratospheric circulations and chemical transports. *Phil. Trans. Roy. Soc. London* **A296**, pp. 129-148.
27. Nielsen, P., 1979. Some basic concepts of wave sediment transport. *Ph. D. Thesis Techn. Univ. of Denmark*, ISVA Series Paper **20**, 160 pp.
28. Pedlosky, J., 1979. *Geophysical Fluid Dynamics.* Springer Verlag, New York etc., 624 pp.
29. Phillips, O.M., 1966. *The Dynamics of the Upper Ocean.* Cambridge University Press, 261 pp.
30. Phillips, O.M., 1977. *The Dynamics of the Upper Ocean.* Second Edition. Cambridge University Press, 336 pp.
31. Pipes, L.A. and Harvill, L.R., 1970. *Applied Mathematics for Engineers and Physicists.* Third Edition. McGraw-Hill Kogakusha, Ltd., Tokyo, 1015 pp.
32. Radder, A.C., 2001. Wind stress induced by sea waves. Report RIKZ/OS/2001.133X, Rijkswaterstaat/RIKZ, the Netherlands, 33 pp.
33. Radder, A.C., 2002. Wind stress induced by sea waves; Part 2: Effects of thermal stratification on wave growth by wind. Report RIKZ/OS/2002.125X, Rijkswaterstaat/RIKZ, the Netherlands, 33 pp.[j]
34. Soulsby, R.L., 1995. Bed shear-stress due to combined waves and currents. *Mast G8M Proceedings*, Gdansk, September 1995, pp. 4.20-4.23.
35. Soulsby, R.L., 1997. *Dynamics of Marine Sands; a manual for practical applications*, Thomas Telford Publ., London, 249 pp.
36. Soulsby, R.L., Hamm, L., Klopman, G., Myrhaug, D., Simons, R.R. and Thomas, G.P., 1993. Wave-current interaction within and outside the bottom boundary layer. *Coastal Engineering* **21**, pp. 41-69.
37. Stuart, J.T., 1958. On the non-linear mechanics of hydrodynamic stability. *J. Fluid Mech.* **4**, pp. 1-21.
38. Truesdell, 1953. Two measures of vorticity. *J. Rational Mechanics and Analysis* **2**, pp. 173-217.
39. Truesdell, C., 1954. *The Kinematics of Vorticity.* Indiana University Publications, Science Series No. 9, 232 pp.
40. Uittenbogaard, R. and Klopman, G., 2001. Numerical simulation of wave-current driven sediment transport. *Coastal Sediments* 2001, Lund, 11 pp.

[j]The reports of Radder (2001, 2002) and Dingemans (1999, 2001) and the paper of Dingemans and Radder (2000) can be sent as a pdf-file; requests to be sent to Dingemans.

HYDRAULIC PERFORMANCE OF A SUBMERGED WAVE ABSORBER FOR COASTAL PROTECTION

H. OUMERACI

Leichtweiss-Institute for Hydraulic Engineering, Technical University Braunschweig and Coastal Research Centre (FZK), Hannover, Germany
E-mail: h.oumeraci@tu-bs.de

G. KOETHER

formerly Leichtweiss-Institute for Hydraulic Engineering
Technical University Braunschweig, Germany

The hydraulic performance of a new reef structure for coastal protection made of submerged thin walls with progressively decreasing porosity (progressive wave absorber concept) has been studied experimentally and theoretically within a more extensive research programme on the hydraulic performance and wave loads of submerged and surface piercing wave absorbers, including numerical modelling. The paper primarily addresses the results of comprehensive large-scale model tests and the theoretical model which has been developed from these results for the prediction of the wave reflection, wave transmission and energy dissipation performance of the new reef structure based on the wave absorber concept.

Before embarking into the more complex multiple-wall reef system, the hydraulic functioning and the associated processes are first systematically investigated for a single submerged wall. As a main result, empirical formulae and a potential flow model, together with a new structure parameter which is required for the matching conditions to account for the energy losses (drag, inertia and vortex losses) at the wall, are developed for the prediction of the hydraulic performance.

Using a two- and a three-wall reef system, the basic difference in the hydraulic functioning as compared with a single-wall reef is demonstrated. In the latter case, the energy dissipation capability is limited and a strong link exists between the reflection and transmission performance, meaning that a reduction of wave transmission necessarily implies an increase in wave reflection. Unlike the single-wall reef, a much higher energy dissipation can be achieved by a multiple-wall reef system. Both wave transmission and reflection can be significantly decreased by a proper selection of the spacing, porosity and submergence depth of the constitutive permeable walls.

Finally, the potential flow model, including a new structure parameter, is extended for multiple-wall reef systems and irregular waves. Example calculations to illustrate the model capability are also discussed. An outlook for the application of the new reef system for the protection against tsunami is also briefly given.

1. Motivation, Objective and Methodology

More than 45% of the 500,000 km world's coastline is affected by permanent erosion. Of the 100,000 km sandy coasts, 70% have been characterized by net erosion over the last decades. One of the most difficult tasks in coastal engineering is to maintain these coasts for recreational purposes without affecting the marine landscape and environment. Therefore, conventional and innovative reef concepts are increasingly getting more attractive for coastal protection, especially at non-tidal coastlines.

An interesting alternative to the existing reef concepts [12], which consists of a submerged progressive wave absorber composed of two or more submerged permeable vertical screens with predetermined porosity and spacing, has been investigated theoretically and experimentally (Fig. 1).

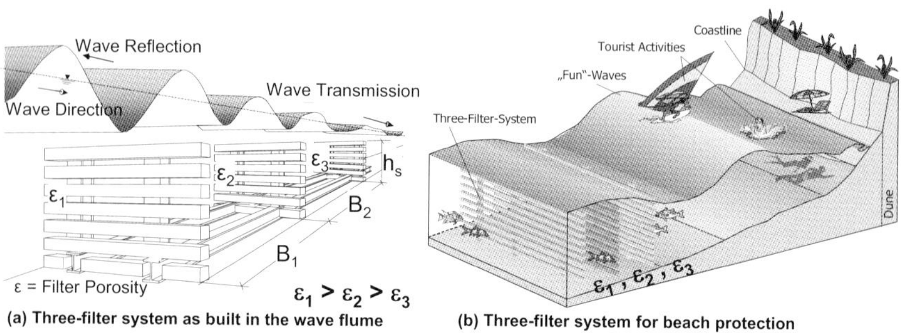

Figure 1. Three-Wall Reef Based on the Progressive Wave Absorber Concept

These investigations were performed in the framework of an extensive joint research programme of both Technical Universities of Braunschweig and Berlin, supported by the German Federal Ministry for Education and Research (BMBF) and including small- and large-scale model testing, theoretical investigations and numerical modelling to study all aspects associated with the hydraulic performance and the wave loading of this new reef concept [11].

In this paper, focus will be put only on the hydraulic performance and the associated processes. The results related to the wave loading are also available [11] and will be published in a forthcoming paper. Numerical modelling, using a commercially available RANS-VOF model, has also been addressed [3].

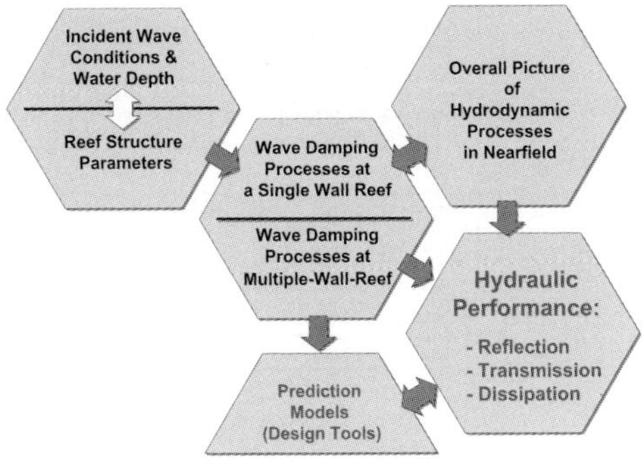

Figure 2. Main Objectives of the Study

Given the wave conditions and the reef structure parameters the main objectives of this paper are therefore (Fig. 2):

(i) to describe the key results of large-scale model tests related to the hydraulic performance of the new reef concept, including wave reflection, wave transmission and energy dissipation;

(ii) to develop an analytical model as a design tool for the prediction of the hydraulic performance which depends on the incident wave parameters as well as on reef parameters such as the submergence depth, the number, porosity and spacing of the constitutive walls. For this purpose, a detailed knowledge of the hydrodynamic processes associated with the wave damping induced by the single walls as well as by the entire reef system will be needed. This overall picture of the processes involved and the way they affect the hydraulic performance will be obtained primarily from large-scale model tests;

(iii) to achieve these objectives a methodology has been adopted which is essentially based on (Fig. 3):
- large-scale model testing in the Large Wave Flume (GWK) of the Joint Coastal Research Centre (FZK) of both Universities Hannover and Braunschweig;
- theoretical investigations and modelling.

First, a submerged single wall with a porosity of 0 to 20% is investigated in the flume using regular waves. Based on the experimental results, an analytical model for the hydraulic performance is developed and validated for various incident wave conditions and structure parameters (number of walls, spacing, porosity and submergence depth).

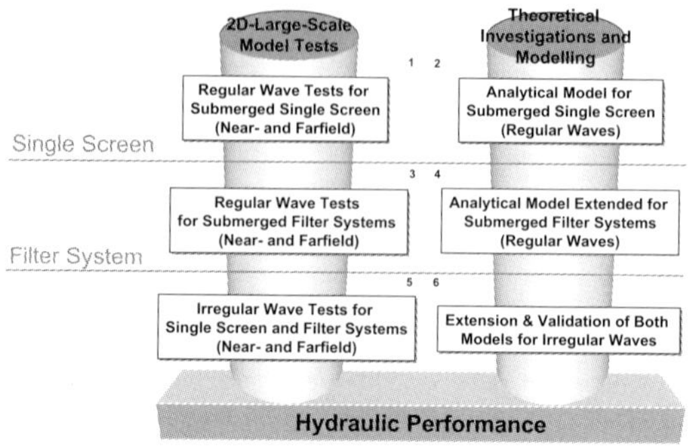

Figure 3. Methodology of the Study

Second, submerged two- and three-wall reefs are then tested using the same regular waves as in the first test series. Based on the experimental results, the single-wall model is extended to account for a wave absorber made of successive submerged walls with progressively decreasing porosity.

Third, irregular wave tests are performed for submerged single walls and progressive wave absorbers. The analytical model is then extended to account for irregular waves.

Making use of the results of previously performed model tests on beach profile development under storm waves [9], the same wave conditions are reproduced for the same beach protected by a two-wall reef system. Comparing the beach behaviour with and without protection, the efficiency of the reef system against beach erosion has been demonstrated. The results of this demonstration were summarized in a previous paper [10] and will not be duplicated here. Further details are reported in [6] and [11].

2. Experimental Set-Up, Test Programme and Procedure

The main scale model tests were performed in the Large Wave Flume (GWK) of the Joint Coastal Research Centre of both Universities of Hannover and Braunschweig (FZK) with a width of 5 m, a depth of 7 m and a length of 325 m. To allow systematic variations of the structure parameters (number, submergence depth, spacing and porosity of the constitutive walls) and of the wave conditions, and to determine a set of optimised model configurations to be finally tested in the Large Wave Flume (GWK), about 900 tests were performed as a preparatory phase in the smaller wave flume of the Technical University Berlin [3], [11]. The programme of the large-scale model tests was performed in two phases:

- First, the hydraulic performance and the wave loads on a set of optimised model configurations were investigated for a broad range of wave conditions, including regular and irregular waves as well as solitary waves and "freak" waves (transient wave packets);
- Second, the behaviour of a beach profile under storm wave conditions was comparatively studied with and without a protection by a reef structure in order to demonstrate the efficiency of the latter.

Only the first phase of the large-scale model tests related to the hydraulic performance is described below. For the results of the second phase and those related to the wave loads see [10], [11] and [6].

2.1. *Model Set-Up and Measuring Techniques*

The first phase includes three types of submerged structures with a relative submergence depth varying from $R_c/H \approx 0$ to 2, H being the local incident wave height:

- Single wall with a porosity $\varepsilon = 0.$ 5, 11 and 20% (Fig. 4a);
- Two-wall system with porosities $\varepsilon = 11\%$ and 5% and a spacing B = 10.3 m (corresponding to a relative chamber width of $B/L \approx 0.25$ with L = local wave length) (Fig. 4b);
- Three-wall system with porosities $\varepsilon = 20\%$, 11% and 5% and a spacing $B_1 = 10.3$ m and $B_{total} = 20.6$ m (Fig. 4c).

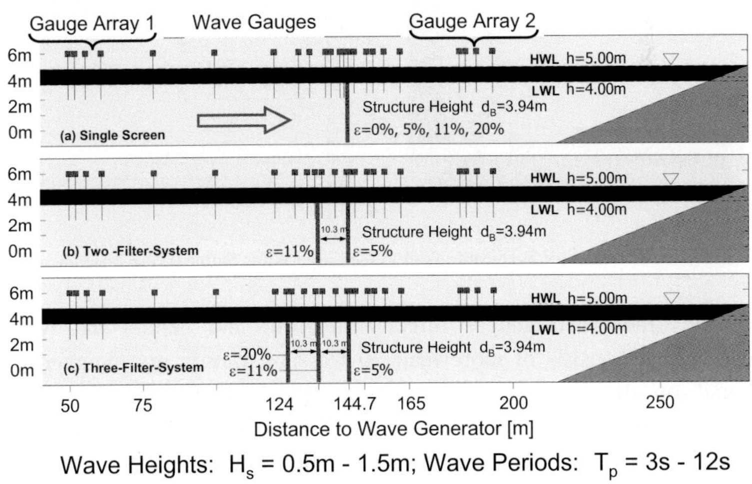

Figure 4. Model Set-Up in Large Wave Flume (GWK), Hannover

The model structures were installed in the middle of the flume in order to allow proper measurement of both wave reflection and transmission. For wave reflection analysis, gauge array 1 was used, and for wave transmission, gauge array 2 (Fig. 4a). A total of 24 gauges were installed (Fig. 4). The height of the walls was about $d_B = 3.93$ m and the water depth h = 4.0 m, 4.25 m and 5.0 m corresponding to relative wall height of $d_B/h = 0.79 - 0.98$.

With significant wave heights of $H_s \approx 0.5$ to 1.5 m the relative submergence depth is in the order of $R_c/H \approx 0$ to 2.0.

Together with the 24 wave gauges a total of 120 measuring devices with a sampling rate of 200 Hz were used simultaneously for a three wall model set-up (Fig. 4c). The instrumentation of a single wall with a porosity $\varepsilon = 5\%$ is exemplarily shown in Fig. 5, which also illustrates the unusually large dimensions of the model wall in the wave flume.

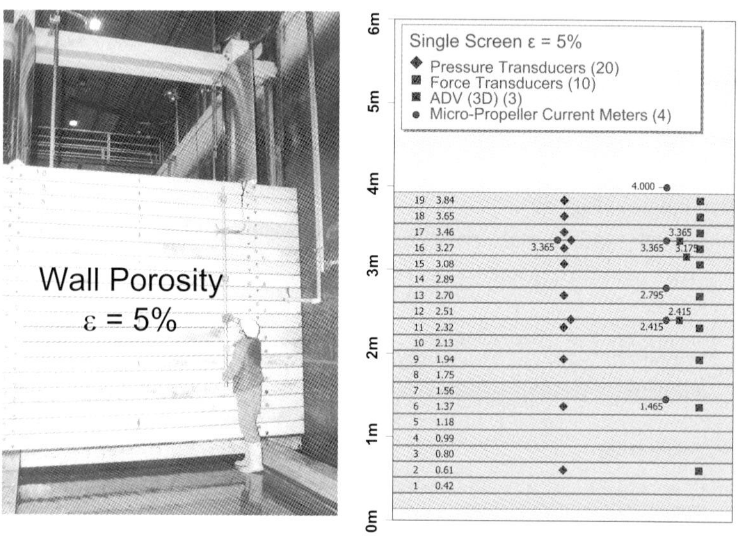

Figure 5. Instrumented Permeable Wall in the Large Wave Flume (GWK), Hannover

Besides the wave gauges directly in front and behind the wall, the instrumentation consists of more than 20 pressure transducers mounted on the front and on the back side of the wall (Piezo-electric PDCR 830 Druck Messtechnik GmbH), 10 forces transducers (strain gauges to record integrated pressure on each horizontal element of the wall), 7 current meters to record the inflow and outflow in the gaps between the horizontal wall elements (4 micro propellers and 3 Acoustic Doppler Velocimeters).

2.2. *Test Programme and Procedure*

In order to achieve a complete understanding of the processes involved in the hydraulic performance and the loading of the entire structure, including their reliable quantification, a total of about 800 tests were performed using regular waves (300), irregular waves (240), transient wave packets (180) and solitary waves (48). The regular wave trains are mainly used to get a better understanding of the processes as well as for the development of the prediction formulae while the irregular wave tests are primarily used for validation. The transient wave packet tests are mainly used for the validation of a commercially available RANS-VOF model [3], [11]. The solitary waves are used to check the validity of the developed formulae for shallow water waves [11].

In this paper the results of regular and irregular wave tests will mainly be addressed. The programme for these tests is summarized in Table 1. With a relative water depth h/L = 0.054 to 0.363 all tests are within the transition between shallow and deep water (h/L = 0.05-0.5) with a wave steepness H/L ≈ 0.01 to 0.06.

Table 1. Test Programme with Regular Waves and Irregular Waves

Water Depth h (m)	**Wave Period T or T_p(s)**	**Wave Height H or H_{mo}**				
		0.50	0.75	1.00	1.25	1.5
4.00	3.0	-	R	-	-	-
	3.5	R, I	R	R	-	-
	4.5	R, I	R, I	R, I	R	R
	6.0	R, I	R, I	R, I	R, I	R
	8.0	-	R, I	R, I	R, I	-
	12.0	-	R, I	R	-	-
4.25	3.0	-	R	-	-	-
	3.5	R, I	R	R	-	-
	4.5	R, I	R, I	R, I	R	R
	6.0	-	R, I	R, I	R, I	R
	8.0	-	I	R, I	I	-
	12.0		I	R	-	-
5.00	3.0	-	R	-	-	-
	3.5	R, I	R	R	-	-
	4.5	R, I	R, I	R, I	R	R
	6.0	R, I	R, I	R, I	R, I	R
	8.0	-	I	R, I	-	-
	12.0	-	-	-	-	-

R = regular wave; I = irregular wave

3. Experimental Results for a Single-Wall Reef and a Multiple-Wall Reef

The hydraulic functioning of a submerged single wall to damp the incident waves is basically different from that of a two-, three- or more walls reef system, but also much more complex. Therefore, and since the processes observed at a single wall are nearly similar to those occurring at each constitutive wall of the entire multiple-wall reef system, the simpler case of a single wall was first systematically investigated, before embarking into the processes associated with the more complex case of a two- and three-wall reef system. Moreover, the hydraulic functioning and the associated processes are examined from a farfield and a nearfield perspective. Particular focus was put on the processes affecting the hydraulic performance, but empirical prediction formulae have also been derived.

The ultimate result, however, remains the development of a theoretical model for the prediction of the hydraulic performance of a single wall and a multiple wall reef system which is principally based on the results of the regular wave tests. The extension of this model for irregular waves was performed using two different approaches, and the results of the irregular wave tests have been used for validation of the theoretical model (see Section 4).

3.1. *Hydraulic Performance of a Submerged Single Wall*

3.1.1. *Farfield Processes and Analyses for Single Wall*

Although the processes in the farfield and the associated hydraulic performance are clearly the results of the local processes occurring directly at the constitutive walls, a global approach (farfield) based on the energy conservation relationship

$$E_d = E_i - (E_r + E_t) \tag{1}$$

$$\text{with } E_i = \frac{1}{8}\rho_w g H_i^2 \tag{2}$$

has first been adopted before analysing the local processes (nearfield); i.e. the incident (E_i), the reflected (E_r) and the transmitted (E_t) wave energy components are determined from the analysis of the associated wave height measurements by using Eq. (2) (H_i is the wave height, ρ_w the water density, g the gravity acceleration and E_i the incident wave energy associated with H_i and corresponding to E_i, E_r, E_t or E_d). The dissipated wave energy E_d is then calculated according to Eq. (1). By defining the reflection, transmission and dissipation coefficient as:

$$C_r = \left(\frac{E_r}{E_i}\right)^{1/2},\ C_t = \left(\frac{E_t}{E_i}\right)^{1/2} \text{ and } C_d = \left(\frac{E_d}{E_i}\right)^{1/2} \tag{3}$$

the unknown dissipation coefficient C_d follows from Eqs. (1) and (3) as:

$$C_d = \sqrt{1-\left(C_r^{\,2}+C_t^{\,2}\right)} \tag{4}$$

C_r, C_t and C_d according to Eqs. (3) are called in the following "energy coefficients".

(a) Influencing Parameters

A detailed parameter study has shown that the effect of wave height and that of the wave period on the energy coefficients are in the same range [7], [6]. The most influencing parameter for both permeable and impermeable walls is, however, the relative submergence depth R_c/H_i ($R_c = d_B - h$, d_B= wall height, h = water depth). This is illustrated by Fig. 6, showing that:

- The reflection and transmission coefficients $C_r = C_r(R_c/H_i)$ and $C_t = C_t(R_c/H_i)$ represent continuous functions over the entire range of "submergence depths", including negative R_c – values (submergence) and positive R_c - values (emergence). Only near the transition between submergence and emergence ($R_c = 0$), a larger scatter of the data is observed.
- The maximum dissipation occurs for $C_r = C_t$ (see Eq. (4)). For impermeable walls ($\varepsilon = 0\%$), max C_d already occurs for $R_c/H_i = -0.2$. It decreases at a higher rate with increasing positive R_c/H_i-values, so that $C_d = 0$ is reached rapidly. For permeable walls, however, max C_d occurs at positive R_c/H_i-values (surface piercing wall). The higher the porosity of the wall, the higher is the corresponding R_c/H_i - value at which max C_d occurs. For $\varepsilon = 5\%$, max C_d occurs at $R_c/H_i = -0.3$, for $\varepsilon = 11\%$ at $R_c/H_i = 0.9$ and for $\varepsilon = 20\%$ at $R_c/H_i > 1$.

To better illustrate the effect of wall porosity ε on the dissipation coefficient C_d, Fig. 7 is drawn, showing that for a submerged wall ($R_c/H_i < 0$), C_d first slightly increases, then decreases with increasing porosity. However, the effect of the porosity remains relatively small, even for $\varepsilon = 20\%$.

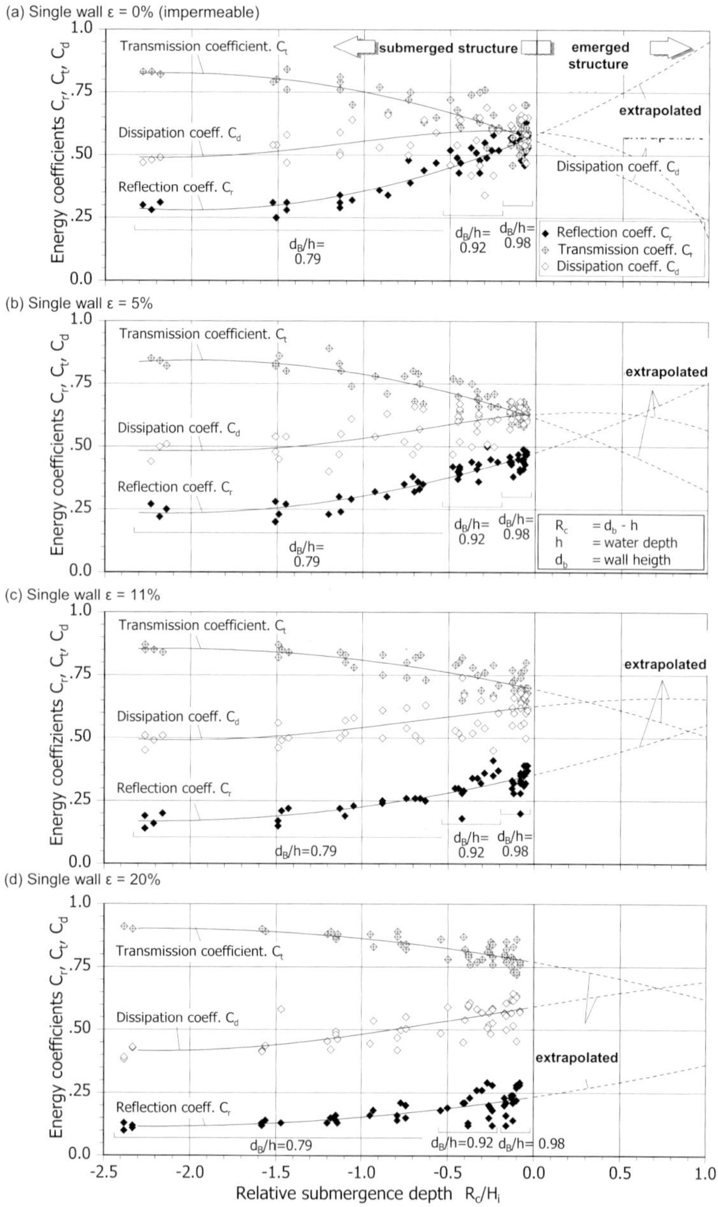

Figure 6. Energy Coefficients C_r, C_t and C_d versus Relative Submergence Depth for Single Wall with Porosity ε = 0%, 5%, 11% and 20%

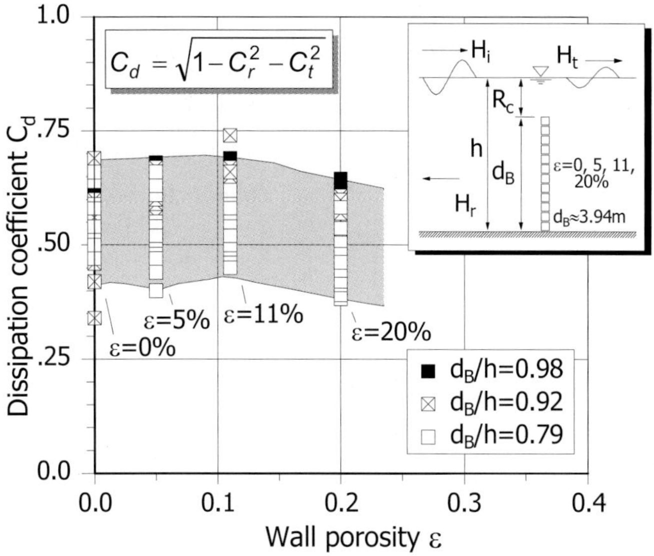

Figure 7. Effect of Wall Porosity on Energy Dissipation for Submerged Single Walls with Porosity ε = 0-20%

(b) Empirical Formulae for Prediction of Energy Coefficients

In the case of a submerged permeable single wall (Fig. 8) both flow and wave transmission occur over the structure (Domain A = permeability over submergence depth $R_c + h_0$) as well as through the structure (Domain B = permeability over the entire wall height d_B). Since the structure porosity $\varepsilon = s/e$ describes only the porosity of the wall with the height d_B and since there is a close relationship between structure porosity and hydraulic permeability, a parameter for the "overall permeability" which is more appropriate to describe the transmission and reflection properties of the structure over the entire water depth $(h + h_0)$ is needed. By defining the permeability domains as follows for $R_c \leq 0$ (Fig. 8):

- Domain A: $R^*_{c_1} = R_c + h_0$ with $h_0 = \text{MWL} - \text{SWL}$
- Domain B: $R^*_{c_2} = \varepsilon^* d_B$ with ε^* = wall permeability over entire wall height d_B [6]:

$$\varepsilon^* = \tan h^{1.4}(2.8\varepsilon) \tag{5}$$

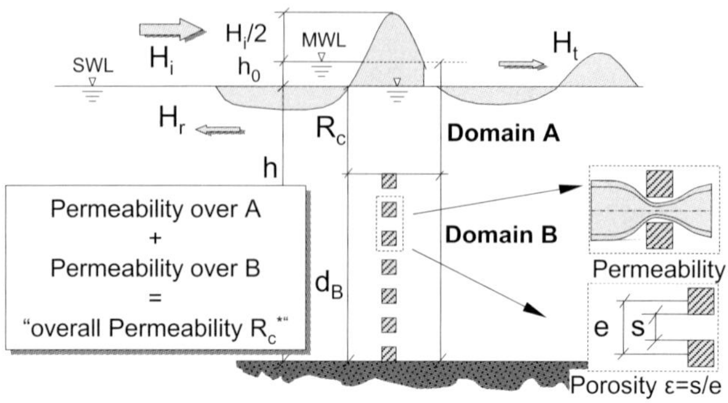

Figure 8. Overall Permeability of Single Submerged Wall (Definition Sketch)

The overall permeability domain over the entire water depth ($h + h_0$) is then:

$$R_c^* = (R_c + h_0) - \varepsilon^* d_B \quad \text{for } R_c \le 0 \tag{6}$$

Approximating the maximum run-up at a surface piercing impermeable wall by [1]:

$$\eta^+_{max,0} = 2\eta^+_{max} \tag{7}$$

With η^+_{max} = max. upward (positive) surface elevation of the incident wave, the following "Overall Permeability Parameter (OPP)" is obtained [6]:

$$OPP = \frac{R_c^* - \eta^+_{max,0}}{\eta^+_{max}} = \frac{R_c^*}{\eta^+_{max}} - 2 \tag{8}$$

for which the boundary condition OPP → 0 for $R_c^* \to \eta^+_{max,0}$ is fulfilled.

To demonstrate that the new proposed parameter is physically appropriate to describe both reflection and transmission properties of the structure, Fig. 9 is given, showing that there is a very good correlation between the energy coefficients and the new "Overall Permeability Parameter OPP":

$$C_t = \frac{8}{\pi^3} \arctan^3 \left(1.4|OPP|\right) \tag{9}$$

$$C_r = 1 - \frac{8}{\pi^3} \arctan^3 \left(1.4|OPP|\right) \tag{10}$$

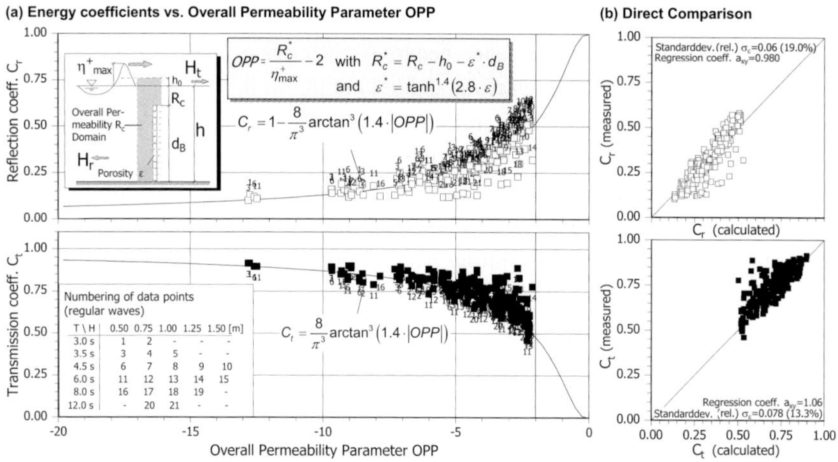

Figure 9. Reflection and Transmission Coefficients for All Tested Single Walls

Summarizing the hydraulic functioning of a submerged single wall from a global perspective it can be concluded, that the wave asymmetry (non-linearity) must necessarily be accounted for, since the wave phase corresponding to the wave crest η^+_{max} directly at the wall is determinant for the flow and wave transmission behind the wall.

The new suggested "Overall Permeability Parameter (OPP)" in Eq. (8) is very appropriate to describe the interaction between wave transmission over and wave transmission through the structure, and thus also both transmission and reflection properties. For a submerged single wall with relatively low porosity ($\varepsilon < 10\%$), the most important influencing parameter is the relative submergence depth (R_c / η^+_{max}). However, the importance of the latter rapidly decreases with increasing wall porosity. The maximum wave energy dissipation which can be achieved by a submerged single wall is limited to 40-70%. The wave transmission and wave reflection are strongly linked by the relation $C_t \approx 1 - C_r$; i.e. a decrease of wave transmission can only be achieved by increased wave reflection [6].

3.1.2. *Nearfield Processes and Analysis for Single Wall*

An extensive study of the local processes observed immediately in front, at and behind a submerged single wall has been performed [6], including:

(i) the maximum surface elevations immediately in front and behind the wall as well as their phase shift,

(ii) the phase shift between reflected and transmitted waves,

(iii) the maximum positive and negative difference between the surface elevations immediately in front and behind the wall,

(iv) the velocity of the flow through the openings of the wall and the contraction coefficient as a function of the incident wave field,

(v) the drag and inertia coefficients resulting from the flow through the permeable wall,

(vi) the energy dissipation resulting from the flow through the permeable wall as compared to that resulting from the vortices induced by flow separation at the submerged crest of the wall.

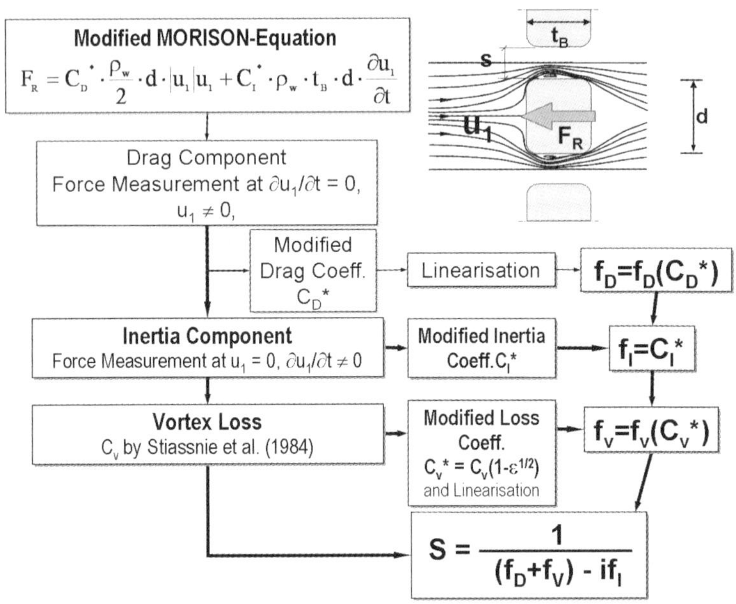

Figure 10. New Structure Parameter S for Submerged Permeable Walls

The results are thoroughly reported by [6]. In the following, only a very brief summary of the results related to the last two items (v) and (vi) will be given, as these will be needed for the development of the analytical model in Section 4. In fact, the existing "concept of permeability parameter G" as commonly suggested for permeable surface piercing walls [13] cannot be applied for submerged permeable walls, for which the flow and wave transmission occur not only through, but also over the submerged wall. Therefore, a new structure parameter S has been developed which is based on a modified MORISON-equation (Fig. 10). The total horizontal reaction force F_R to the flow with velocity u_l and acceleration $\partial u_l / \partial t$ consists of a drag component F_D^* and an inertia component F_I^* with the corresponding modified drag and inertia coefficient C_D^* and C_I^*. The modification is necessary to account for the effect of the neighbouring wall elements. The procedure summarized in Fig. 10 for the development of the new structure parameters is briefly discussed below.

(a) Modified Drag Coefficient and Linearization of Drag Force

The modified drag coefficient C_D^* is determined from the force measured on each wall element at the time where $\partial u_l / \partial t = 0$ and $u_l = u_{l\max}$ (wave crest at the wall). The results are plotted against a modified Reynolds-Number R_e^* as defined in Fig. 11, showing the function $C_D^* = C_D^*(R_e^*)$ with a coefficient of variation of about 16%.

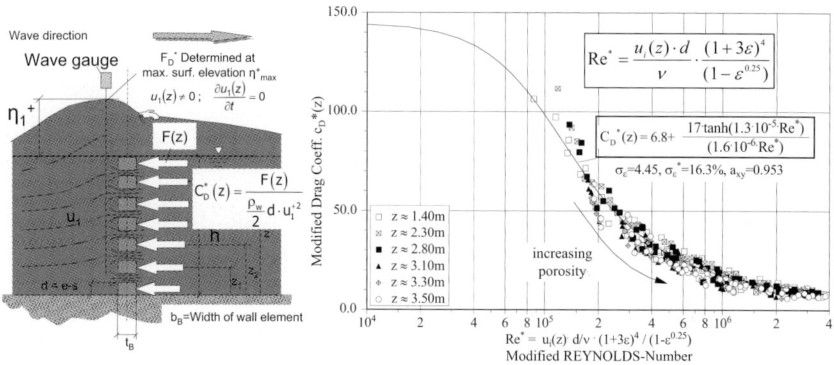

Figure 11. Modified Drag Coefficient versus Modified Reynolds-Number for Submerged Permeable Walls

The incident wave energy flux due to the non-linearized drag of the flow through the permeable submerged wall with height d_B and porosity ε is [4], [6]:

$$F_D = \frac{1}{2}\rho_w \frac{4}{3\pi}(1-\varepsilon)\int_0^{d_B} C_D^*(z)\, u_{1,\max}^3(z)\, dz \qquad (11)$$

with $u_{1,\max}$ = maximum velocity amplitude and ρ_w = water density.

Using Lorentz-Hypothesis for equivalent work (i.e. energy over a wave cycle must be equal for non-linear force and linearized force!) as suggested for instance by [13], the energy flux in Eq. (11) corresponding to the linearized drag force becomes:

$$F_{D,lin} = \frac{1}{2}\rho_w \cdot f_D \cdot \omega(1-\varepsilon)\, t_B \int_0^{d_B} u_{1,max}^2(z)\,dz \tag{12}$$

where t_B = wall thickness; $\omega = 2\pi/T$ = angular wave frequency; f_D = drag coefficient of the linearized force which is then obtained according to Lorentz-Hypothesis by equating Eqs. (11) and (12), i.e. $F_D = F_{D,lin}$:

$$f_D = \frac{1}{\omega}\frac{4}{3\pi\, t_B}\frac{\int_0^{d_B} C_D^*(z)\, u_{1,max}^3(z)\,dz}{\int_0^{d_B} u_{max}^2(z)\,dz} \tag{13}$$

(b) Modified Inertia Coefficient

The modified inertia coefficient C_I^* is directly calculated from the measured force on each element at the time where $u_1 = 0$:

$$C_I^*(z) = \frac{F_I^*\left(t_{u_1=0}, z\right)}{\rho_w d \cdot t_B \cdot \partial u_1(z)/\partial t} \tag{14}$$

with d = width of the wall (see Fig. 10).

Using a modified Keulegan-Carpenter-Number KC^* proposed by [6]:

$$KC^* = \frac{u_{1,max,d_B} \cdot T}{t_B} \cdot \tanh^{0.57}(1.5\varepsilon) \tag{15}$$

(with $u_{1,max,d_B} = u_{1,max}(z = d_B)$ and T = wave period), a simple expression is obtained for the modified inertia coefficient f_I averaged over the entire wall height d_B [6].

$$f_I = \overline{C}_I^* = \frac{L/t_B}{15\, KC^*} \tag{16}$$

where L is the local wave length.

Eq. (16) is plotted in Fig. 12 with the experimental data exemplarily for submerged permeable walls with a relative height $d_B/h = 0.98$ and wall porosities ε = 5, 11 and 20%, showing a coefficient of variation of 27%.

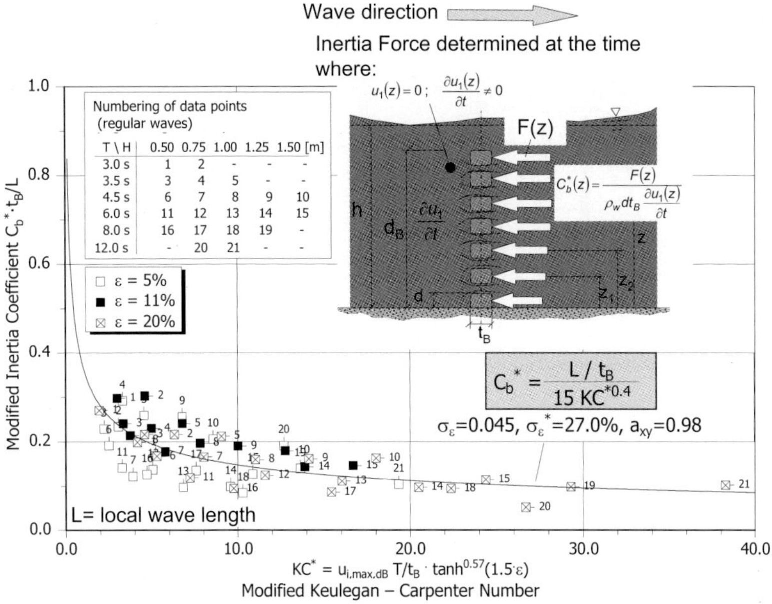

Figure 12. Modified Inertia Coefficient versus Modified Keulegan-Carpenter Number for Submerged Permeable Walls with $d_B/h = 0.98$

(c) Modified Vortex Loss Coefficient due to Flow Separation at Submerged Wall Crest

Due to flow separation at the crest of the submerged wall, vortices are induced twice during a wave cycle (Fig. 13), leading to energy losses which may reach up to 20% of the total dissipated energy by the wall for moderate wave lengths ($kh = 0.75\text{-}1.25$ with $k = 2\pi/L$) and large relative wall height ($d_B/h \rightarrow 1$). For shorter wave periods, the two vortices generated during each wave cycle generally remain close to the wall (Fig. 13a). As a result, the flow and wave field at the structure are strongly affected by the vortices. For longer wave periods, however, this effect becomes less important, as the vortices are transported far from the wall (Fig. 13b).

A theoretical solution for an impermeable wall ($\varepsilon = 0$) proposed by [14] to predict the vortex loss coefficient C_v was used as a starting point (Fig. 14):

$$C_v = \frac{F_v}{F_I} = \frac{2.79\,k\left(\dfrac{H_i}{2}\right)^{2/3}}{\left[k(h-d_B)\left(K_i^2(k(h-d_B)) + \pi^2 I_i^2(k(h-d_B))\right)\right]^{4/3}} \tag{17}$$

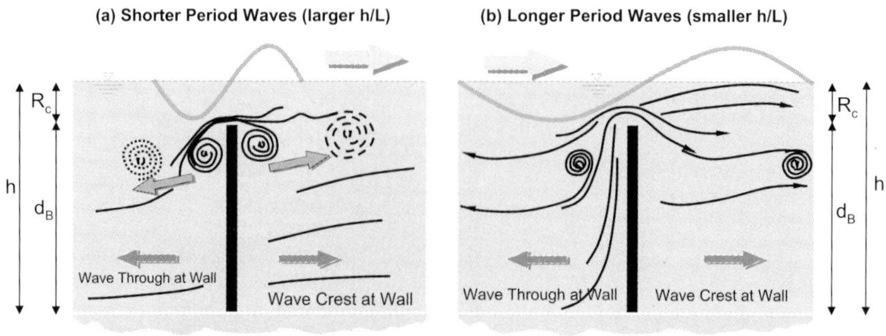

Figure 13. Flow Separation at Submerged Wall Crest and Resulting Vortices During a Wave Cycle

With F_I = incident wave energy flux; F_v = dissipated wave energy flux due to flow separation (vortices) at wall crest; K_i, I_i: First-Order Bessel-functions; $k = 2\pi/L$ = wave number; H_i = incident wave height.

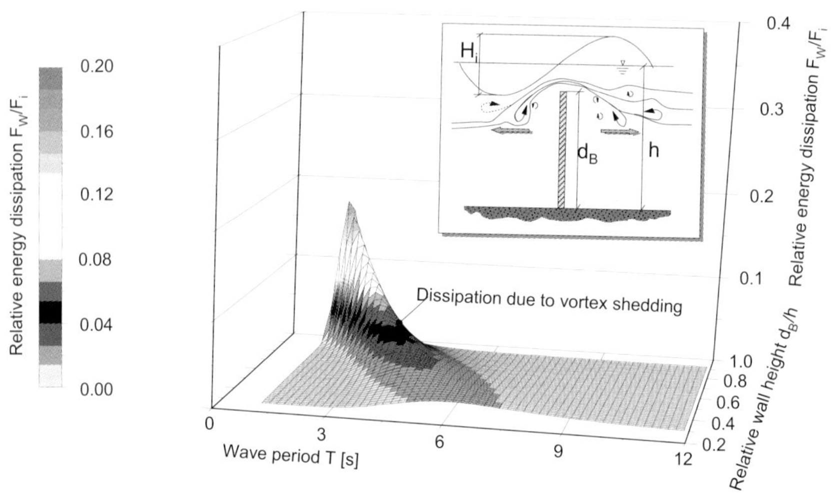

Figure 14. Vortex Loss Coefficient Due to Flow Separation at Impermeable Wall Crest According to Eq. (17)

To account for the reduction of the vortex loss coefficient C_V in Eq. (17) which will be induced by the wall porosity ε, a reduction factor $\left(1-\sqrt{\varepsilon}\right)$ is introduced, yielding the modified vortex loss coefficient C_v^*:

$$C_v^* = \left(1-\sqrt{\varepsilon}\right)C_v \qquad (18)$$

After linearization using the Lorentz-Hypothesis in the same way as for the drag force in Section 3.1.2a, a modified vortex loss coefficient f_v for the linarized case is finally obtained [6]:

$$f_v = \frac{1}{\omega} \frac{2F_I \cdot C_v^*}{\rho_w d_B \cdot t_B \cdot u_{1,max,d_B}^2} \tag{19}$$

(d) New Structure Parameter

Using the modified drag, inertia and vortex loss coefficients f_D, f_I and f_v determined by Eqs. (13), (16) and (19) the following structure parameter S is obtained to describe the matching conditions associated with the drag, inertial and vortex losses at the submerged wall (see also Section 4 and Fig. 20):

$$S = \frac{1}{(f_D + f_v) - i f_I} \tag{20}$$

with $i = \sqrt{-1}$.

The new structure parameter S in Eq. (20) can be used for both permeable and impermeable submerged walls and is physically more appropriate than the commonly used "complex permeability parameter" $G = \varepsilon C_c / (C_D - iC_I)$ with C_c = contraction coefficient [2] which yields G = 0 for $\varepsilon = 0$.

3.2. *Hydraulic Performance of Submerged Progressive Wave Absorbers*

The main drawback of a single wall reef as compared to a submerged progressive wave absorber (multiple-wall reef) is that the dissipation performance of the former is limited (40-70% of incident wave energy) and that both wave transmission and reflection are closely tied together ($C_t \approx 1 - C_r$); i.e. a reduction in wave transmission (C_t) is necessarily accompanied by an increase in wave reflection (C_r). Unlike a single wall reef, a multiple wall reef can dissipate a much higher portion of the incident wave energy, and a decrease of both wave transmission and reflection can simultaneously be achieved by a proper selection of the spacing, porosity and submergence depth of the constitutive walls. By means of variation of these parameters, the wave energy dissipation at each wall and the interference of incident and reflected waves can be much better controlled and tuned. As a result, a hydraulic performance tailored to any specific functional design requirements can be obtained. The wave damping mechanisms and the associated influencing parameters which make this possible are discussed in Sections 3.2.1 and 3.2.2 below. A further discussion, including a comparative analysis of the hydraulic performance of both single-wall and multiple-wall reefs is also given in Section 3.2.3 to illustrate the basic difference between the two types of reef concepts.

3.2.1. *Farfield Processes and Analysis for Multiple Wall Reef*

The processes in the farfield and the associated hydraulic performance have been analysed using the same equations (1)-(4) as for a single wall [11], [6]. The conclusions drawn from this analysis, which have also been confirmed by the results of the analysis in the nearfield (see Section 3.2.2), may be summarized as follows:

(i) The relative spacing B/L (L = local wave length) and the relative submergence depth R_c/H_i (H_i = incident wave height) of the constitutive walls represent the most dominant parameters affecting the hydraulic performance of the entire reef structure. An example result is shown in Fig. 15 for a two-wall reef structure with porosities ε = 11% and 5% and a relative spacing B/L > 0.5. In addition, R_c/H_i is determinant for the local processes at each constitutive wall. Further influencing parameters are the wall porosity ε, the number of walls N, the relative water depth h/L and the relative wall height d_B/h: $C_i = C_i\left(B/L; R_c/H_i; \varepsilon, N, d_B h, h/L\right)$ where C_i represents any of the three energy coefficients C_r, C_t, or C_d.

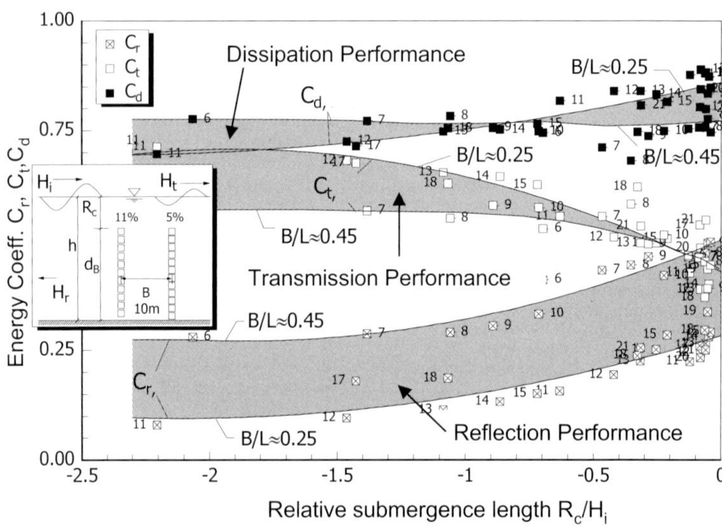

Figure 15. Hydraulic Performance of a Two-Wall Reef as a Function of the Relative Submergence Depth

(ii) The wave reflection of a multiple-wall reef is essentially determined by the reflection properties of the first wall (see submerged single wall in Section 3.1.1). A further contribution is provided by the interference effects which are induced by the next wall(s) and are therefore mainly governed by the relative spacing B/L (see Fig. 16 and Section 3.2.2 below).

(iii) A partial standing wave field develops in front of and inside the reef, while behind the reef a progressive wave field is transmitted (see Fig. 16). A minimum wave transmission is obtained when the combination of both reflected and dissipated wave energy is maximum (see Eq. (4)). For higher dB/h-values, dissipation represents the dominant process, so that minimum wave transmission is obtained near the peak of energy dissipation ($B/L \approx 0.25$). For lower dB/h-values, the influence of the local processes at the first wall on the performance of the entire reef structure decreases and the minimum transmission will shift to $B/L \approx 0.4$.

To confirm these results an analysis of the processes in the nearfield has been conducted which also highlights the relative contribution of each constitutive wall to the wave damping performance of the entire reef structure.

3.2.2. *Nearfield Processes and Analysis for Multiple Wall Reef*

The local processes at each constitutive wall of the submerged progressive wave absorber have been analysed in the same way as for a single-wall reef (see Section 3.1.2). In the following, however, particular focus will be put only on those processes which influence the relative contribution of each wall to the performance of the entire reef system. Further details and results are reported by [11] and [6].

As already mentioned above (Section 3.2.1iii), a partial standing wave field develops in front of and inside the reef system shown in Fig. 16a for a two-wall reef structure. Depending on the phase shift between the incident and reflected waves which is essentially determined by the relative spacing B/L, the "node" of the partial standing wave just behind the first wall varies. The closer the "node" approaches the backside of the first wall, the larger the difference of the surface elevations at the front and backside of the first wall; i.e. the pressure gradient driving the flow through and over the wall (and thus energy dissipation) will increase. The maximum η_{12}^{+} and minimum η_{12}^{-} surface elevation at the backside of the first wall of a two-wall reef ($\varepsilon = 11\%$ and 5%) are related to the corresponding values η_{2SW}^{+} and η_{2SW}^{-} for a single wall reef and plotted in Fig. 16b for three relative wall height d_B/h as a function of the relative spacing B/L. The results show that both positive ($\eta_{12}^{+}/\eta_{2SW}^{+}$) and negative ($\eta_{12}^{-}/\eta_{2SW}^{-}$)

extreme values oscillate around +1 and -1, respectively. This indeed confirms that the wave reflection of the entire reef system is essentially determined by the reflection behaviour of the first wall with, in addition, a contribution of the interference effects induced by the next wall(s) which are primarily determined by the relative spacing B/L. The lower the relative wall height d_B/h, the smaller is this contribution (Fig. 16b). This may be explained by the lower reflection associated with lower d_B/h-values.

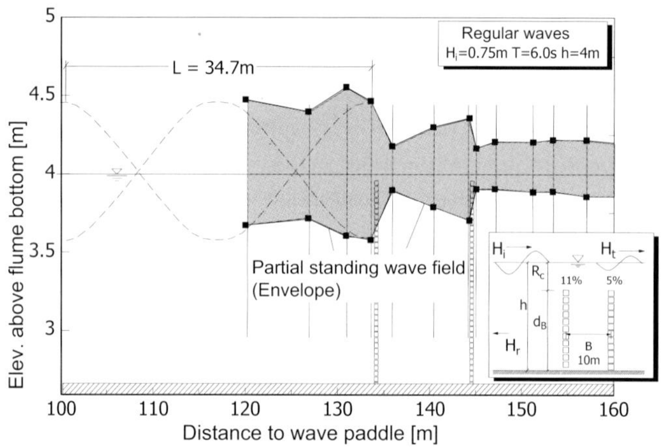

(a) Partial Standing Wave Inside and In front of the Structure

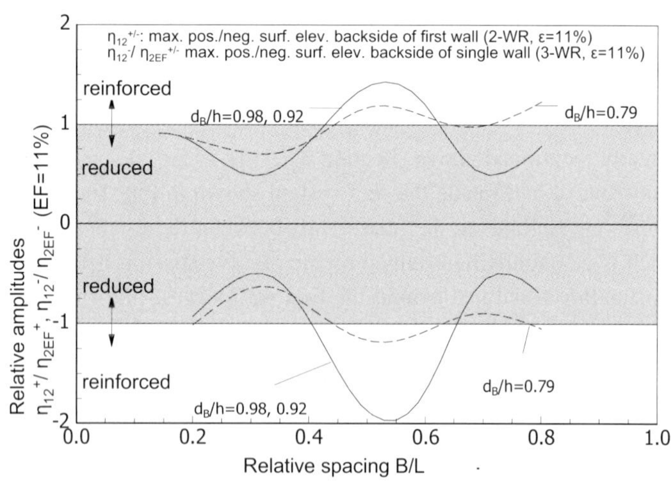

(b) Relative Surface Elevation at Backside of first Wall

Figure 16. Contribution of First Wall and of Interference Effects to the Reflection Performance of a Two-Wall Reef Structure for Different Wall Heights d_B/h

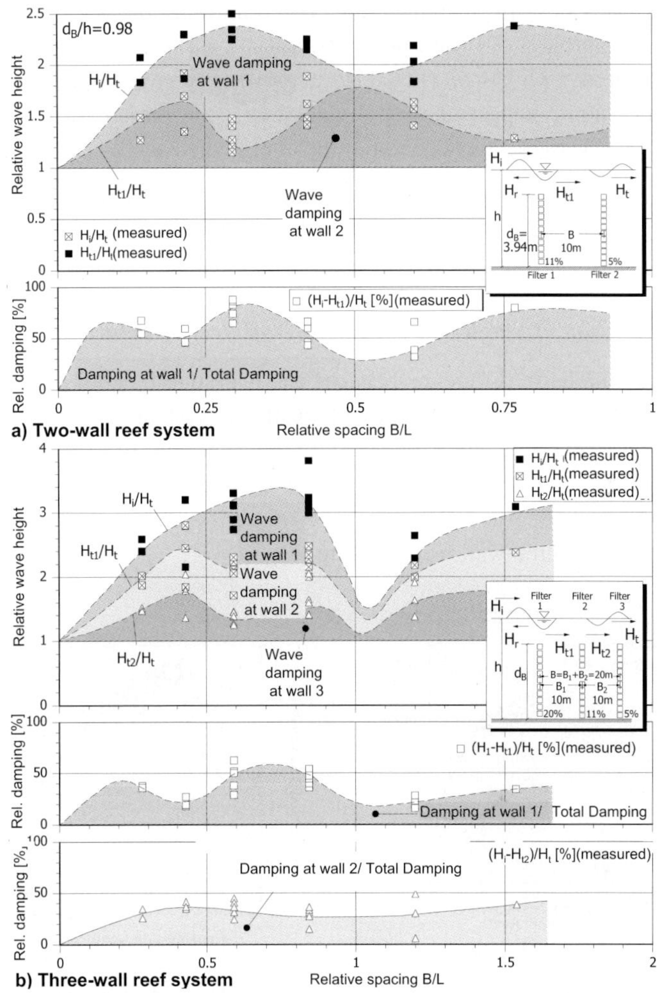

Figure 17. Relative Contribution of Each Wall to the Wave Damping of a Two- and Three-Wall Reef for a Relative Wall Height $d_B/h = 0.98$

The relative contribution of each constitutive wall to the wave damping as a function of the spacing B/L is exemplarily illustrated in Fig. 17 for a two- and three-wall reef system with a relative wall height of $d_B/H = 0.98$:

- *For the two-wall reef system* (Fig. 17a), the contribution of the first wall varies between 30% (B/L ≈ 0.5) to 85% (B/L ≈ 0.25) and is much larger than that of the second wall, but the latter is necessary to generate the interference effects, thus ensuring the hydraulic functioning of the reef structure as a submerged progressive wave absorber.

- *For the three-wall system* (Fig. 17b), the overall wave damping performance is significantly improved, but the relative contribution of the first wall decreases due to the introduction of the intermediate wall.

In fact, the contribution of the latter amounts up to 30-40% and remains almost unchanged for B/L > 0.25. Like in Fig. 17a, the contribution of the last wall is relatively small, but it is required to ensure additional interference effects.

3.3. *Comparison of Single- and Multiple-Wall Reef Structure*

To illustrate the basic difference between the hydraulic functioning of a single-wall reef structure and that of a multiple-wall reef system, a comparison of the reflection, transmission and dissipation performance is given in Fig. 18. The direct comparison in Fig. 18c illustrates how significantly the energy dissipation increases while both wave reflection and transmission decrease when moving from a single to a three-wall reef structure.

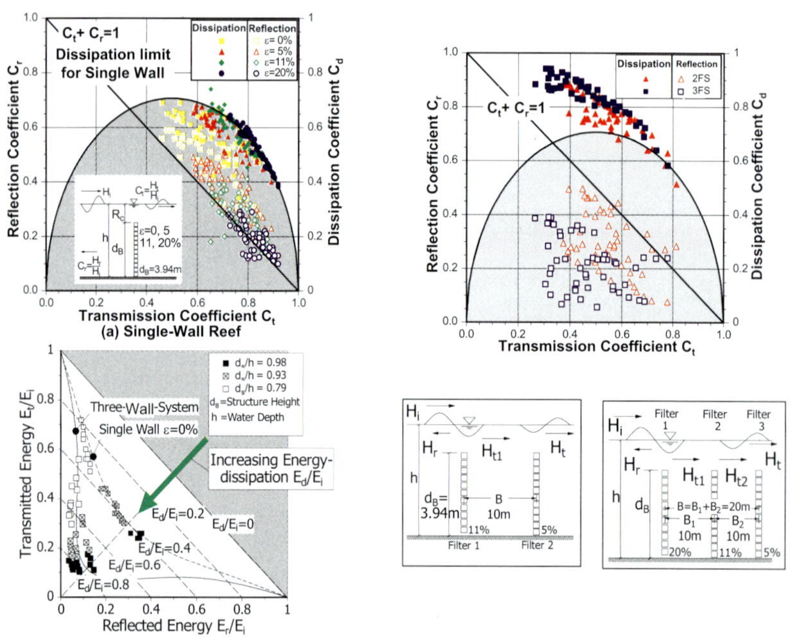

Figure 18. Basic Difference Between Hydraulic Functioning of Single-Wall Reef and Multiple-Wall Reef

In Fig. 18a,b the midline $C_r + C_t = 1$ and the dissipation limit $C_{d\max} = \sqrt{1 - C_r^2 - C_t^2}$ of a single-wall reef are also drawn in both figures to better highlight the basic difference between the two cases.

It is seen that for

- *A single-wall reef* (Fig. 18a), the measured C_r- and C_t-values are generally within the lower bound (midline $C_r + C_t = 1$) and the upper bound (dissipation limit $C_{d\max} = \sqrt{1 - C_r^2 - C_t^2}$). To overcome these two limits a new reef concept is required which consists of successive submerged walls with progressively decreasing porosity in the direction of incident waves (progressive wave absorber).
- *A two- and three-wall reef* (Fig. 18b), the dissipation coefficient C_d is clearly over the aforementioned dissipation limit and may reach 90% while both wave transmission and reflection coefficients C_t and C_r strongly decrease below the midline $C_t + C_r = 1$.

4. Theoretical Model for the Prediction of the Hydraulic Performance

4.1. *Basic Requirements for the Prediction Model*

The theoretical model to be developed should be an engineering model which can be applied as a design tool to optimise the hydraulic performance by varying the number of submerged walls N, their porosity ε, their spacing B and their submergence depth R_c as a function of the incident wave parameters and depth conditions. It should therefore:

(i) be simple, allowing parameter studies to be performed in shorter time and at lowest costs;

(ii) be versatile in the sense that it can easily be adapted for any number and configuration of the submerged walls, including a single wall reef alternative;

(iii) account in an integrated manner (Fig. 19) for all relevant processes and energy losses, such as local losses (drag and inertia) of the flow through the permeable walls and the vortex losses caused by flow separation at the submerged wall crest which will induce a water level difference $\Delta\eta$ between the front and the back side of the wall (i.e. wave damping). Particular focus should be put on the phase shift which is essentially determined by the spacing B/L, but also by the aforementioned losses. Other influencing parameters such as submergence depth R_c/H_i and wall porosity ε will also affect the phase shift. The resulting partial standing wave field will then affect the water level difference $\Delta\eta$ which again affects the local and vortex losses, etc. This means that all these processes are strongly interdependent, and the hydraulic performance of a progressive wave absorber cannot be analysed by separately addressing the wave damping induced by each constitutive wall (Fig. 19);

(iv) account for irregular wave trains.

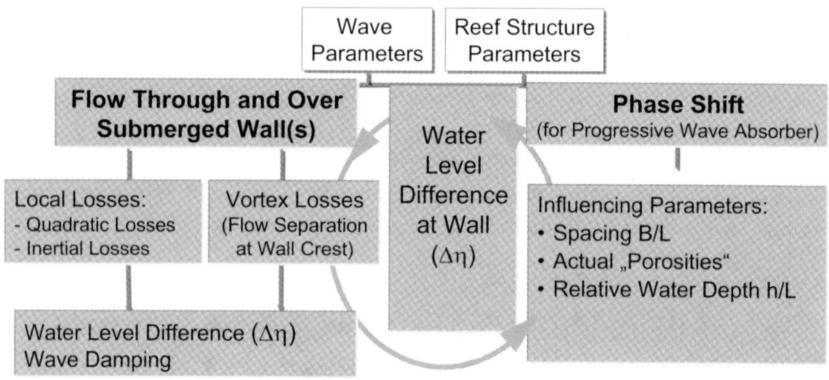

Figure 19. Overview of Most Relevant Processes and Interactions for the Hydraulic Performance of Reef Structures

4.2. *Potential Flow Model Formulation*

The theoretical model which can best fulfill the aforementioned requirements is a potential flow model in which the energy losses are introduced through the matching conditions at the wall by using the new structure parameter S described in Eq. (20). Since the principle is the same for each constitutive wall of the entire reef system, it is much simpler to illustrate for the case of a single wall (Fig. 20). Further details are provided in the PhD-thesis of Koether [6].

The flow field is subdivided in fluid domains 1 and 2 in front and behind the submerged permeable wall, respectively. The two fluid domains are again subdivided in an upper zone A (flow over the wall) and a lower zone B (flow through the wall).

For fluid domains 1 and 2 the velocity potentials φ_1 describes the partial standing wave field, φ_2 the transmitted wave field, $\varphi_i = \varphi_i(x,z) = \cosh(kz)\exp(ikx)$ the incident waves and μ_m the complex wave number for the waves of order $m \geq 1$ (with $\mu_m \tanh(\mu_m h) = -\omega^2$ and $\mu_o = ik$ and $k = 2\pi / L$).

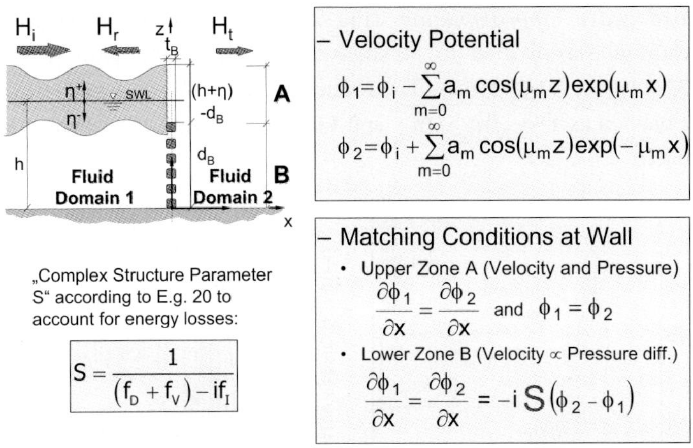

Figure 20. Potential Flow Model: Exemplarily for the Case of a Single Submerged Permeable Wall

The matching conditions at the wall for the upper zone A (overflow) are trivial (continuity of horizontal velocity and pressure). More difficult are the matching conditions for the lower part B (flow through the wall). In addition to the continuity of horizontal velocity which must be fulfilled, a proportionality of this velocity to the pressure difference between the front and backside of the wall is assumed. This assumption is justified for a thin wall [15], [5], [2] and the proportionality factor is provided by the complex structure parameter S which accounts for all energy losses (see Fig. 20 and Eq. (20)).

Substituting φ_1 and φ_2 into the matching conditions, integrating the latter over the lower and upper zones A and B, and adding the results to obtain the matching conditions over the entire water column $(h+\eta)$, a system of linear equations with the unknown complex amplitudes a_m is obtained. Given the incident waves (φ_i), the partial standing waves (φ_1 = incident + reflected) and the transmitted waves (φ_2) can then be determined. Further details on the deviations of the analytical solutions for a single, a two- and three-wall reef are given in [6].

The potential flow model has first been developed for regular waves and then extended for irregular waves by using two different approaches (Fig. 21):

- *Regular wave approximation* (Fig. 21a): The incident irregular wave spectrum is described by the representative wave height (H_{mo}) and period (T_{o1}), so that the reflection and transmission coefficient are calculated in the same way as for regular waves with $H = H_{mo}$ and $T = T_{o1}$ (Frequency-independent approach).

- *Spectral wave approximation* (Fig. 21b): The incident irregular wave spectrum is sub-divided in 50 sine-wave components with $H_i = 2\sqrt{S(f_i)}$ and $T_i = 1/f_i$ $(i = 1, 2, ..., 50)$. The reflection and transmission coefficients are then obtained as $C = \sqrt{m_{or}/m_{oi}}$ and $C_r = \sqrt{m_{ot}/m_{oi}}$ (Frequency-dependent approach).

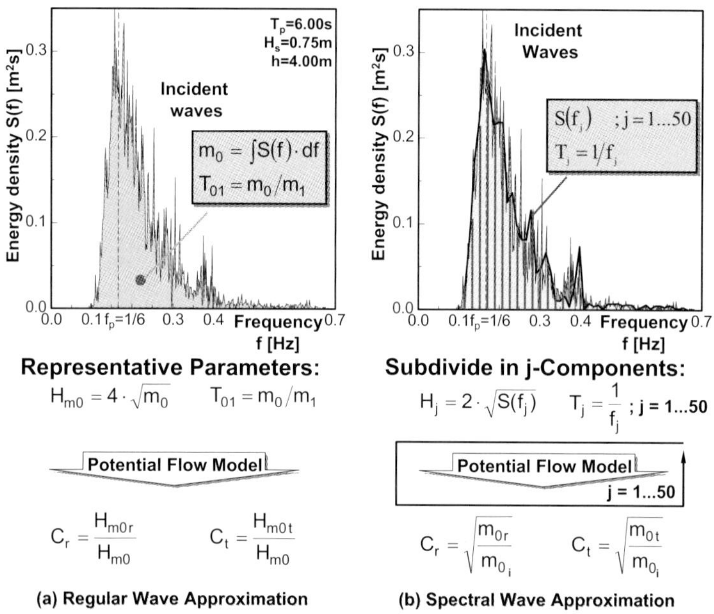

Figure 21. Model Extension Approaches for Irregular Waves

4.3. *Model Validation*

(a) Comparison with Regular Wave Tests

An extensive comparative analysis of the calculated and measured reflection, transmission and dissipation coefficients C_r, C_t, and C_d has been performed for single-wall reefs and multiple-wall reef systems with different submergence depths. An example for a two-wall reef with porosity $\varepsilon = 11\%$ and 5% and spacing B = 10 m is given in Fig. 22. The energy coefficients C_r, C_t and C_d are plotted as a function of the relative spacing B/L for three submerged depths d_B/h (Fig. 22a). A direct comparison of the calculated and measured values is also given (Fig. 22b), showing only a slight underestimation of the reflection coefficient C_r and a slight overestimation of the transmission coefficient C_t.

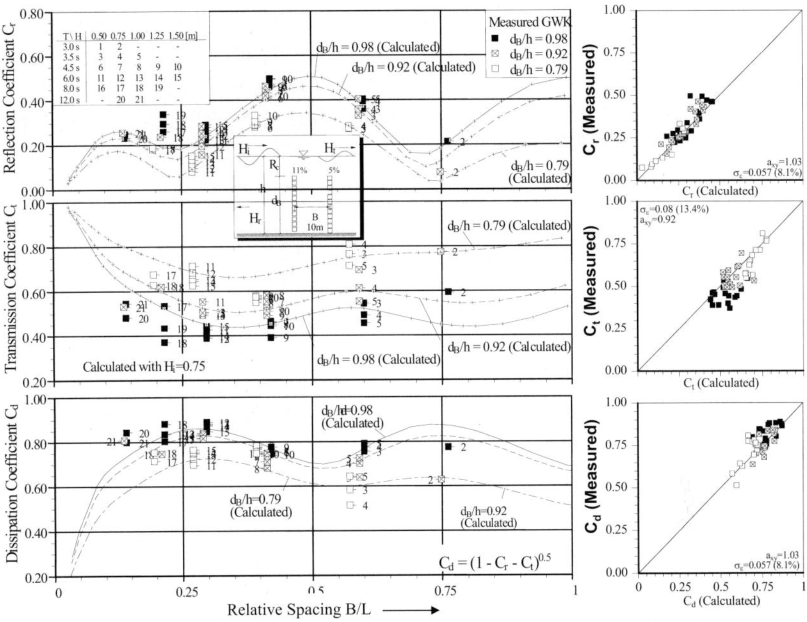

(a) Energy Coefficients vs. Relative Spacing **(b) Direct Comparison**

Figure 22. Example Model Verification for a Two-Wall Reef Subject to Regular Waves

(b) Comparison with Irregular Wave Tests

A comparison of the measured and calculated refection coefficient as a function of the incident wave frequency is exemplarily shown in Fig. 23a for a two-wall reef system with spacing B = 10 m, porosity 11% and 5% and height $d_B = 3.9$ m in a water depth h = 4.0 m which is subjected to irregular waves with $H_s = 0.50$ m and $T_p = 3.5$ s. The oscillations in the reflected wave spectrum are due to the strong effect of the relative spacing B/L. This effect is also responsible for the strong variation of the reflection coefficient $C_r(f)$ with the wave frequency (f).

The agreement between measured and calculated C_r-values is satisfactory within the frequency range of practical interest (fundamental mode), i.e. where most incident wave energy in the spectrum is concentrated. The disagreements in the higher frequency range are less relevant for engineering practice related to coastal protection. Using the "spectral wave approximation" (like in Fig. 23a), a direct comparison of the measured and calculated energy coefficients C_r, C_t and C_d for a two- and three-wall reef systems is drawn in Fig. 23b, showing a relatively good agreement. Using "regular wave approximation" slightly higher coefficients of variation σ' are obtained [6].

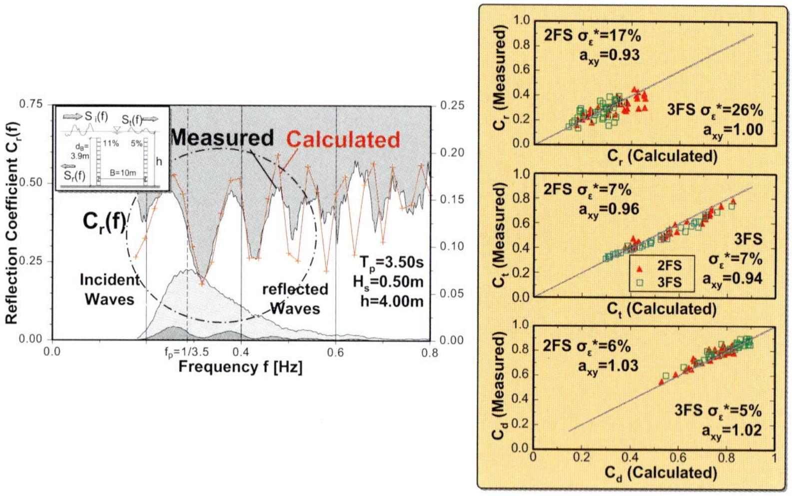

(a) Frequency Dependent Reflection Coefficient for a Two-Wall Reef

(b) Direct Comparison of Measured and calculated Energy Coefficients for Two-Wall (2FS) and Three-Wall (3FS) Reef

Figure 23. Example Model Verification for a Two-Wall Reef (2FS) and Three-Wall Reef (3FS) Subject to Irregular Waves

4.4. *Prediction Capability of the Model: Example Results*

After validation of the model using the data obtained from large-scale model tests (see Section 2), an extensive parameter study has been performed [6]. Here only few example results are given to illustrate the prediction capability of the model.

For any submerged single-wall reef and any submerged multiple-wall system, the reflection, transmission and dissipation coefficients C_r, C_t and C_d can be calculated as a function of the incident wave parameters and the structure parameters. For a wave height H = 1 m and a water depth h = 4 m, C_r, C_t and C_d are calculated as a function of the wave length L for (i) a single-wall with porosity $\varepsilon = 11\%$ (Fig. 24a) and (ii) a two-wall reef with porosity $\varepsilon = 11\%$ and 5% and spacing B = 20.6 m (Fig. 24b).

The comparison of the two results in Fig. 24a and Fig. 24b well-illustrates the basic difference of the hydraulic functioning of the two reef concepts. In the first case (Fig. 24a), the maximum level of energy dissipation is limited (here $C_d \approx 60\%$), so that a further decrease of wave transmission will necessarily lead to a corresponding increase of wave reflection. In the second case (Fig. 24b), the maximum dissipation level is much higher (80%) with typical oscillations due to the interference effects.

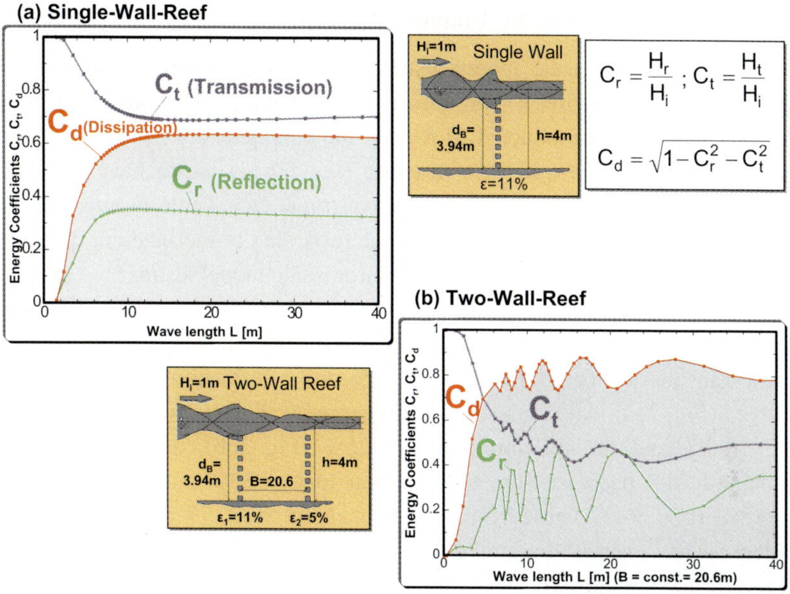

Figure 24. Calculated Hydraulic Performance for Regular Waves

For any multiple wall-reef system consisting of two- or more submerged walls, the contribution of each wall to the overall wave damping can also be calculated (Fig. 17).

Given any incident wave spectrum, the reflected and transmitted wave spectrum can be calculated for any submerged progressive wave absorber system and structure configuration. This is illustrated by Fig. 25 for a two- and three-wall reef system, showing that by adding an intermediate wall to the two-wall reef will decrease both wave reflection and transmission.

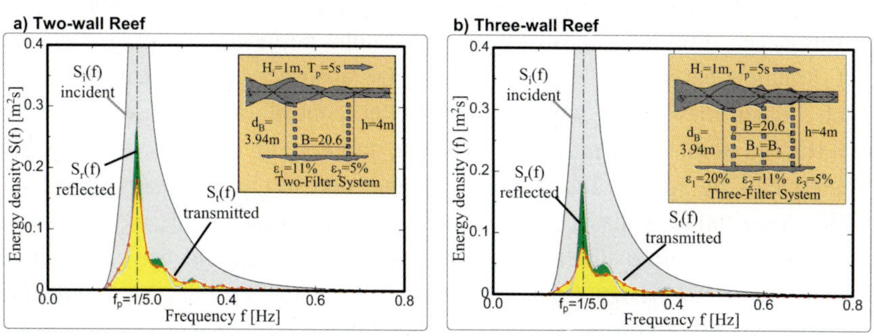

Figure 25. Calculation of Reflected and Wave Spectra for a Two- and Three-Wall Reef System

5. Potential Application for Tsunami Protection

In the course of the scale model testing programme performed in the Large-Wave Flume (see Section 2), preliminary tests using solitary waves up to 1 m height have also been conducted. However, the parameters of the single-wall, the two-wall and the three-wall reef structure (see Fig. 4) which have been fixed according to the regular and irregular wave trains (see Table 1) have not been modified and adapted for the solitary wave tests, due to the inherent difficulties (time, costs, man-power) associated with large-scale model testing.

Nevertheless, the results of the preliminary tests unexpectedly show that a relatively high transmission performance has been achieved. For a single-wall reef, a transmission coefficient of $C_t = 54\%$ and a reflection of $C_r = 38\%$ were achieved (Fig. 26a). Adding a second submerged wall decreases the transmission to $C_t = 40\%$ and the reflection to $C_r = 26\%$ (Fig. 26b), and adding a third submerged wall decreases the transmission to $C_t = 33\%$ and the reflection to $C_r = 18\%$ (Fig. 26c). These correspond to about 56%, 77%, and 86% of the incident wave energy dissipated by the single-wall, the two-wall and the three-wall reef structure, respectively. These encouraging results suggest that an appropriate numerical model should be developed to investigate in more details the peculiarities associated with the application of this new reef concept for the protection against tsunami. A hybrid model, including BOUSSINESQ-model for the farfield and a RANS-VOF model for the nearfield [8] would probably represent an appropriate candidate for this purpose.

6. Concluding Remarks

An interesting cost effective and soft alternative solution for coastal protection against erosion has been systematically investigated, showing that unlike conventional artificial reef concepts which have serious drawbacks (limited wave damping performance, strong link between wave reflection and transmission performance, difficult control of the hydraulic performance, etc.), the new reef structure based on the progressive wave absorber concept can provide a much higher dissipation with a much better control of the wave reflection and transmission. The latter can even be tailored to any functional design requirements by varying the number, spacing, porosity and submergence depth of the constitutive walls. The proposed theoretical (linearized) model can be used to optimise these structure parameters given the incident wave conditions and any functional design criteria for coastal protection. Further development of the model is required to account for the generation of higher harmonics in the transmitted wave field.

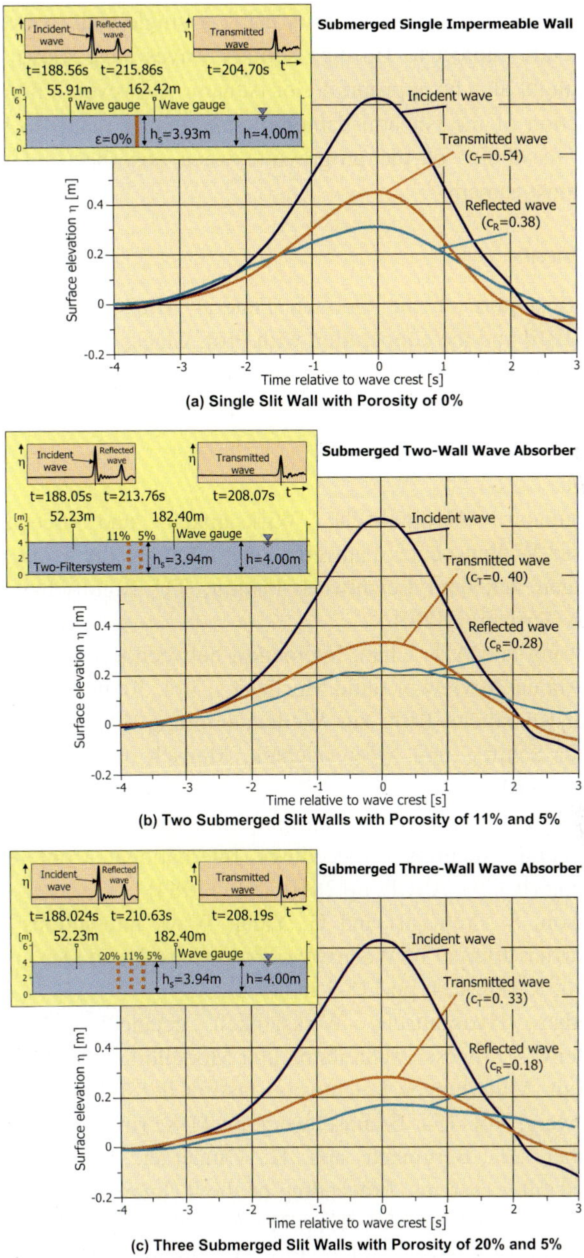

Figure 26. Incident, Reflected and Transmitted Solitary Wave by Submerged Slit Wall Systems–Results from Large-Scale Model Tests

Preliminary large-scale model tests using a single-wall, a two- and three-wall reef structure subject to 1m high solitary waves have shown that the new reef concept might also be appropriate for tsunami. However, a numerical model for the prediction of the tsunami damping performance needs to be developed to check the feasibility of the progressive wave absorber reef concept as a protection against tsunami.

Acknowledgments

The financial support of the German Federal Ministry for Education and Research (BMBF) and the cooperation with Prof. Clauss and Dr. Habel from the Technical University Berlin within the Joint Research Programme 03KIS 001/1 are gratefully acknowledged.

References

1. H. Bergmann, Hydraulische Wirksamkeit und Seegangsbelastung senkrechter Wellenschutz- bauwerke mit durchlässiger Front. PhD-Thesis, Mitteilungen aus dem Leichtweiss-Institut, TU Braunschweig, No. 147, 234p., (in German) (2000).
2. A.T. Chwang and A.T. Chan, Interaction between porous media and wave motion. Annual Review of Fluid Mechanics, Vol. 30, pp.53-84 (1998).
3. R. Habel, Künstliche Riffe zur Wellendämpfung. PhD-Thesis, TU Berlin, Institut für Schiffs- und Meerestechnik, Mensch und Buch Verlag, 112p, (in German) (2001).
4. K. Hagiwara, Analysis of upright structure for wave dissipation using integral equation. Proceedings International Conference Coastal Engineering (ICCE), ASCE, pp.2810-2826 (1984).
5. M. Isaacson, S. Premasiri and G. Yang, Wave interactions with vertical slotted barrier. Journal of Waterways, Port, Coastal and Ocean Engineering, ASCE, No. 3, Vol. 124, pp.118-126 (1998).
6. G. Koether, Hydraulische Wirksamkeit getauchter Einzelfilter und Filtersysteme – Prozessbeschreibung und Modellbildung für ein innovatives Riffkonzept. Mitteilungen Leichtweiss-Institut der Technischen Universität Braunschweig, Vol. 154, Braunschweig, pp.41-82 (in German) (2007).
7. G. Koether, H. Bergmann and H. Oumeraci, Wave attenuation by submerged filter systems. Proceeding of the 4. International Conference on Hydrodynamics (ICHD), Vol. II, pp.711-716, Yokohama, Japan (2000).
8. P. Lynett, K. Sitanggang and P. Liu, Development of a Boussinesq and RANS-VOF Hybrid wave model. Proceeding ICCE'06, pp.24-35 (2007).

9. J. Newe, Methodik für großmaßstäbliche 2D-Experimente zum Strandverhalten unter Sturmflutbedingungen. PhD-Thesis, TU Braunschweig, Leichtweiss-Institute für Wasserbau, (in German) (2004).
10. H. Oumeraci, Nonconventional wave damping structures, Handbook of Coastal and Ocean Engineering, Ed. Y.C. Kim, Chapter 12, World Scientific, Singapore, 29pp. (2008) (in print).
11. H. Oumeraci, G.F. Clauss, R. Habel and G. Koether, Unterwasserfiltersysteme zur Wellendämpfung. Abschlussbericht zum BMBF-Vorhaben "Unterwasserfiltersysteme zur Wellendämpfung". Final Research Report, (in German) (2001).
12. K.W. Pilarczyk, Design of low-crested (submerged) structures – an overview. Proceedings of the 6th International Conference on Coastal and Port Engineering in Developing Countries, Columbo, Sir Lanka (2003).
13. C.K. Sollitt, and R.H. Cross, Wave transmission through permeable breakwaters. Proceedings International Conference Coastal Engineering (ICCE), ASCE, part II, pp.1828-1846. (1972).
14. M. Stiassnie, E. Naheer, and I. Boguslavsky, Energy losses due to vortex shedding from the lower edge of a vertical pate attacked by surface waves. Proceedings of the Royal Society of London, Vol. 396, pp.131-142 (1984).
15. X. Yu, Diffraction of water waves by porous breakwaters. Journal of Waterways, Port, Coastal and Ocean Engineering, ASCE, No. 6, Vol. 121, pp.275-282 (1995).

EFFICIENT WAVE AND CURRENT MODELS FOR COASTAL STRUCTURES AND SEDIMENTS

NOBUHISA KOBAYASHI

Center for Applied Coastal Research, University of Delaware, Newark Delaware, 19716 USA

A computationally-efficient model based on time-averaged continuity, momentum and energy equations coupled with a probabilistic runup model is developed to predict the mean and standard deviation of the free surface elevation and horizontal fluid velocities above and inside a porous layer. The developed model is calibrated and verified using laboratory experiments on irregular breaking wave transmission over submerged porous breakwaters, irregular wave runup on permeable slopes, and irregular wave seepage and overtopping on wide permeable crests. Furthermore, the developed model coupled with a model for sand suspension and transport is used to predict irregular wave transformation and sediment transport from outside the surf zone to the lower swash zone on sand breaches. The coupled model is compared with laboratory experiments in which detailed measurements were made of the free surface elevations, velocities and sand concentrations. A bedload model is developed and added to this numerical model for the prediction of beach profile evolution including berm and dune erosion.

1. Introduction

The prediction of irregular wave runup, overtopping and transmission is necessary for the design of a coastal structure and in determining the landward extent of wave action on a beach. A large number of studies were performed to understand wave dynamics on inclined structures and beaches, as reviewed by Kobayashi (1999). The prediction of wave runup, overtopping and transmission was initially based on experiments and empirical formulas because of the complexity involved in wave breaking, runup, overtopping and transmission. Empirical formulas for coastal structures have been improved gradually to account for various factors (e.g., van der Meer and Janssen 1995; van Gent 2001; Pozueta *et al.* 2004) but are not versatile enough to deal with various combinations of different beaches and structures. Furthermore, the empirical formulas require the input of the representative height and period of incident waves at the toe of the structure which is normally located inside the surf zone during a severe storm. Consequently, a wave model will be necessary to predict the wave transformation from offshore to the toe of the structure.

As reviewed by Kobayashi (1995, 2003), time-dependent numerical models have been developed to predict the interactions of breaking and nonbreaking waves with inclined structures and beaches since Kobayashi *et al.* (1987) and Kobayashi and Otta (1987) showed the feasibility and utility of such a numerical model. The time-dependent numerical models predict the detailed temporal and spatial variations of the free surface elevation and fluid velocities which are needed to understand the complicated hydrodynamics. However, these models require significant computational effort and experience to run computer programs and obtain quantities of practical importance. Consequently, the time-dependent numerical models have not been applied routinely for practical applications. On the other hand, numerical models based on time-averaged momentum and energy equations such as that of Battjes and Stive (1985) are widely used to predict irregular wave breaking and wave setup on an impermeable beach of an arbitrary profile. However, these time-averaged models have not been extended to the swash zone on a beach because the swash zone is inherently time-dependent.

The brief review above indicates the need of a computationally-efficient, time-averaged model for predicting the irregular wave transformation from offshore to the swash zone on coastal structures and beaches. The time-averaged model is more empirical than the corresponding time-dependent model but can be calibrated using available data sets. The time-averaged models for coastal structures and beaches developed by the author with his graduate students and visiting scientists are described concisely in the following.

2. Time-Averaged Model for Permeable Slopes

Fig. 1 depicts a relatively steep permeable slope on a gently sloping beach where alongshore uniformity and normally incident irregular waves are assumed. The cross-shore coordinate x is positive onshore. The vertical coordinate z is positive upward with $z = 0$ at the still water level (SWL). The upper and lower boundaries of the permeable layer are located at $z = z_b$ and z_p, respectively, where the lower boundary is assumed impermeable to simplify the permeable structure. The beach is assumed to be impermeable and $z_b = z_p$ on the beach. Fig. 1 corresponds to an emerged permeable slope. For a submerged porous structure, $z_b < 0$ and a beach also exists landward of the submerged structure. The input profiles of $z_b(x)$ and $z_p(x)$ are arbitrary in the numerical model. The instantaneous water depth and free surface elevation are denoted by h and η, respectively, and $h = (\eta - z_b)$. The horizontal fluid velocity u is the

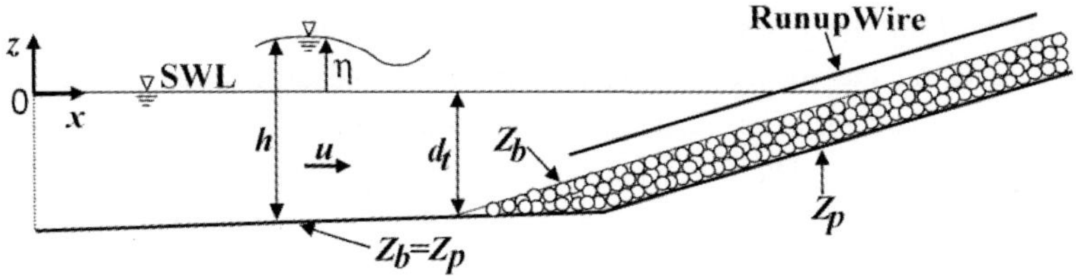

Fig. 1. Definition sketch for time-averaged model for irregular waves on permeable slope.

depth-averaged velocity. The still water depth at the toe of the emerged permeable slope is denoted by d_t.

The time-averaged continuity, momentum and energy equations given by Kobayashi *et al.* (2007a) are summarized in the following. These equations are applicable for the case of negligible reflected waves. The time averaged momentum and energy equations are expressed as

$$\frac{dS_{xx}}{dx} = -\rho g \bar{h} \frac{d\bar{\eta}}{dx} - \tau_b \ ; \quad \frac{dF}{dx} = -D_B - D_f - D_r \tag{1}$$

where S_{xx} = cross-shore radiation stress; ρ = fluid density, g = gravitational acceleration; $\bar{h}$ = mean water depth with the overbar denoting time averaging; $\bar{\eta}$ = wave setup or setdown; τ_b = time-averaged bottom shear stress; F = wave energy flux per unit width; and D_B, D_f and D_r = time-averaged energy dissipation rate per unit horizontal area due to wave breaking, bottom friction, and porous flow resistance, respectively.

Linear wave theory for onshore progressive waves is used to estimate S_{xx} and F where the root-mean-square wave height H_{rms} used by Battjes and Stive (1985) is defined as $H_{rms} = \sqrt{8}\ \sigma_\eta$ with σ_η = standard deviation of η.

$$S_{xx} = \rho g \sigma_\eta^2 (2n - 0.5) \ ; \quad F = \rho g C_g \sigma_\eta^2 \tag{2}$$

where $n = C_g/C_p$ with C_g and C_p = group velocity and phase velocity in the mean water depth $\bar{h}$ corresponding to the spectral peak period T_p of incident waves. The representative wave period of irregular waves is taken as T_p, but C_g and C_p equal $(g\bar{h})^{0.5}$ in shallow water.

The bottom shear stress τ_b and the corresponding dissipation rate D_f are expressed using the formulas based on the quadratic drag force based on the horizontal velocity u. The mean and standard deviation of u are denoted by $\bar{u}$ and σ_u, respectively. The Gaussian distribution of u and the equivalency of the time and probabilistic averaging are assumed to express τ_b and D_f in terms of $\bar{u}$ and σ_u

$$\tau_b = \frac{1}{2}\rho f_b \sigma_u^2 G_2(u_*); \quad D_f = \frac{1}{2}\rho f_b \sigma_u^3 G_3(u_*); \quad u_* = \frac{\bar{u}}{\sigma_u} \tag{3}$$

where f_b = bottom friction factor which is taken as $f_b = 0$ on the smooth slope and $f_b = 0.01$ on the stone slope, as calibrated by Kobayashi *et al.* (2007a). The analytical functions $G_2(r)$ and $G_3(r)$ for the arbitrary variable r are given by Kobayashi *et al.* (2005) and can be approximated as $G_2 \simeq 1.64\ r$ and $G_3 \simeq (1.6 + 2.6\ r^2)$ for $|r| < 1$.

The standard deviation σ_u is estimated using the relationship between σ_u and σ_η based on linear shallow-water wave theory (Kobayashi *et al.* 1998)

$$\sigma_u = \sigma_* (g\bar{h})^{0.5}; \quad \sigma_* = \sigma_\eta / \bar{h} \tag{4}$$

The mean $\bar{u}$ is estimated using the time-averaged, vertically-integrated continuity equation $(\sigma_u\sigma_\eta + \bar{u}\bar{h} + \bar{v}h_p) = 0$ with the condition of no net landward water flux in the absence of wave overtopping. In this equation, $\sigma_u\sigma_\eta$ is the onshore flux due to linear shallow-water waves (Kobayashi *et al.* 1998), $\bar{u}\bar{h}$ is the offshore flux due to the return current $\bar{u}$, and $\bar{v}h_p$ is the water flux inside the permeable layer of vertical height h_p due to the time-averaged horizontal discharge velocity $\bar{v}$. Substitution of Eq. (4) into the continuity equation yields

$$\bar{u} = -\left[\sigma_*^2 (g\bar{h})^{0.5} + \bar{v}h_p/\bar{h}\right]; \quad h_p = z_b - z_p \tag{5}$$

where $h_p = 0$ on the impermeable beach.

The energy dissipation rate D_r in Eq. (1) is estimated using the formula by Wurjanto and Kobayashi (1993) based on the discharge velocity v whose probability distribution is assumed to be Gaussian

$$D_r = \rho h_p \left[\alpha\sigma_v^2(1 + v_*^2) + \beta\sigma_v^3 G_3(v_*)\right]; \quad v_* = \bar{v}/\sigma_v \tag{6}$$

where σ_v = standard deviation of v; G_3 = same function as in Eq. (3) except for $r = v_*$; and α and β = laminar and turbulent flow resistance coefficients.

Kobayashi *et al.* (2007a) modified the formulas of α and β by van Gent (1995) for irregular waves in the form

$$\alpha = \alpha_o \left(\frac{1-n_p}{n_p}\right)^2 \frac{\nu}{D_{n50}^2}; \quad \beta = \left(\beta_1 + \frac{\beta_2}{\sigma_v}\right); \quad \beta_1 = \frac{\beta_o(1-n_p)}{n_p^3 D_{n50}}; \quad \beta_2 = \frac{7.5\,\beta_o(1-n_p)}{\sqrt{2}\,n_p^2 T_p} \tag{7}$$

where α_o and β_o = empirical parameters taken as α_o = 1,000 and β_o = 5, as explained by Kobayashi *et al.* (2007a); n_p = porosity of the stone; D_{n50} = nominal stone diameter defined as $D_{n50} = (M_{50}/\rho_s)^{1/3}$ with M_{50} = median stone mass and ρ_s = stone density; ν = kinematic viscosity of water ($\nu \simeq$ 0.01 cm^2/s); and T_p = spectral peak period. The mean $\overline{v}$ and standard deviation σ_v are estimated assuming the local force balance between the horizontal gradient of hydrostatic pressure and the flow resistance inside the permeable layer

$$(\alpha + 1.64\,\beta\sigma_v)\overline{v} = -g\frac{d\overline{\eta}}{dx}; \quad \alpha\sigma_v + 1.9\,\beta\sigma_v^2 = gk_p\overline{h}\sigma_* \tag{8}$$

where k_p = linear wave number based on $\overline{h}$ and T_p. Eq. (8) can be solved analytically to obtain σ_v and $\overline{v}$ for known $k_p, \overline{h}, \sigma_*$ and $d\overline{\eta}/dx$.

The energy dissipation rate D_B due to wave breaking in Eq. (1) is estimated using the formula by Battjes and Stive (1985) which is modified by Kobayashi *et al.* (2007a) as

$$D_B = \frac{\rho g a Q H_B^2}{4T_p}; \; a = \frac{T_p S_b}{3}\left(\frac{g}{\overline{h}}\right)^{0.5} \geq 1; \; \frac{Q-1}{\ln Q} = \left(\frac{H_{rms}}{H_m}\right)^2; \; H_m = \frac{0.88}{k_p}\tanh\left(\frac{\gamma k_p \overline{h}}{0.88}\right) \tag{9}$$

where a = empirical coefficient; Q = fraction of breaking waves with $Q = 0$ for no wave breaking and $Q = 1$ for all waves break; H_B = wave height used to estimate D_B; S_b = local bottom slope defined as $S_b = dz_b/dx$, which is positive for an upward slope; H_m = local depth-limited wave height with $H_m = \gamma\overline{h}$ in shallow water; and γ = breaker ratio parameter. The coefficient a with its lower limit of unity is the ratio between the wavelength and the horizontal length scale $(3\overline{h}/S_b)$ imposed by the small depth $\overline{h}$ and the bottom slope S_b near the shoreline or on the steep seaward slope of a submerged breakwater. This coefficient a increases D_B in the region of $a > 1$ and improves the formula by Battjes and Stive (1985). The requirement of $0 \leq Q \leq 1$ implies $H_{rms} \leq H_m$ but H_{rms} becomes larger than H_m in very shallow water. When $H_{rms} > H_m$, use is made of $Q = 1$ and $H_B = H_{rms}$ to estimate D_B instead of $H_B = H_m$ for $H_{rms} \leq H_m$. The values of γ calibrated by Battjes and Stive (1985) were in the range of 0.6 – 0.8. For the subsequent comparisons with experiments, γ was calibrated to obtain fair agreement between the measured and predicted cross-shore variations of σ_η, indicating the empirical nature of Eq. (9).

Eqs. (1) – (9) are solved using a finite difference method with constant nodal spacing Δx of a sufficient resolution near the shoreline. The bottom elevation $z_b(x)$ and the impermeable boundary $z_p(x)$ are specified as input. The stone is characterized by its nominal diameter D_{n50} and porosity n_p. The measured values of T_p, $\overline{\eta}$ and $H_{rms} = \sqrt{8}\ \sigma_\eta$ are specified at the seaward boundary $x = 0$ outside the surf zone. The landward-marching computation is continued until the landward boundary is reached or until the computed value of $\overline{h}$ or σ_η becomes of the order of 0.1 cm. The computation time is of the order of one second using a workstation. This computational efficiency allows the calibration of several empirical parameters and comparisons with a number of data sets.

The time-averaged model based on Eqs. (1) – (9) neglects reflected waves. Kobayashi *et al.* (2007b) estimated the degree of wave reflection in the following crude manner. The onshore energy flux F in Eq. (1) decreases landward due to wave breaking, bottom friction and flow resistance inside the permeable slope. The residual energy flux F_{sws} at the still water shoreline located at $z = 0$ is assumed to be reflected from the slope and propagate seaward. This assumption neglects the fact that the landward-marching computation is made without regard to wave reflection. The root-mean-square wave height $(H_{rms})_r$ due to the reflected wave energy flux is obtained as

$$(H_{rms})_r = \left[8F_{sws}/(\rho g C_g)\right]^{0.5} \tag{10}$$

Kobayashi *et al.* (2007b) also developed a probabilistic model for irregular wave runup using the computed $\overline{\eta}(x)$ and $\sigma_\eta(x)$ on the permeable slope. A runup wire was used in the subsequent experiments to measure the shoreline oscillations above the slope as shown in Fig. 1. The vertical height δ_r of the wire above the average stone surface was given in each experiment. The wire measures the instantaneous elevation $\eta_r(t)$ above SWL of the intersection between the wire and the free surface unlike a wave gauge that measures $\eta(t)$ at given x. Fig. 2 depicts an intuitive method used to estimate the mean $\overline{\eta_r}$ and standard deviation σ_r of $\eta_r(t)$. The probabilities of η exceeding $(\overline{\eta} + \sigma_\eta)$, $\overline{\eta}$ and $(\overline{\eta} - \sigma_\eta)$ are assumed to be the same as the probabilities of η_r exceeding $(\overline{\eta_r} + \sigma_r)$, $\overline{\eta_r}$ and

($\overline{\eta_r}$ − σ_r), respectively. The elevations of Z_1, Z_2 and Z_3 of the intersections of ($\overline{\eta}$ + σ), $\overline{\eta}$ and ($\overline{\eta}$ − σ) with the runup wire are obtained using the computed $\overline{\eta}(x)$ and $\sigma(x)$ together with the wire elevation $[z_b(x) + \delta_r]$. The obtained elevations are assumed to correspond to Z_1 = ($\overline{\eta_r}$ + σ_r), Z_2 = $\overline{\eta_r}$ and Z_3 = ($\overline{\eta_r}$ − σ_r). The mean and standard deviation of $\eta_r(t)$ are estimated as

$$\overline{\eta_r} = (Z_1 + Z_2 + Z_3)/3; \quad \sigma_r = (Z_1 - Z_3)/2 \tag{11}$$

where the use of Z_1, Z_2 and Z_3 to estimate $\overline{\eta_r}$ is slightly more reliable than $\overline{\eta_r}$ = Z_2 because the elevation Z_2 is somewhat sensitive to the detailed spatial variation of $\overline{\eta}(x)$.

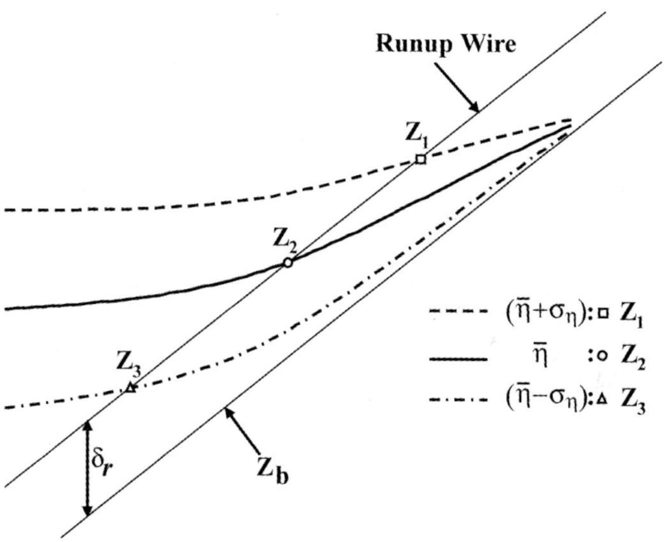

Fig. 2. Elevations Z_1, Z_2 and Z_3 of intersections of ($\overline{\eta}$ + σ_η), $\overline{\eta}$ and ($\overline{\eta}$ − σ_η) with runup wire where $\overline{\eta}$ and σ_η are the mean and standard deviation of free surface elevation.

The runup height R is defined as the crest height above SWL of the temporal variation of η_r. The measured time series of [$\eta_r(t)$ − $\overline{\eta_r}$] is analyzed using a zero-upcrossing method to identify the crests in the time series. This procedure is the same as that used for the analysis of the wave crests in the time series of $\eta(t)$ except that the wave crest is defined as the height above the mean water level. The probability distribution of linear wave crests is normally given by the

Rayleigh distribution. As a first approximation, the runup height ($R - \overline{\eta_r}$) above the mean level $\overline{\eta_r}$ is given by the Raleigh distribution in the absence of wave overtopping.

$$P(R) = \exp\left[-2\left(\frac{R-\overline{\eta_r}}{R_{1/3}-\overline{\eta_r}}\right)^2\right] \tag{12}$$

where $P(R)$ = exceedance probability of the runup height R above SWL; and $R_{1/3}$ = significant runup height defined as the average of 1/3 highest values of R. The mean $\overline{\eta_r}$ related to wave setup is normally neglected in Eq. (12) for the prediction of irregular wave runup on steep coastal structures [e.g., van der Meer and Janssen (1995)]. However, wave setup on gentler slopes is not negligible as was shown by Kobayashi *et al.* (2007b). It is necessary to express $R_{1/3}$ in terms of $\overline{\eta_r}$ and σ_r estimated using Eq. (11). If the probability distribution of η_r is approximately Gaussian, use may be made of ($R_{1/3} - \overline{\eta_r}$) $\simeq 2\sigma_r$ (Goda 2000). Kobayashi *et al.* (2007b) estimated $R_{1/3}$ as

$$R_{1/3} = \overline{\eta_r} + (2 + \tan\theta)\sigma_r \tag{13}$$

to improve the agreement where the slope $\tan\theta$ = 1/5 and 1/2 in their experiments.

2.1. *Wave Transmission*

Kobayashi *et al.* (2007a) conducted a reef breakwater experiment in a wave flume that was 33 m long, 0.6 m wide and 1.5 m high. Typical reef breakwaters in Japan are sufficiently submerged for their aesthetics and very wide to attenuate storm waves. An impermeable smooth beach with a 1/35 slope was placed in the flume. A submerged breakwater was constructed of angular stone on the 1/35 slope. The measured porosity and nominal diameter were n_p = 0.5 and D_{n50} = 3.4 cm. The seaward and landward slopes of the breakwater were 1/2.28 and 1/1.40, respectively. The width of the horizontal crest was 164 cm, that is, 48 D_{n50} and much wider than the crest width of a conventional low-crested rubble mound structure. The still water depth on the crest was in the range of 4 – 10 cm. The spectral peak period and root-mean-square wave height at $x = 0$ outside the surf zone were T_p = 2.32 s and H_{rms} = 10.28 – 10.81 cm. The number of test runs was 12.

The cross-shore variations of the wave setup $\overline{\eta}$ and the standard deviation $\sigma_\eta = H_{rms}/\sqrt{8}$ computed using Eq. (1) were compared with the measured values of $\overline{\eta}$ and σ_η at eight cross-shore locations. The numerical model predicted the landward increase of $\overline{\eta}$ and the landward decrease of σ_η due to wave breaking on the steep seaward slope and wide crest of the breakwater fairly well. The agreement for the horizontal velocity measured at three cross-shore locations was marginal partly because Eqs. (4) and (5) are based on the depth-averaged horizontal velocity, whereas the velocity was measured at a specified elevation. The numerical model will be useful in designing the optimal geometry of a reef breakwater at a specific field site.

A reef breakwater with a wide crest is normally designed for no or little damage during a design storm. This practice increases the cost of the structure. Ota *et al.* (2006) conducted an experiment and showed that a deformed reef breakwater did not increase wave transmission because the crest height of the deformed breakwater increased due to the landward stone movement on its wide crest. They also showed that the present numerical model could predict the wave transformation above the deformed reef breakwater as well.

2.2. *Wave Runup*

Meigs *et al.* (2004) compared this time-averaged model with the corresponding time-dependent model by Wurjanto and Kobayashi (1993) which was verified using three tests for irregular wave runup on a 1/3 slope with a thick gravel layer. The computed cross-shore variations of $\overline{\eta}$, σ_η, $\overline{u}$, σ_u, $\overline{v}$ and σ_v on the 1/3 slope were shown to be in reasonable agreement for the three tests except that the time-averaged model does not include the region where the free surface is inside the permeable layer. This comparison may be regarded to verify Eq. (8) for the mean and standard deviation of the discharge velocity v in the permeable layer which was not measured in any experiment described here.

Kobayashi *et al.* (2007b) compared the time-averaged model with the irregular wave runup experiments on 1/2 and 1/5 permeable slopes conducted by Kearney and Kobayashi (2000) and de los Santos *et al.* (2005), respectively. The number of tests was 27 and 30 for the 1/2 and 1/5 slopes, respectively. The numerical model was shown to predict the cross-shore variations of $\overline{\eta}$, σ_η, $\overline{u}$ and σ_u fairly accurately. Furthermore, the numerical model coupled with Eqs. (11) – (13) predicted the significant and 2% runup heights within an error of about 20%. The wave reflection coefficient estimated using Eq. (10) was less accurate but is useful in estimating the order of magnitude of wave reflection from permeable slopes.

2.3. *Wave Overtopping*

Kobayashi *et al.* (2007c) extended the time-averaged model based on Eqs. (1) – (13) to include the landward water flux q_{os} in Eq. (5) due to wave seepage and overtopping. The flux q_{os} is the sum of the overtopping rate q_o and the seepage rate q_s computed after the landward marching computation. Since q_{os} is unknown before the computation, the landward marching computation was repeated until the computed q_{os} converged after several iterations. The formulas for q_o and q_s based on the computed water flux and wave setup on the permeable slope were developed using the 12 seepage tests and 10 overtopping/seepage tests conducted on a 1/5 slope by de los Santos *et al.* (2006) as well as the 12 overtopping tests conducted on a 1/2 slope by Kobayashi and Raichle (1994). The 22 tests on the 1/5 slope were performed by modifying the crest geometry of the permeable slope in the wave runup experiment by de los Santos *et al.* (2005).

The measured wave runup distributions were fitted to the Weibull distribution whose shape parameter increased with the increase of the wave overtopping probability. For the case of no or little wave overtopping, this Weibull distribution reduces to the Rayleigh distribution given by Eq. (12). The wave overtopping rate q_o normalized by the wave-induced water flux at the still water shoreline was shown to depend on the wave overtopping probability and the horizontal number of stones above the maximum wave setup. A simple formula for the seepage rate q_s was proposed by analyzing the seepage flow driven by the wave setup on the seaward slope. The extended model predicted the significant and 2% runup height within an error of 20% and the combined overtopping and seepage rate q_{os} within a factor of two. However, the model will need to be evaluated using more extensive data sets because the wave overtopping and seepage rates are very sensitive to the detailed crest geometry and the degree of infiltration above the maximum wave setup.

3. Time-Averaged Model for Sediment Transport

The time-averaged model developed here is an extension of the Dutch models by Battjes and Stive (1985), Reniers and Battjes (1997), and Ruessink *et al.* (2001). Fig. 3 shows obliquely incident irregular waves on a straight shoreline where the cross-shore coordinate x is positive onshore and the longshore coordinate y is positive in the downwave direction. The depth-averaged cross-shore and longshore velocities are denoted by U and V, respectively. Incident waves are assumed to be unidirectional with θ = incident angle relative to the shore normal and uniform in the longshore direction. The time-averaged cross-shore momentum, longshore momentum and energy equations for the case of longshore uniformity are expressed as

$$\frac{dS_{xx}}{dx} = -\rho g \bar{h} \frac{d\bar{\eta}}{dx} - \tau_{bx}; \quad \frac{dS_{xy}}{dx} = -\tau_{by}; \quad \frac{dF_x}{dx} = -D_B - D_f \tag{14}$$

where S_{xx} = cross-shore radiation stress; ρ = fluid density; g = gravitational acceleration; $\bar{h}$ = mean water depth given by $\bar{h}$ = $(\bar{\eta} - z_b)$ with $\bar{\eta}$ = mean free surface elevation and z_b = bottom elevation; τ_{bx} = cross-shore bottom stress; S_{xy} = shear component of the radiation stress; τ_{by} = longshore bottom stress; F_x = cross-shore energy flux; and D_B and D_f = energy dissipation rates due to wave breaking and bottom friction, respectively. The terms τ_{bx} and D_f in Eq. (14) are normally neglected but included here because of the importance of bottom friction for sediment transport. On the other hand, the effects of wind, tide and lateral mixing are neglected in Eq. (14) for simplicity.

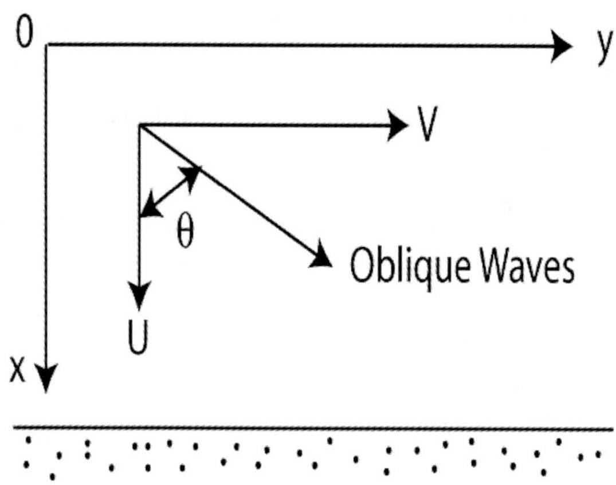

Fig. 3. Obliquely incident irregular waves on straight shoreline.

Linear wave theory for onshore progressive waves are used to estimate S_{xx}, S_{xy} and F_x

$$S_{xx} = (En + M_r)\cos^2\theta + E\left(n - \frac{1}{2}\right); \quad S_{xy} = (En + M_r)\cos\theta\sin\theta; \quad F_x = EC_g\cos\theta \tag{15}$$

with

$$E = \rho g \sigma_\eta^2; \quad n = C_g / C_p; \quad \sin\theta / C_p = \alpha; \quad M_r = \rho C_p q_r \tag{16}$$

where E = specific wave energy; σ_η = standard deviation of the free surface elevation η related to the root-mean-square wave height $H_{rms} = \sqrt{8}\,\sigma_\eta$; C_g and C_p = group velocity and phase velocity, respectively, in the mean water depth $\overline{h}$ corresponding to the spectral peak period T_p; α = constant in the region $x \geq 0$ based on Snell's law; and q_r and M_r = volume and momentum fluxes of a roller propagating with the speed C_p. It is noted that the roller energy density used by Ruessink *et al.* (2001) corresponds to $M_r/2$. The roller effect has been represented by its area or energy (Svendsen 1984) but the roller volume flux has been used by Kobayashi *et al.* (2005) who have found that the roller effect is the most apparent in the increase of undertow current for normally incident waves. Kobayashi *et al.* (2007d) found the roller effect necessary to predict the cross-shore distributions of the longshore current and sediment transport rate. The energy equation for the roller may be expressed as (Ruessink *et al.* 2001)

$$\frac{d}{dx}(\rho C_p^2 q_r \cos\theta) = D_B - D_r; \quad D_r = \rho g \beta_r q_r \tag{17}$$

where the roller dissipation rate D_r is assumed to equal the rate of work to maintain the roller on the wave-front slope β_r of the order of 0.1.

The time-averaged bottom shear stresses and dissipation rate are expressed as

$$\tau_{bx} = \frac{1}{2}\rho f_b \overline{UU_a}; \quad \tau_{by} = \frac{1}{2}\rho f_b \overline{VU_a}; \quad D_f = \frac{1}{2}\rho f_b \overline{U_a^3}; \quad U_a = (U^2 + V^2)^{0.5} \tag{18}$$

where f_b = bottom friction factor taken as $f_b = 0.015$ for sand beaches (Kobayashi *et al.* 2005); and the overbar indicates time averaging. Linear shallow-water wave theory has been used to find the approximate local relationships between the free surface elevation and the depth-averaged velocity in the direction of wave propagation [e.g., Kobayashi *et al.* (1998)]. The velocities U and V in Eq. (18) are assumed to be expressed as

$$U = \overline{U} + U_T \cos\theta; \quad V = \overline{V} + U_T \sin\theta; \quad U_T = (g/\overline{h})^{0.5}(\eta - \overline{\eta}) \tag{19}$$

where $\overline{U}$ and $\overline{V}$ = depth-averaged cross-shore and longshore currents; and U_T = oscillatory horizontal velocity with zero mean. Eq. (19) yields

$$\sigma_U = \sigma_T \cos\theta; \quad \sigma_V = \sigma_T \sin\theta; \quad \sigma_T = (g\overline{h})^{0.5}\sigma_*; \quad \sigma_* = \sigma_\eta / \overline{h} \tag{20}$$

where σ_U, σ_V and σ_T = standard deviations of U, V and U_T, respectively.

Assuming the equivalency of the time and probabilistic averaging as well as the Gaussian distribution of U_T, Eq. (18) is approximated as

$$\tau_{bx} = \frac{1}{2}\rho f_b \sigma_T^2 G_{bx}; \quad \tau_{by} = \frac{1}{2}\rho f_b \sigma_T^2 G_{by}; \quad D_f = \frac{1}{2}\rho f_b \sigma_T^3 G_f \tag{21}$$

with

$$G_{bx} = \int_{-\infty}^{\infty} F_U F_a f(r) dr; \quad G_{by} = \int_{-\infty}^{\infty} F_V F_a f(r) dr; \quad G_f = \int_{-\infty}^{\infty} F_a^3 f(r) dr \tag{22}$$

$$F_U = \frac{\overline{U}}{\sigma_T} + r\cos\theta; \; F_V = \frac{\overline{V}}{\sigma_T} + r\sin\theta; \; F_a = (F_U^2 + F_V^2)^{0.5}; \; f(r) = \frac{1}{\sqrt{2\pi}}\exp\left(-\frac{r^2}{2}\right) \tag{23}$$

where the mean and standard deviation of the Gaussian variable $r = U_T/\sigma_T$ is zero and unity, respectively. The numerical integration with respect to r in Eq. (22) is performed in the range of $-5 < r < 5$ where $f(r) = 1.5 \times 10^{-6}$ for $r^2 = 25$.

The longshore momentum equation in Eq. (14) is used to obtain the value of G_{by} and the corresponding longshore current $\overline{V}$ using a bisection method. It is more efficient computationally to adopt the following explicit relationship between G_{by} and $\overline{V}$ obtained by Feddersen *et al.* (2000) using field data and probabilistic analyses:

$$G_y = \frac{\overline{V}}{\sigma_T}\left[1.16^2 + \left(\frac{\overline{V}}{\sigma_T}\right)^2\right]^{0.5} \tag{24}$$

The difference between the values of G_{by} computed using Eqs. (22) and (24) is less than about 20% for the computations made by Kobayashi *et al.* (2007d). This difference is less than the uncertainty of the bottom friction factor f_b. As a result, Eq. (24) is adopted in the following. On the other hand, the cross-shore momentum and energy equations in Eq. (14) can be solved numerically to obtain $\overline{\eta}$ and σ_η in the same way as in Section 2. The depth-integrated continuity equation of water on the beach, in the absence of wave overtopping and overwash, is expressed as $(\overline{hU} + q_r \cos\theta) = 0$ with $h = (\eta - z_b)$. Using Eqs. (19) and (20), the continuity equation yields

$$\overline{U} = -\sigma_U \sigma_* \left(1 + \sqrt{\frac{\overline{h}}{g}} \frac{q_r}{\sigma_\eta^2} \right) \tag{25}$$

The energy dissipation rate D_B due to wave breaking in Eq. (14) is estimated using Eq. (9). The effect of the local bottom slope $S_b = dz_b/dx$ is included in the roller slope β_r in Eq. (17)

$$\beta_r = (0.1 + S_b) \geq 0.1 \tag{26}$$

where the increase of β_r due to S_b is introduced to reduce the increase of q_r in Eq. (17) resulting from the increased D_B in the region of small $\overline{h}$ and large S_b as explained below Eq. (9). The formula in Eq. (26) may need to be improved in the future because the slope effects on D_B and β_r have been examined very little.

Eqs. (14) – (17) and (20) – (26) are solved using a finite difference method of constant grid spacing Δx of the order of 1 cm to resolve the detail in the swash zone where Δx was not increased seaward because the computation time was of the order of 10 s. The measured bottom elevation $z_b(x)$ is specified in the computation domain $x \geq 0$ where $x = 0$ is the seaward boundary outside the surf zone. The measured values of T_p, $\overline{\eta}$, $H_{rms} = \sqrt{8}\,\sigma_\eta$ and θ as well as $q_r = 0$ at $x = 0$ are specified as input. The landward-marching computation is continued until the computed value of $\overline{h}$ is of the order 0.1 cm. The computation is made with and without the roller effect. For the case of no roller, $q_r = 0$, $D_r = D_B$ and Eq. (17) is not used.

After the landward-marching computation, the suspended sediment volume $\overline{V_c}$ per unit horizontal area is estimated using the sediment suspension model by Kobayashi and Johnson (2001)

$$\overline{V_c} = \frac{e_B D_r + e_f D_f}{\rho g(s-1)w_f} \tag{27}$$

where s and w_f = specific gravity and fall velocity of the sediment, respectively, and e_B and e_f = suspension efficiencies associated with breaking waves and bottom friction, respectively. Use has been made of $e_f = 0.01$ and $e_B = 0.002$ or 0.005. the longshore suspended sediment transport rate q_s is simply estimated as $q_s = \overline{V}\,\overline{V_c}$ where $\overline{V}$ = computed longshore current.

3.1. *Cross-Shore Suspended Sand Transport*

Kobayashi *et al.* (2005) conducted three tests in a wave flume and investigated time-averaged suspended sediment transport processes under normally-incident irregular breaking waves on equilibrium beaches consisting of fine sand. Free surface elevations were measured at ten locations for each test. Velocities and concentrations were measured in the vicinity of the bottom at 94 elevations. The relations among the three turbulent velocity variances were found to be similar to those for the bottom boundary layer flow. The vertical variation of the mean concentration $\overline{C}$ was fitted by the exponential and power-form distributions equally well. The vertical variation of the mean horizontal velocity, which causes offshore transport, was fitted by a parabolic profile fairly well. The correlation coefficient between the horizontal velocity and concentration, which results in onshore transport, was of the order of 0.1 and decreased upward linearly. The offshore and onshore transport rates of suspended sediment were expressed as

$$q_{\text{off}} = 0.9\overline{U}\,\overline{V_c}; \quad q_{\text{on}} = 0.8\sigma_* \sigma_U \overline{V_c} \tag{28}$$

where $\sigma_* = \sigma_\eta / \overline{h}$; and $\overline{U}$ and σ_U = mean and standard deviation of the depth-averaged velocity U predicted using Eqs. (20) and (25); and $\overline{V_c}$ = suspended sediment volume predicted by Eq. (27).

The time-averaged numerical model for the normally-incident waves with $\cos\theta = 1$ and no longshore current was compared with the measured cross-shore variations of $\overline{\eta}$, σ_η, $\overline{U}$, σ_U, $\sqrt{k}$ and $\overline{V_c}$ for the three tests where k = time-averaged turbulent kinetic energy per unit mass. The agreement was fair but $\overline{U}$ was difficult to predict accurately. The measured $\sqrt{k}$ was found to be more related to the turbulent velocity estimated from the energy dissipation rate D_f due to bottom friction. The suspended sediment volume $\overline{V_c}$ could be predicted only within a factor of about 2. The roller effect represented by the roller volume flux q_r did not necessarily improve the agreement for the three tests.

3.2. *Longshore Suspended Sand Transport*

Kobayashi *et al.* (2007d) and Agarwal *et al.* (2006) compared the time-averaged model for obliquely incident waves with the corresponding time-dependent model by Kobayashi and Karjadi (1996) who compared their model with field data. The measured and predicted cross-shore variations of σ_η and the longshore current $\overline{V}$ were in fair agreement. The time-averaged model is limited to the lower swash zone unlike the time-dependent model which can predict the shoreline movement in the swash zone.

The time-averaged model was also compared with the spilling and plunging wave tests conducted in the Large-Scale Sediment Transport Facility at the US Army Engineer Research and Development Center (Wang *et al.* 2002). The measured and predicted cross-shore variations of $\overline{\eta}$, σ_η, $\overline{U}$, σ_U, $\overline{V}$, σ_V, $\overline{V_c}$, q_{off} and q_s were in fair agreement where the offshore suspended sediment transport rate q_{off} was predicted using Eq. (28) and the longshore suspended sediment transport rate q_s was estimated as $q_s = \overline{V}\,\overline{V_c}$. However, the degree of the agreement for q_s was difficult to pinpoint because the longshore sediment transport rate measured using downdrift bottom traps included bedload, whereas no bedload was included in the numerical model.

3.3. *Bedload and Suspended Sand Transport and Profile Evolution*

Schmied *et al.* (2006) expanded the time-averaged model to predict the net cross-shore sediment transport rates of suspended sediment and bedload. The suspended sediment model is based on Eq. (28) which predicts the offshore suspended sediment transport due to undertow current and the onshore suspended sediment transport due to the correlation of the time-varying sediment concentration and horizontal velocity (more sediment suspension under breaking waves with the onshore fluid velocity). The net suspended sediment transport is offshore but relatively small. The probability of sediment suspension is included so that this suspended sediment model may also be applied to coarse sediments for which sediment suspension may be limited by their large settling velocities. A new formula for the net bedload transport rate is developed by synthesizing available formulas and data. The probability of sediment movement is included so that this bedload formula may be applied to coarse sediments that may not move continuously under wave action. The new formula can predict the measured net onshore transport rates on rippled and sheet-flow beds in a water tunnel and a large wave flume. The combined bedload and suspended sediment transport model predicts the equilibrium profile proposed by Dean (1991) under certain assumptions. The model also predicts the terraced and barred equilibrium profiles under different spectral peak periods produced by Kobayashi *et al.* (2005). Furthermore, this time-averaged model coupled with the continuity equation of bottom sediment was compared with erosional and accretional beach profile evolution tests in a wave flume. The numerical profile evolution model did not always predict the fairly subtle profile changes of the order of 5 cm in the small-scale tests.

Payo *et al.* (2006) conducted two berm erosion tests in a wave basin where four still water levels in the basin were used for each test to produce the sequence of foreshore erosion, berm overwash, berm erosion, and minor dune overtopping. The test was terminated when the berm erosion reached the toe of the dune. Erosion was relatively rapid after the berm was overwashed. The numerical profile evolution model underpredicted the berm erosion considerably. They neglected the onshore bedload transport rate and included the offshore suspended sediment transport rate only to reproduce the observed berm erosion reasonably. This simplified approach can predict erosion only.

To remedy this shortcoming, Kobayashi *et al.* (2007e) added the effect of a steep bottom slope to the formulas for bedload and suspended load. The proposed bottom slope function for bedload depends on the local bottom slope relative to the limiting slope of approximately 0.6. This empirical function reduces the onshore bedload transport rate on a steep upward slope and reverses the bedload transport direction on the upward slope exceeding about 0.3. The effect of a steep slope on the offshore suspended sand transport rate is taken into account by the use of the actual slope length exposed to wave action but is relatively small. Kobayashi *et al.* (2007e) compared the improved profile evolution model with seven small-scale tests including the two berm erosion tests by Payo *et al.* (2006) and the seven large-scale tests including dune erosion tests reported by Roelvink and Reniers (1995). The numerical model predicted the significant dune and berm erosion within a factor of about 2 and the minor profile changes within errors of 5 cm and 20 cm in the small-scale and large-scale tests, respectively.

4. Conclusions

A time-averaged model for irregular waves is developed to predict wave transmission, runup and overtopping of porous coastal structures as well as to predict wave transformation and sediment transport on beaches. The time-averaged model is more empirical than the corresponding time-dependent model (Kobayashi 1999, 2003) but the computation time is reduced by a factor of 10^{-3}. Furthermore, no numerical difficulty is encountered unlike the time-dependent model with a moving shoreline. This computational efficiency allows one to calibrate the model using a number of data sets for specific applications.

The numerical model is relatively simple and versatile. This may allow one to extend the model for various applications such as damage progression of rubble mound structures, toe erosion, sediment transport on gravel beaches, dune overwash and breaching. However, each extension and application will require significant efforts and reliable data. The alongshore variability of damage and erosion is always present in nature but such variability is very difficult to predict deterministically.

References

Agarwal, A., Kobayashi, N., and Johnson, B.D. (2006). "Longshore suspended sediment transport in surf and swash zones." *Coastal Engineering 2006, Proc. 30th Coastal Engineering Conf.*, World Scientific, Singapore.

Battjes, J.A., and Stive, M.J.F. (1985). "Calibration and verification of a dissipation model for random breaking waves." *J. Geophys. Res.*, 90(C5), 9159 – 9167.

Dean, R.G. (1991). "Equilibrium beach profile: Characteristics and applications." *J. Coastal Res.*, 7(1), 53 – 84.

de los Santos, F.J., Kobayashi, N., Meigs, L.E., and Losada, M.A. (2005). "Irregular wave runup on porous structures and cobble beaches." *Proc. Waves'2005 Conf.*, ASCE, Paper No. 30, 1 – 10.

de los Santos, F.J., Kobayashi, N., and Losada, M. (2006). "Irregular wave runup and overtopping on revetments and cobble beaches." *Coastal Engineering 2006, Proc. 30th Coastal Engineering Conf.*, World Scientific, Singapore.

Feddersen, F., Guza, R.T., Elgar, S., and Herbers, T.H.C. (2000). "Velocity moments in alongshore bottom stress parameterization." *J. Geophys. Res.*, 105(C4), 8673 – 8686.

Goda, Y. (2000). *Random Seas and Design of Maritime Structures.* World Scientific, Singapore.

Kearney, P.G., and Kobayashi, N. (2000). "Time-averaged probabilistic model for irregular wave runup on coastal structures." *Coastal Engineering 2000*, Proc. 27th Coastal Engineering Conf., ASCE, Reston, Va., 2004 – 2017.

Kobayashi, N., and Otta, A.K. (1987). "Hydraulic stability analysis of armor units." *J. Waterw., Port, Coastal, Ocean Eng.*, 113(2), 171 – 186.

Kobayashi, N., Otta, A.K., and Roy, I. (1987). "Wave reflection and runup on rough slopes." *J. Waterw., Port, Coastal, Ocean Eng.*, 113(3), 282 – 298.

Kobayashi, N., and Raichle, A.W. (1994). "Irregular wave overtopping of revetments in surf zones." *J. Waterw., Port, Coastal, Ocean Eng.*, 120(1), 56 – 73.

Kobayashi, N. (1995). "Numerical models for design of inclined structures." *Wave Forces on Inclined and Vertical Wall Structures*, ASCE, 118 – 139.

Kobayashi, N., and Karjadi, E.A. (1996). "Obliquely incident irregular waves in surf and swash zones." *J. Geophys. Res.*, 101(C3), 6527 – 6542.

Kobayashi, N., Herrman, M.N., Johnson, B.D., and Orzech, M.D. (1998). "Probabilistic distribution of surface elevation in surf and swash zones." *J. Waterw., Port, Coastal, Ocean Eng.*, 124(3), 99 – 107.

Kobayashi, N. (1999). "Wave runup and overtopping on beaches and coastal structures." *Advances in Coastal and Ocean Engineering*, World Scientific, Singapore, 5, 95 – 154.

Kobayashi, N., and Johnson, B.D. (2001). "Sand suspension, storage, advection and settling in surf and swash zones." *J. Geophys. Res.*, 106(C5), 9363 – 9376.

Kobayashi, N. (2003). "Numerical modeling as a design tool for coastal structures." *Advances in Coastal Structure Design*, ASCE, Reston, Va., 80 – 96.

Kobayashi, N., Zhao, H., and Tega, Y. (2005). "Suspended sand transport in surf zones." *J. Geophys. Res.*, 110, C12009, doi:10.1029/2004JC002853.

Kobayashi, N., Meigs, L.E., Ota, T., and Melby, J.A. (2007a). "Irregular breaking wave transmission over submerged porous breakwaters." *J. Waterw., Port, Coastal, Ocean Eng.*, 133(2), 104-116.

Kobayashi, N., de los Santos, F.J., and Kearney, P.G. (2007b). "Time-averaged probabilistic model for irregular wave runup on permeable slopes." *J. Waterw., Port, Coastal, Ocean Eng.*, (in press).

Kobayashi, N., and de los Santos, F.J. (2007c). "Irregular wave seepage and overtopping of permeable slopes." *J. Waterw., Port, Coastal, Ocean Eng.*, 133(4).

Kobayashi, N., Agarwal, A., and Johnson, B.D. (2007d). "Longshore current and sediment transport on beaches." *J. Waterw., Port, Coastal, Ocean Eng.*, 133(4).

Kobayashi, N., Payo, A., and Schmied, L.D. (2007e). "Cross-shore suspended sand and bedload transport on beaches." *J. Geophys. Res.*, (submitted).

Meigs, L.E., Kobayashi, N., and Melby, J.A. (2004). "Cobble beaches and revetments." *Coastal Engineering 2004*, Proc. 29th Coastal Engineering Conf., World Scientific, Singapore, 3865 – 3877.

Ota, T., Kobayashi, N., and Kimura, A. (2006). "Irregular wave transformation over deforming submerged structures." *Coastal Engineering 2006, Proc. 30th Coastal Engineering Conf.*, World Scientific, Singapore.

Payo, A., Kobayashi, N., and Kim, K.H. (2006). "Beach nourishment strategies." *Coastal Engineering 2006, Proc. 30th Coastal Engineering Conf.*, World Scientific, Singapore.

Pozueta, B., van Gent, M.R.A., and van den Boogaard, H. (2004). "Neural network modeling of wave overtopping at coastal structures." *Coastal Engineering 2004*, Proc. 29th Coastal Engineering Conf., World Scientific, Singapore, 4275 – 4287.

Reniers, A.J.H.M., and Battjes, J.A. (1997). "A laboratory study of longshore currents over barred and non-barred beaches." *Coastal Eng.*, 30, 1 – 21.

Roelvink, J.A., and Reniers, A.J.H. (1995). "LIP 11 D Delta Flume experiments." Data Rep. H2130, Delft Hydraulics, Delft, The Netherlands.

Ruessink, B.G., Miles, J.R., Feddersen, F., Guza, R.T., and Elgar, S. (2001). "Modeling the alongshore current on barred beaches." *J. Geophys. Res.*, 106(C10), 22,451 – 22,463.

Schmied, L.D., Kobayashi, N., Puleo, J.A., and Payo, A. (2006). "Cross-shore suspended sand transport on beaches." *Coastal Engineering 2006, Proc. 30th Coastal Engineering Conf.*, World Scientific, Singapore.

Svendsen, I.A. (1984). "Mass flux and undertow in a surf zone." *Coastal Eng.*, 8, 347 – 365.

van der Meer, J.A., and Janssen, P.F.M. (1995). "Wave run-up and wave overtopping at dikes." *Wave Forces on Inclined and Vertical Wall Structures*, ASCE, Reston, Va., 1 – 27.

van Gent, M.R.A. (1995). "Porous flow through rubble-mound materials." *J. Waterw., Port, Coastal, Ocean Eng.*, 121(3), 176 – 181.

van Gent, M.R.A. (2001). "Wave runup on dikes with shallow foreshores." *J. Waterw., Port, Coastal, Ocean Eng.*, 127(5), 254 – 262.

Wang, P., Ebersole, B.A., Smith, E.R., and Johnson, B.D. (2002). "Temporal and spatial variations of surf-zone currents and suspended sediment concentration." *Coastal Eng.*, 46, 175 – 211.

Wurjanto, A., and Kobayashi, N. (1993). "Irregular wave reflection and runup on permeable slopes." *J. Waterw., Port, Coastal, Ocean Eng.*, 119(5), 537 – 557.

TOWARDS AN ENGINEERING APPLICATION OF COBRAS (CORNELL BREAKING WAVE AND STRUCTURES)

INIGO J. LOSADA, JAVIER L. LARA, RAUL GUANCHE and JOSE M. GONZALEZ-ONDINA

Ocean and Coastal Research Group, Universidad de Cantabria, Instituto de Hidráulica Ambiental "IHCantabria". Avda. De los Castros s/n Santander, 39005 Spain

In this paper the capability of the numerical model named COBRAS, based on the Reynolds Averaged Navier-Stokes (RANS) equations, to simulate the most relevant hydrodynamic near-field processes that take place in the interaction between waves and coastal structures is presented. Initially developed by Lin and Liu (1998), the Ocean and Coastal Research Group at the University of Cantabria has been working on the improvement of the model capabilities in order to transform it into a tool useful for engineering applications. The model considers wave reflection, transmission, overtopping and breaking due to transient nonlinear waves including turbulence in the fluid domain and in the permeable regions for any kind of geometry and number of layers. The model is now able to deal with long irregular incident wave time series impinging on almost any structure typology. The new version COBRAS-UC has been extensively validated against small and large-scale laboratory data as well as field data providing excellent results. This model represents a substantial improvement in the numerical modelling of wave and coastal structures interaction.

1. Introduction

Coastal structures are built to protect coastal areas from erosion and flooding, as well as to shelter ports and marinas from wave action. The wave and structure interaction process depends on incident wave conditions and structure typology. In the past, in order to provide design guidelines for the hydraulic response of coastal structures, many semi-empirical formulations based on flume and basin experiments have been developed. Such formulae try to consider and parameterize the most relevant variables ruling the process taking part in the wave and structure interaction phenomenon, mainly, reflection, run-up, rundown, overtopping and transmission. An extensive number of very successful semi-empirical formulations are available in the literature and can be found, as part of the state of the art in coastal structures design in several books and manuals (e.g. Losada, 2005; Burharth and Hughes, 2006).

These empirical formulations have been the main tool for coastal structures design and have proved to be successful. However, they are based on flume or wave basin experiments covering a limited number of typologies and setups. Furthermore, many of them are based on a reduced set of incident wave conditions including mostly regular waves or narrow-banded spectra. Their use out of range, a problem often faced for design, may require extrapolations increasing uncertainties, and/or lead to important errors. To overcome these limitations, computational modeling of wave interaction with coastal structures has grown as a serious complementary approach during the last decade (Losada, 2003), especially when pre-designing the hydraulic response of some typologies of coastal structures.

The most relevant numerical models currently available for wave and coastal structures interaction are different versions of models solving the nonlinear shallow equations (NSWE) (e.g. Kobayashi and Wurjanto, 1989; Mingham and Causson, 1998; Hu *et al.*, 2000; Hubbard and Dodd, 2002; Stansby and Feng, 2004); models based on Reynolds Average Navier-Stokes equations including any kind of free surface-tracking or -capturing technique (e.g. Troch and de Rouck, 1998; Kawasaki, 1999; Li *et al.*, 2004) and, more recently, particle methods like the Smoothed Particle Hydrodynamics (SPH) (e.g. Dalrymple *et al.*, 2001; Gotoh *et al.*, 2004; Shao, 2006), in which the governing partial differential equations of continuum fluid dynamics, are transformed into particle form by integral equations through the use of an interpolation function that gives kernel estimation of the field variables at a point. Turbulence is usually treated with a Large Eddy Simulation approach.

The use of NSWE places severe restrictions to real applications inherent, essentially, to the intrinsic requirement to satisfy the shallow water limit. This restriction is especially crude when considering high frequency components in the incident wave spectrum and also requires that the offshore boundary condition of the numerical model be located close to the structure. Other restrictions are associated with the semi-empirical introduction of breaking, porous flow modeling or the difficulty in simulating complicated free surfaces.

However, they are computationally very efficient providing the chance to simulate wave trains including about 1000 waves very rapidly, which may be of importance, for example, to analyze extreme statistics of wave overtopping.

Regarding the SPH technique, even if the weakly incompressible or the incompressible method is used, this technique has several disadvantages. The high number of particles needed to have a reliable solution; the use of fixed particle spacing; the limited validation available and a very low computational efficiency are some of the issues that need to be solved before considering SPH for practical purposes.

The numerical model's ability to analyze wave and structure interaction problems successfully depend mainly on the equations and solving technique used, Losada (2003). Moreover, the leap to its application to case studies requires robustness and efficiency in terms of computation as well as a thorough validation process only available today for a limited set of models.

2. Cornell Wave Breaking and Structures (COBRAS) model

2.1. *COBRAS*

As said above, a third alternative is the numerical modelling based on the Navier-Stokes equations. The main advantage of these models is that they overcome the limitations associated with using a given wave theory and include wave breaking thanks to incorporating a turbulence model and considering Reynolds Average (RANS) equations. Those including a Volume of Fluid technique to track the free surface are becoming very powerful since they are able to consider large free surface deformations.

Liu *et al.* (1999), based on a previously existing model called RIPPLE (Kothe *et al.*, 1991), presented a RANS model, nicknamed COBRAS (Cornell Breaking Waves and Structures) to simulate breaking waves and wave interaction with coastal structures. The model calculated the mean flow in the fluid region solving the Reynolds Averaged Navier-Stokes equations and turbulence including a $k-\varepsilon$ model. The flow in porous structures was described by the spatially averaged Navier-Stokes equations. One of the main advantages of this kind of models is the use of a Volume of Fluid (VOF) technique, Hirt and Nichols (1981), to track the free surface. This technique can be used to consider large free surface deformations.

Due to model and computational limitations, COBRAS was validated for a vertical caisson protected by an armoured layer made of tetrapods in a very small computational domain (7.348 m x 0.43 m) and for a very short simulation time (t = 18.2 s) and therefore for a limited number of regular waves (13).

In a second paper Hsu *et al.* (2002) extended the original COBRAS model, introducing the Volume-Averaged/Reynolds Averaged Navier-Stokes (VARANS). In the VARANS equations, the volume-averaged Reynolds stress is modelled by adopting the nonlinear eddy viscosity assumption. The volume-averaged turbulent kinetic energy and its dissipation rate are derived by taking the volume-average of the standard $k-\varepsilon$ equations, therefore introducing the turbulence flow in the porous media. In this case, most of the validation focuses on the regular wave field in front of the structure.

2.2. *COBRAS-UC*

Even though the initial validations described in the aforementioned works have shown excellent and promising results, the initial code required extensive additional modifications to carry out a detailed validation and to make it useful for engineering applications.

Therefore, a number of modifications have been introduced into the original COBRAS code at the University of Cantabria in order to overcome some of the initial limitations and especially to convert it into a tool for practical application. Most of these modifications in the new version COBRAS-UC, have been founded on the extensive validation work carried out for low-crested structures (Garcia *et al.*, 2004; Losada *et al.*, 2005; Lara *et al.*, 2006a) and wave breaking on permeable slopes (Lara *et al.*, 2006b). Wave generation has been improved by developing a set of pre-processing tools to define target regular and irregular wave conditions based on either theoretical formulations, laboratory or field information. Irregular wave generation has been implemented considering TMA and JONSWAP-type input spectra as well as wave groups with a free number of components. Optionally, sponge layers can be defined in any region of the domain to absorb wave propagating offshore the generation region.

Most of the initial COBRAS code is directly based on the RIPPLE code (Kothe *et al.*, 1991) strongly modified and extended by several researchers from multiple institutions resulting in a complex and cryptic structure making it difficult to find problems in the code. Consequently the initial code has been converted to FORTRAN 90 and most of the code has been rewritten and compiled to run on standard PCs. Furthermore, one of the main subroutines in RIPPLE, devoted to the resolution of the Poisson pressure equation, has been modified and the iterative process used to solve the matrix improved, resulting in a remarkable reduction in the number of iterations.

One of the major limitations in the use of COBRAS as an operative engineering tool lays on the mesh generation process and definition of the elements of the flume or prototype to be modeled (solid obstacles, porous media, etc.). The obstacles and porous media were defined in COBRAS with a series of conic sections that might overlap one another to form arbitrarily complex boundaries and whose equations were to be defined and introduced by the user. This can result in an extremely tedious task for large and complicated structures. Furthermore, this method could give rise to problems on the border between obstacles, where a false fluid area might be generated. In order to overcome these problems COBRAS-UC includes a graphical user interface to define the initial parameters necessary for the simulation including the generation of elements with irregular shapes to describe the structures. The interface is based on a new technique to generate obstacles and avoids previously undetected sources of errors.

The aforementioned modifications have reduced the computational cost of the model by 40% compared to the initial code for a typical run. Several execution problems have been eliminated and the memory requirements reduced opening the possibility to consider larger and more complex computational domains and longer simulation times.

Although other RANS models are available in literature (e.g. Troch and de Rouck, 1998; Kawasaki, 1999; Li *et al.*, 2004), COBRAS and COBRAS-UC are probably the most extensively validated, especially for wave interaction with permeable low-crested structures (Garcia *et al.*, 2004; Losada *et al.*, 2005; Lara *et al.*, 2006a) and wave breaking on permeable slopes (Lara *et al.*, 2006b). The new version of the code, COBRAS-UC allows the possibility to consider larger and more complex computational domains and longer simulation times, bridging the gap between the initial developments and practical applications due to the production of reliable statistical information in the short term.

3. Validation for low-crested structures

Low-crested structures (hereafter LCS) are increasingly regarded by coastal engineers and planners as a valuable alternative to more classical surface-piercing and/or hard structures. By low-crested structures we mean detached rubble-mound breakwaters built with the crown elevation near the still water level. Thus, these structures may be submerged, emerged, or both alternatively, but are characterized to be strongly overtopped. The relative importance of each one of the physical processes, hereafter described, that take place in the interaction of a wave train with a low-crested structure depends on the wave parameters (height, period, relative depth) and on the rubble mound characteristics (geometry, porosity, permeability).

Validation of COBRAS-UC has been conducted considering the experimental data base on low-crested structures (LCS) described in Garcia *et al.* (2004), which includes both regular and irregular wave tests with different values of water depth and characteristics of incident waves. The water depth considered in the irregular wave tests chosen to be reproduced numerically is h = 40 cm at the wave paddle. The structure freeboard and crest width are F = - 5 cm and b = 100 cm respectively. Peak period is Tp = 2.4 s significant wave height (Hs = 10 cm), wave spectrum type (TMA) and peak enhancement factor ($\gamma = 3.3$). The test includes a flow recirculation system aimed at preventing water piling-up in the leeward side of the low-crested breakwater (see Garcia *et al.*, 2004). The total duration of the simulation is set to 200 s in order to achieve time series long enough to fully define the wave spectra.

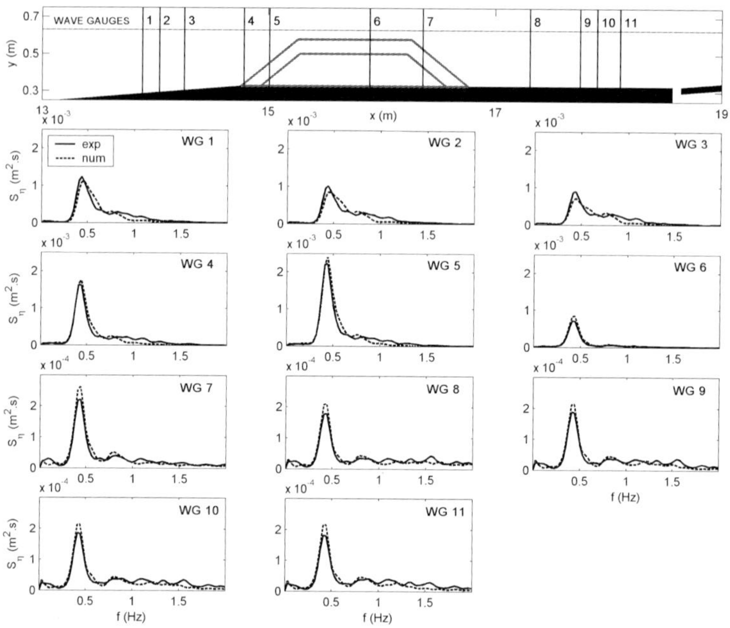

Figure 1. Spatial evolution of the wave spectrum. F = -5 cm, b = 100 cm, Hs = 10 cm, Tp = 2.4 s.

The wave spectrum evolution along the flume is presented in Figure 1. An accurate description of the spectral evolution of the incident waves is essential for a correct assessment of the breakwater performance. As commented in Garcia *et al.* (2004), transmission at low-crested structures induces a reduction of the total incident wave energy, but also a redistribution of the remaining energy in the frequency domain. The knowledge of the transmitted spectrum characteristics is of great interest for design purposes, in order to estimate for instance the morphological response of the beach protected by the considered structure. The figure illustrates the spectral changes taking place as the incident waves interact with the permeable submerged breakwater. The solid and dotted lines represent the measured and computed spectra respectively. In order to allow a better visualisation of the low-energy transmitted spectra, the vertical scale of the graphs corresponding to the seaside sections (WG7 to WG11) is amplified.

The numerical model is able to correctly reproduce the measured spectrum at each one of the considered sections. A good description of the incident wave train spectral evolution is obtained in terms of peak energy location and spectral shape. The numerical model adequately reproduces the main characteristics of the wave spectrum transformation over the submerged porous breakwater. Wave spectra at sections 4 and 5 display a significant enhancement of the energy peak, corresponding to the reflection-induced standing wave pattern and shoaling effect over the seaward slope of the breakwater. A considerable reduction of the incident wave energy is achieved past the submerged structure, due to wave breaking over the structure and dissipation inside the porous media. At section 7 and following sections, the energy is carried out at a wider range of frequencies as a result of energy transfer to higher frequencies as the incident irregular wave train propagates over the submerged breakwater. The numerical model is seen to reproduce very well the broadening of the wave spectrum past the submerged structure.

It is interesting to note that, though the high frequency part of the incident spectrum is poorly described due to the original internal wavemaker formulation, the numerical model is able to simulate the spectrum widening past the breakwater related to energy transfers to high frequency components.

4. Validation for vertical breakwaters

4.1. *Introduction*

In order to classify the structures' typology and associated response Kortenhaus and Oumeraci (1998) introduced an extended parameter map which allows a wave load-based classification in terms of three simple non-dimensional parameters. The relative height, h_b^*, determines the type of structure: a simple vertical wall, a composite structure with a low mound, a composite structure with a high mound or a rubble mound with a crown wall. Relative wave height, H_s^*, and relative berm width, B^*, are used to decide which loading case will be used. In this section we will concentrate our numerical modelling on wave interaction with low-mound breakwaters, i.e. $0.3 < h_b^* < 0.6$ and mostly small waves, $0.1 < H_s^* < 0.2$, therefore avoiding the existence of impact loads.

In order to validate the numerical model a set of experiments on wave interaction with a low mound breakwater were carried out in the wave flume of the University of Cantabria. The flume is 60.5 m long, 2 m wide and 2 m high. The breakwater, see Figure 2, is built of an impermeable caisson installed on a rubble mound foundation. The impermeable caisson is 1.04 m long, 0.3 m high and 2 m wide. The still water level was 0.8 m and the structure freeboard was 0.2 m. A 0.3 m high, D50 = 0.035 cm and 1V/2H gravel foundation was placed below the breakwater. The foundation layer was covered with an external layer of gravel with D50 = 3.5 cm. At the seaside face of the structure, a 10 cm berm was built. The porosity was 0.48 for the foundation layer and 0.50 for the outer layers.

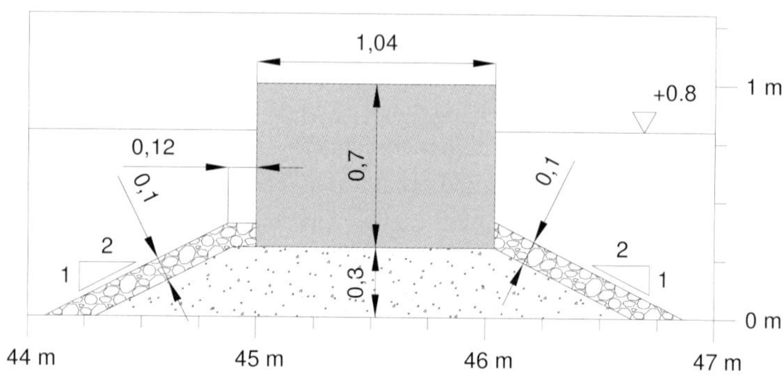

Figure 2. Experimental set-up. Data in meters.

4.2. *Free surface*

Figure 3 corresponds to the measurements and numerical solutions for the free surface time series at 7 different positions for irregular waves, Hs = 0.21 m and Tp = 3 s. Slight variations between measured and predicted free surface already appear in WG7 but, in general, the agreement is good. One source of these slight deviations could be the high reflection induced by the structure, almost 90%, which may increase the differences between the numerical absorption based on a sponge layer and the way active absorption in the flume deals with high reflection. As waves propagate towards the structure, the combined effect of reflection and shoaling results in a general increase of wave amplitudes in the time series. In WG8, WG9 and WG10, the model is able to predict the free surface time history with reasonable accuracy. In WG10 it can be seen that the highest waves, reaching wave heights in the range of 60 cm, attack the structure without breaking. In front of the structure, the model tends to overestimate measured wave heights, which results in an overestimation of the number and magnitude of overtopping events as can be seen in WG11 (t = 70 s – 85 s and t = 170 s – 180 s). This problem is also evident in WG12 and WG13. It can be said that the general agreement between experimental and numerical free surface time series is very good.

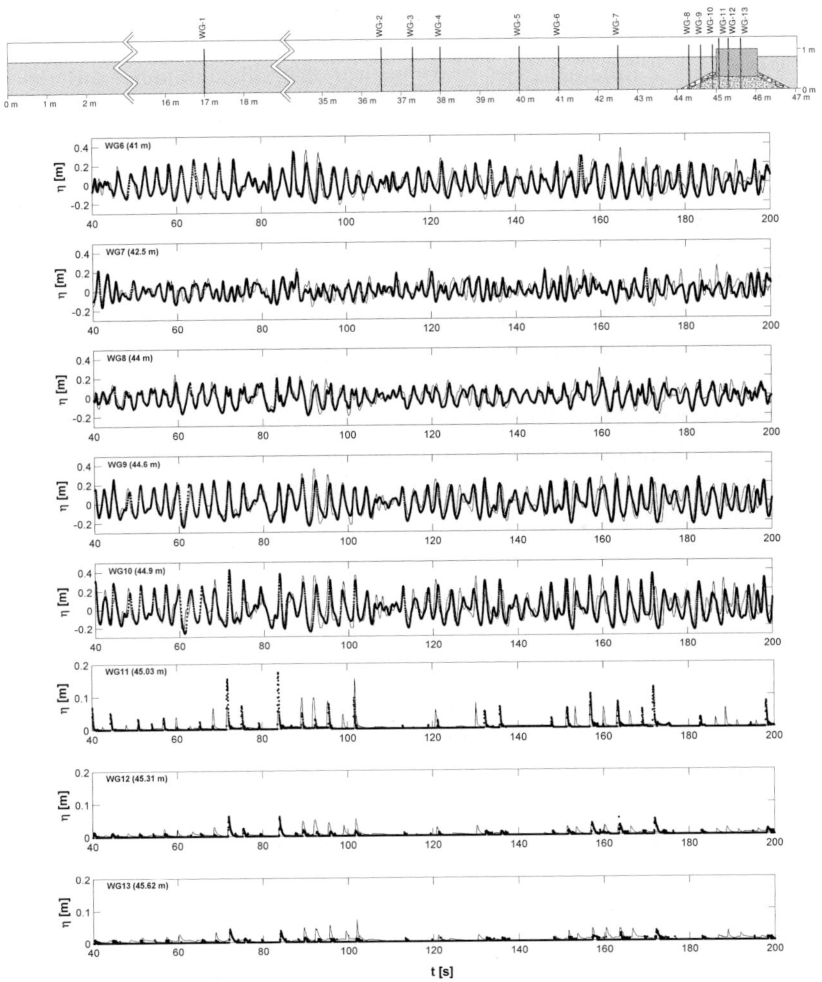

Figure 3. Free surface evolution, solid line: laboratory measurements, dotted line: numerical results. Case ia35: Hs = 0.21 m, Tp = 3 s.

4.3. *Wave overtopping*

A crucial potential application of the model is the evaluation of wave overtopping as one of the most relevant hydraulic parameters in the design of coastal structures. In this section results obtained from the numerical model are compared with experimental results and existing wave overtopping formulae.

The accumulated overtopping discharge and the average discharge during the simulation period, QA, is shown in the Figure 4 for three irregular wave cases ia35 (Hs = 0.21 m, Tp = 3 s), ia55 (Hs = 0.18 m, Tp = 5 s) and ia66 (Hs = 0.18 m, Tp = 6 s).

For the three different cases it can be seen that although minor differences exist, there is a very good agreement between the numerical and experimental accumulated overtopping discharge time history. Only during short interval periods, some differences appear in the time series of wave overtopping volumes, due to the fact that slight differences in wave height deviations in front of the structure induce significant changes in the overtopped amount of water. The best agreement is for the highest waves and shorter wave periods.

Experimental and numerical results show that since the difference in wave height is small, larger overtopping is due to longer wave periods and smaller wave height. The comparison between the numerical and experimental average discharge over the simulation period shown in the plots indicates an excellent agreement with almost negligible errors, proving the abilities of COBRAS-UC to simulate with a high degree of accuracy the wave overtopping process in complicated structures.

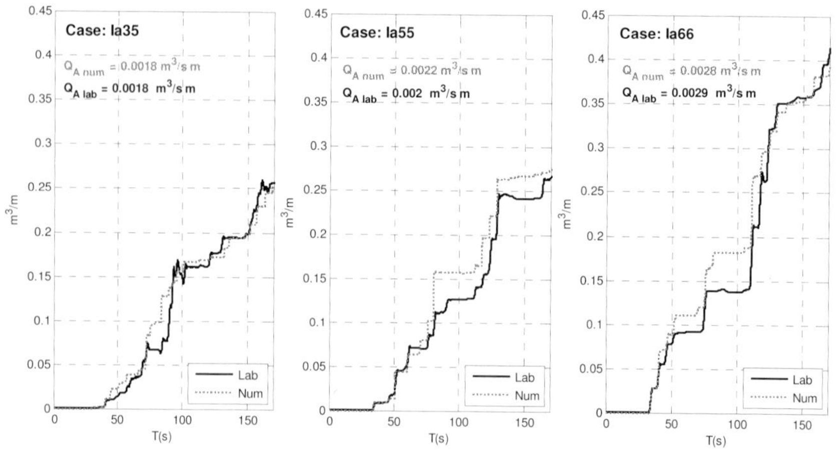

Figure 4. Wave overtopping as a function of time.

4.4. *Stability*

In order to show some additional applications of the numerical model, in this section the stability of the vertical structure is considered. Several semi-empirical formulations are available in literature to evaluate structure's stability. The total horizontal force on the vertical front wall of the structure can be evaluated using linear theory or the formulae by Sainflou (1928), Nagai (1973) and Takahashi *et al.* (1994).

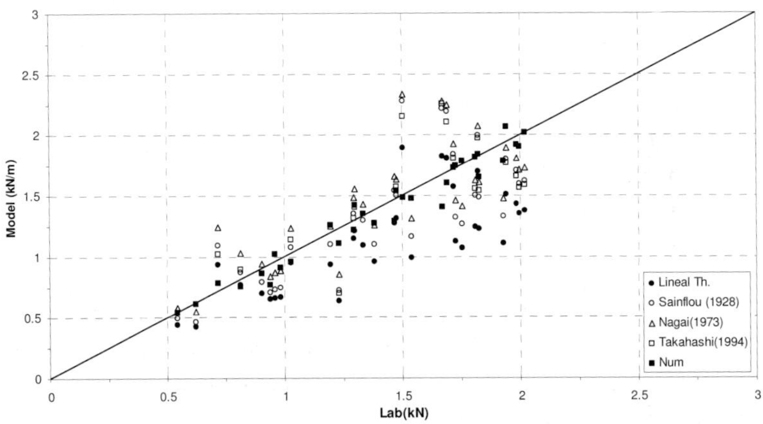

Figure 5. Total horizontal force on the structure calculated using linear theory, Sainflou (1928), Nagai (1973), Takahashi (1994) and the numerical model versus experiments for all the experimental cases considered.

Figure 5 and Table 1 show a summary of the results obtained for the analysis of the stability of the vertical breakwater and incident wave conditions considerd in the experimental work describe above. It is important to point out that linear theory, Nagai (1973) and Sainflou (1928) have been applied to all the experimental cases considered, regular and irregular waves. Takahashi *et al.* (1994) has been only applied for the irregular wave cases.

Results based on formulations and on the numerical model show a similar fit to experimental data, except for linear theory. Even of the incident wave conditions correspond to linear wave theory range (H/h < 0.5) and breaking does not take place, the presence of the berm generates a transformation of the wave train and a small dissipation that may induce the important deviations when applying linear wave theory. Furthermore, linear theory assumes perfect reflection, which is non-existing in the experimental results. Nagai (1973), Goda

(1994) and Sainflou (1928) show similar results. Sainflou underestimates the total force in a 5%, while the two other formulations show a slight overprediction. However, standard deviation is similar (±22.77%), contributing to predict similar results for all three formulae. It is important to note that the three formulations show very good agreement even if they do not take into account overtopping. However, they are not as good as the results obtained using the numerical model. COBRAS-UC shows a small underestimation of the total force (-2.76%), but the standard deviation is three times smaller, which results in a higher confidence when evaluating the total force.

Table 1. Experimental and calculated total horizontal force on the caisson using several formulations and COBRAS-UC. Mean relative error and standard deviation.

	Mean relative error	Standard deviation
Linear theory	-19.26%	±18.52%
Sainflou (1928)	-5.66%	±22.77%
Nagai (1973)	4.25%	±22.79%
Takahashi (1994)	2.88%	±23.35%
COBRAS-UC	-2.76%	±6.77%

5. Validation for rubble-mound breakwaters

5.1. *Introduction*

In order to consider a different structure typology, a set of experiments on wave interaction with a rubble mound breakwater was carried out in the same wave flume of the University of Cantabria. In this case, the breakwater, see Figure 6, is built of an impermeable caisson installed on a rubble mound foundation. The impermeable caisson is 1.04 m long, 0.3 m high and 2 m wide. The still water level was 0.8 m and the structure freeboard was 0.2 m. A 0.7 m high, D50 = 0.01 cm and 1V/2H gravel core was placed below the breakwater. The core was covered with three external layers; two of them were gravel with D50 = 3.5 cm and one external layer of gravel with D50 = 13.5 cm. At the seaside face of the structure, a 14 cm berm was built. The porosity was 0.48 for the core and 0.50 for the outer layers. A reservoir was built at the crown of the structure in order to collect the water which overtops the structure. With this new experimental setup, run-up, breaking and wave transmission due to overtopping become the most relevant processes.

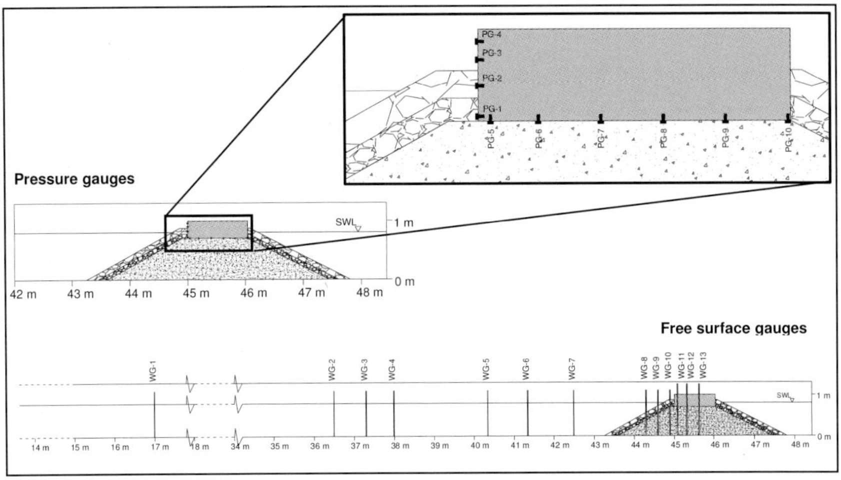

Figure 6. Instrumentation set-up. Pressure gauges (PG) and free surface gauges (WG).

5.2. *Pressure, forces and moments time series*

In this case the validation process is based on the comparison of experimental and numerical pressure time series underneath and on the seaward face of the caisson. The upper panel of Figure 7 shows the distribution of the 10 pressure gauges on the caisson. The lower panels plot comparisons between predicted and experimental pressure time series for approximately 50 waves, in case ib66 (Hs = 0.18 m, Tp = 6 s).

In general, the model gives a very accurate prediction of the recorded pressure time series. It is important to note that the prediction of pressure time series, especially in gauges PG3 to PG10 is very difficult since it requires an excellent simulation of the flow across the different porous layers. Gauges PG1 to PG4 in the front face of the caisson show a good agreement in phase and amplitude. Gauges PG2 to PG4 present a discontinuous record since, due to their location, they only register the pressure induced by the larger waves overtopping the structure. Regarding the gauges underneath the caisson, the damping induced by the porous material is clearly visible in both the numerical and the experimental records. But for minor discrepancies, the model is able to reproduce the measured time series very well in all the gauges. Please, note that the pressure time series in gauges 9 and 10 are of the same order of magnitude and non-zero.

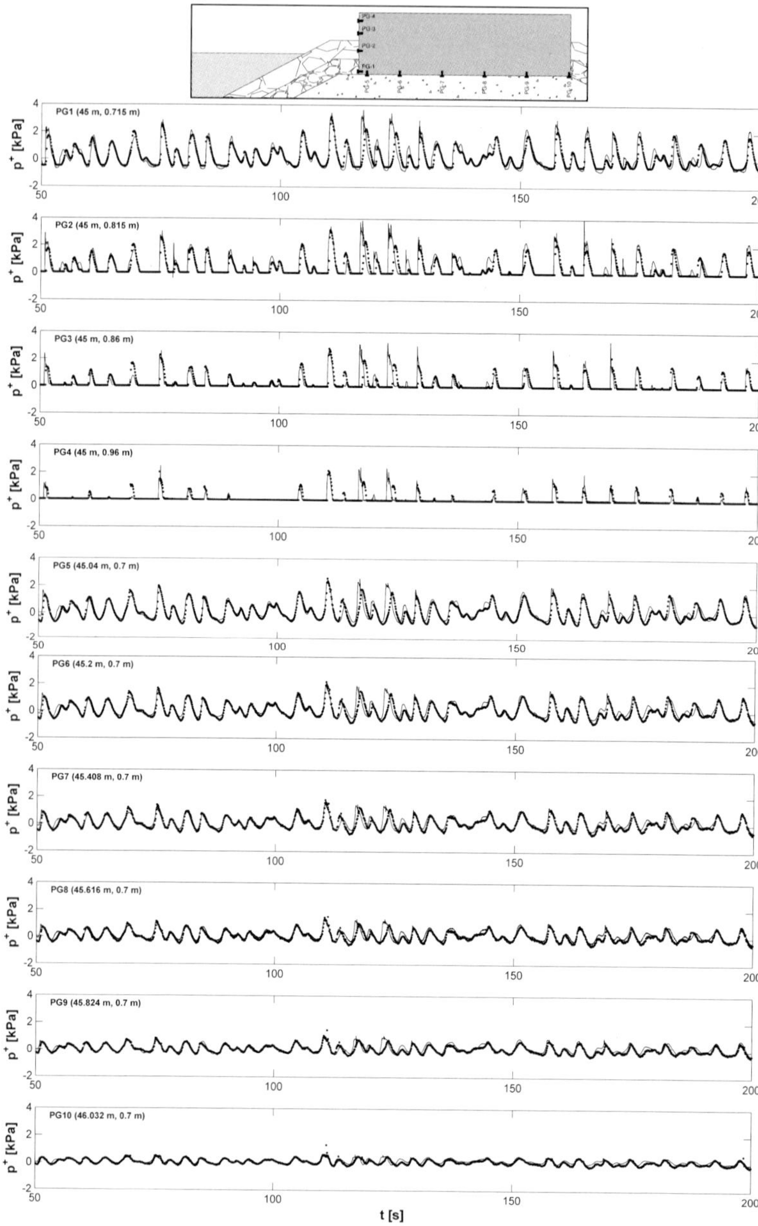

Figure 7. Pressures time evolution, solid line: laboratory measurements, dotted line: numerical results. Case ib66: Hs = 0.18 m, Tp = 6 s.

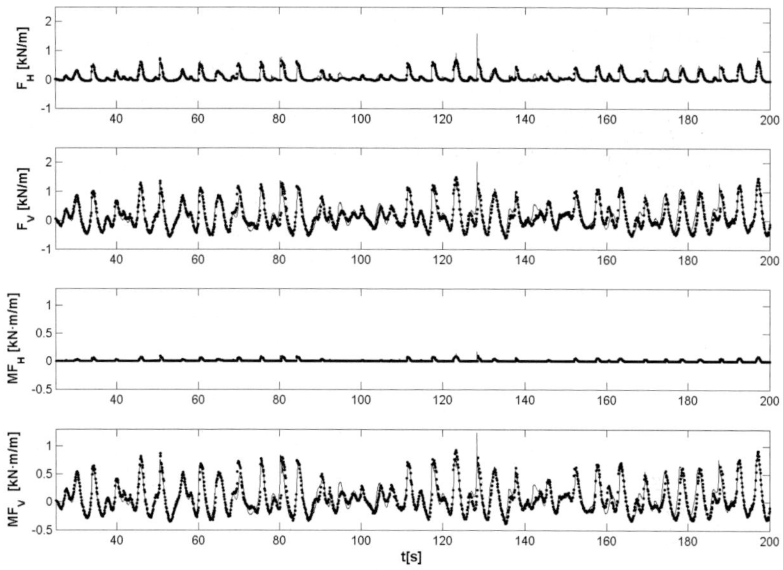

Figure 8. Experimental (dots) and numerical (solid line) time series of horizontal forces (FH), pressure underneath the caisson (FV), moment induced by horizontal forces (MFH), moment induced by vertical forces (MFV). Hs = 0.18 m, Tp = 5 s.

Figure 8 shows a comparison between the experimental and numerical time series of horizontal forces (FH), pressure underneath the caisson (FV), moment induced by horizontal forces (MFH), moment induced by vertical forces (MFV) for a different case, with a shorter peak period, Hs = 0.18 m, Tp = 5 s. The agreement is very good with minor discrepancies (t = 130 s) where the model is not able to predict impulsive forces induced by wave breaking on the crown wall, due to the sampling frequency.

6. Application to a real case: Castro Urdiales breakwater

Castro Urdiales (Spain) breakwater, Figure 9, is an existing composite breakwater originally constructed as a vertical breakwater later protected by a rubble mound protection layer. Due to wave action the current geometry corresponds to a typology in between rubble mound and composite breakwaters for which empirical stability formulae are not available in literature. Therefore, the analysis of structure's stability can be carried out only based on a pure experimental approach or using a numerical model.

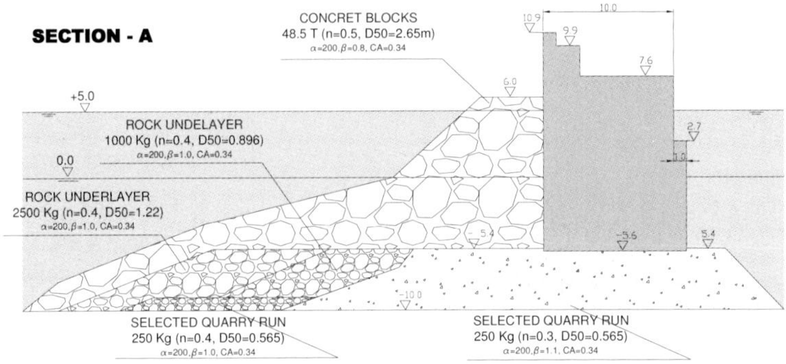

Figure 9. Castro Urdiales breakwater section.

Based on the deep water wave information, the hindcast numerical wave data base is propagated to 25 m water depth, where the mean and extreme wave height distributions are defined. At this water depth, considering a 25 year life expectancy and a failure probability of 0.1, the design conditions were calculated resulting in a 250 return period significant wave height Hs = 5.9 m and peak period Tp = 16 s. Based on a JONSWAP spectrum ($\gamma = 10$), a sea state has been generated numerically. The sea state is long enough to consider the maximum wave height defined by Hmax = 1.9Hs (see Figure 10). The complete sea state is simulated using COBRAS-UC and evaluating the time series of pressure, forces and moments along the sea state.

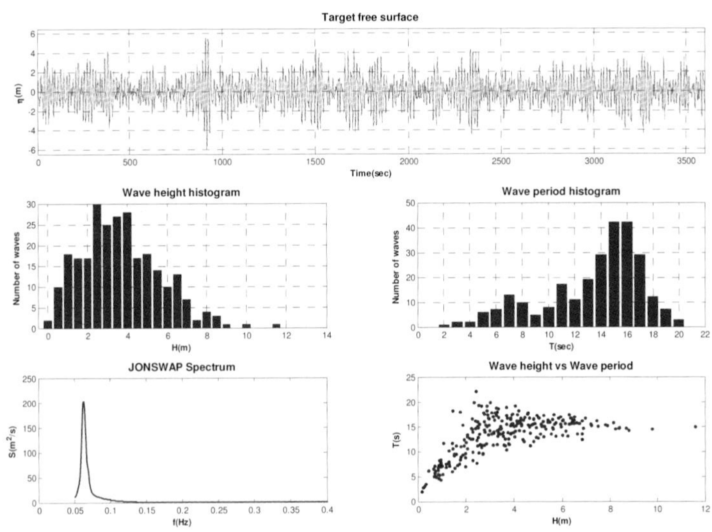

Figure 10. Synthetic generated sea state in 25 m water depth used to feed COBRAS-UC.

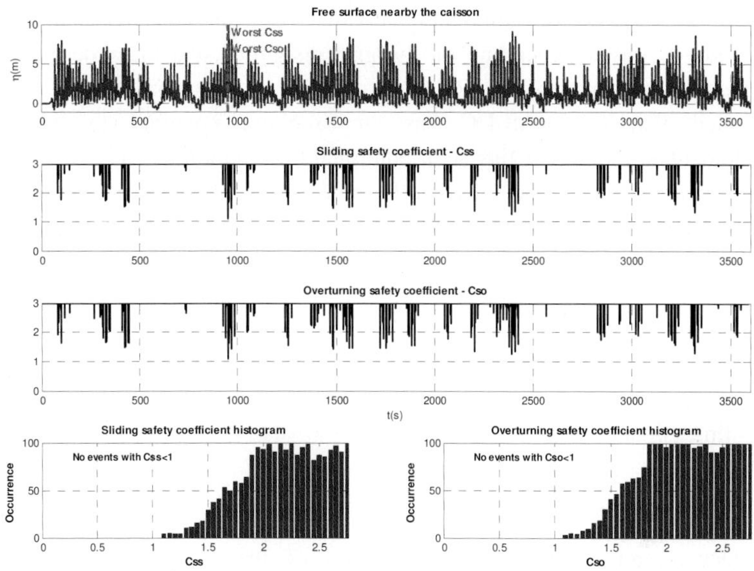

Figure 11. Upper panel: Calculated free surface in front of the breakwater. Sliding and overturning safety coefficients time series during the sea state. Lower panel: Sliding and overturning safety coefficients histograms.

The numerical model provides the opportunity to locate free surface and pressure gauges at any location in the numerical wave flume. Figure 11 shows in the upper panel the free surface time series in a wave gauge in front of the breakwater for the design sea state. The lower panels provide information regarding sliding and overturning safety coefficients time series as well as the corresponding histograms. The correspondence between higher free surface and low safety coefficients is clearly visible, as well as the location of the minimum value of the coefficients along the sea state. The safety coefficients do not take values lower than 1, indicating than the structure is stable for this sea state.

7. Concluding remarks

By improving several aspects of the initial version of the COBRAS model, COBRAS-UC allows the application of a numerical model solving the volume-average RANS equations and corresponding $k-\varepsilon$ equations, to the investigation of irregular wave interaction with several structures' typologies. The model has been validated against several data sets showing an excellent behavior when evaluating functional and stability parameters. The improvements in this new version allow for longer computational times that can vary from hundreds to thousands of waves depending on the grid resolution and the processes to be considered, in reasonable periods of time and on standard PCs. It can be concluded that COBRAS-UC has an excellent potential to become a complementary tool to existing semi-empirical formulations.

Even though the model has performed very well, there is still room for improvement. The computational efficiency of the model must be upgraded. A second issue to be considered is a detailed analysis of the dependency of the empirical coefficients, α and β, associated with the linear and nonlinear drag force in the porous media flow (not discussed in this paper). Finally, including the effect of air on the wave and structure interaction process could also be of interest for the analysis of some specific situations as well as working towards a three-dimensional version of the model.

Acknowledgments

The Ocean and Coastal Research Group of the University of Cantabria acknowledges Prof. Philip Liu for his scientific generosity and friendship, two substantial elements for the development of this work.

References

1. H.F. Burcharth and S.A. Hughes, Fundamentals of Design - Part 1, Chapter 1, 2 and 3. *In: Vincent, L., and Demirbilek, Z. (editors), Coastal Engineering Manual, Part II, Hydrodynamics, Chapter II-2, Engineer Manual 1110-2-1100, U.S. Army Corps of Engineers, Washington, DC.* (2006).
2. R.A. Dalrymple, O. Knio, D.T. Cox, M. Gomez-Gesteira, S. Zou, Using a Lagrangian particle method for deck overtopping. *Proc. Waves ASCE 2001*, 1082-1091 (2001).
3. N. Garcia, J.L. Lara, I.J. Losada, 2-D Numerical analysis of near-field flow at low-crested permeable breakwaters. *Coast. Eng.* **51**, 991-1020 (2004).
4. H. Gotoh, S.D. Shao, T. Memita, SPH–LES model for numerical investigation of wave interaction with partially immersed breakwater. *Coast. Eng. Japan* **46(1)**, 39-63 (2004).
5. C.W. Hirt and B.D. Nichols, Volume of Fluid (VOF) method for dynamics of free boundaries. *J. Comp. Physics* **39**, 201-225 (1981).
6. T.-J. Hsu, T. Sakakiyama, P.L.-F. Liu, A numerical model for wave motions and turbulence flows in front of a composite breakwater. *Coast. Eng.* **46**, 25-50 (2002).

7. K. Hu, C.G. Mingham, M.D. Causon, Numerical simulation of wave overtopping of coastal structures using the non-linear shallow water equations. *Coast. Eng.* **41**, 433-465 (2000).
8. M.E. Hubbard, N. Dodd, A 2D numerical model of wave run-up and overtopping. *Coast. Eng.* **47**, 1-26 (2002).
9. K. Kawasaki, Numerical simulation of breaking and post-breaking wave deformation process around a submerged breakwater. *Coast. Eng. J.* **41**, 201-223 (1999).
10. N. Kobayashi, A. Wurjanto, Wave overtopping on coastal structures. *J. Waterw., Port, Coast., Ocean Eng.*, ASCE, **115**, 235-251 (1989).
11. A. Kortenhaus, H. Oumeraci, Classification of wave loading on monolithic coastal structures, *Proceedings of the 26th International Conference Coastal Engineering (ICCE), ASCE*, **Volume 1**, Copenhagen, Denmark, pp. 867-880 (1998).
12. D.B. Kothe, R.C. Mjolsness, M.D. Torrey, RIPPLE: A computer program for incompressible flows with free surfaces, Los Alamos National Laboratory, **LA-12007-S** (1991).
13. J.L. Lara, N. Garcia, I.J. Losada, RANS modelling applied to random wave interaction with submerged permeable structures. *Coast. Eng.* **53**, 395-417 (2006a).
14. J.L. Lara, I.J. Losada, P.L.-F. Liu, Breaking waves over a mild gravel slope: Experimental and numerical analysis. *J. Geophysical Research*, AGU, **111**, **C11019**; doi: 10-1029/2005 JC003374 (2006b).
15. T.Q. Li, P. Troch, J.D. Rouck, Wave overtopping over a sea dike. *J. Comp. Physics* **198**, 686-726 (2004).
16. P.Z. Lin and P.L.-F. Liu, A numerical study of breaking waves in the surf zone. *J. Fluid Mech.* **359**, 239-264 (1998).
17. P.L.-F. Liu, P.Z. Lin, K.A. Chang, T. Sakakiyama, Numerical modelling of wave interaction with porous structures. *J. Waterw., Port, Coast., Ocean Eng.*, ASCE **125(6)**, 322-330 (1999).
18. I.J. Losada, Advances in modeling the effects of permeable and reflective structures on waves and nearshore flows. *In: Advances in Coastal Modeling. Elsevier Oceanography Series*, **67**, Ed. V. Chris Lakhan (2003).
19. I.J. Losada, Advances in the design and construction of coastal structures. In: *Port and Coastal Engineering. Ed. P. Bruun. Coastal Engineering Research Foundation.* ISBN 1891276506., **Chapter 3**, 55-99 (2005).
20. I.J. Losada, J.L. Lara, E. Damgaard, N. Garcia, Modelling of velocity and turbulence fields around and within low-crested rubble-mound breakwaters. *Coast. Eng.* **52**, 887-913 (2005).

21. C.G. Mingham and D.M. Causon, High-resolution finite-volume method for shallow water flows. *J. Hydraul. Eng.*, ASCE **124(6)**, 605-614 (1998).
22. S. Nagai, Wave Forces on Structures. *Advances in Hydroscience*. **Vol. 9**. Ed. Academic Press (1973).
23. G. Sainflou, Essai sur les digues maritimes verticals. *Annales Ponts e Chausses*, **Vol. 98**, No. 4 (1928).
24. S. Shao, C. Ji, D.I. Graham, D.E. Reeve, P.W. James, A.J. Chadwick, Simulation of wave overtopping by an incompressible SPH model. *Coast. Eng.* **53**, 723-735 (2006).
25. P.K. Stansby and T. Feng, Surf zone wave overtopping a trapezoidal structure: 1-D modelling and PIV comparison. *Coast. Eng.* **51**, 483-500 (2004).
26. S. Takahashi, K. Tanimoto, K. Shimosako, A Proposal of Impulsive Pressure Coefficient for Design of Composite Breakwaters, *Proceedings of the International Conference on Hydro-Technical Engineering for Port and Harbor Construction, Port and Harbour*, Research Institute, Yokosuka, Japan, pp. 489-504 (1994).
27. P. Troch and J. de Rouck, Development of two-dimensional numerical wave flume for wave interaction with rubble mound breakwaters. *26th Int. Coastal Eng. Conf.*, pp. 1638-1649 (1998).

OCEAN-BOTTOM PRESSURE VARIATIONS DURING THE 2003 TOKACHI-OKI EARTHQUAKE

WENWEN LI and HARRY YEH

Department of Civil, Construction and Environmental Engineering
Oregon State University
Corvallis, OR 97331-3212, USA

KENJI HIRATA and TOSHITAKA BABA

Institute for Research on Earth Evolution
Japan Agency for Marine Science and Technology
2-15 Natsuhima-cho, Yokosuka 237-0061, Japan

During the 2003 Tokachi-Oki earthquake, pressure variations were registered by the ocean-bottom observatory located near the epicenter. The acoustic (pressure) waves bouncing up and down between the hard bottom and the sea surface were generated by the co-seismic seafloor displacement. The numerical simulations using the method of characteristics show that the dominant frequency of the simulated acoustic wave is in good agreement with that of the measurement if the soft sediment layers are taken into account. The quantitative analysis suggests that the co-seismic seafloor uplift took place in the duration of 6.8 ~ 7.9 seconds. The amplitude modulation of acoustic waves was found to be regular, which indicates that the neighboring geologic factors control its characteristics. Acoustic-wave attenuations for the main shock and the following aftershocks are consistent, except for one of the aftershocks. Hence, that particular aftershock must have occurred away from the location of the pressure gage.

1. Introduction

On September 26, 2003 at 04:50:06 JST (Japan Standard Time), a large thrust-type earthquake ($Mw = 8.0$) occurred near Tokachi, Japan. The epicenter (41.78°N, 144.08°E) is approximately 60 km east of Cape Erimo (see Fig. 1). Yamanaka and Kikuchi [1] estimated the rupture duration to be 40 seconds and the maximum slip 5.8 meters. This earthquake was an inter-plate earthquake associated with the subduction of the Pacific plate under the Eurasian plate. A number of aftershocks subsequently occurred around the epicenter of the main shock, and the largest took place at 06:08 local time, a little more than one hour after the main shock [2].

This 2003 Tokachi-Oki earthquake generated tsunamis that struck coastal areas in the Hokkaido and Tohoku regions. The maximum tsunami run-up height of about 4 meters was recorded about 70 km northeast of the Cape Erimo [3]. Based on the observed tsunami travel times at 17 Japanese tide gauge stations, Hirata *et al.* [4] estimated the tsunami source area of this earthquake to be approximately 1.4×10^4 km^2 (see Fig. 1).

The seismic waves, ocean-bottom pressures, and other geophysical data were captured by the cabled seafloor observatory located right in the source region of the earthquake: the Japan Agency for Marine-Earth Science and Technology (JAMSTEC) deployed the observatory in 1999, just four years prior to the earthquake [5]. The locations of ocean-bottom seismometer (KOBS) and ocean-bottom pressure gages (PGs) are shown in Fig. 1. Note that the pressure gages (PG1 and PG2) are located within the tsunami source area. In this paper, we focus on the analyses of the pressure data obtained at PG1 (41.70°N, 144.44°E) that is located the closest to the epicenter.

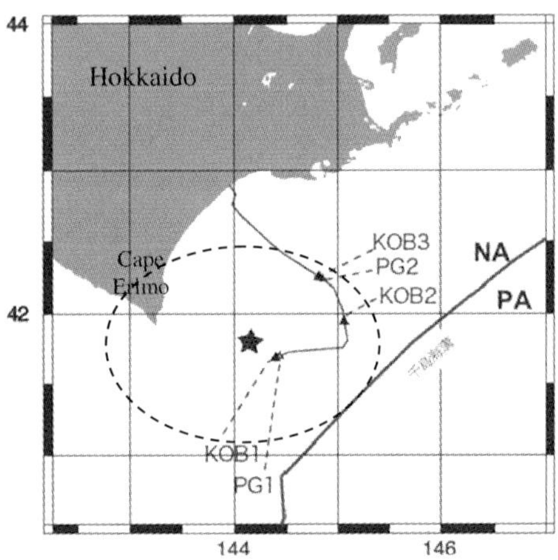

Figure 1. The 2003 Tokachi-Oki earthquake epicenter (star) and locations of ocean-bottom seismometers KOBS and pressure gauges PGs. The estimated tsunami source area is enclosed by the dashed line (JAMSTEC, 2003).

The pressure gage PG1 is placed at the depth of 2283 m and the pressure data are plotted in Fig. 2a. The sampling rate was 10 Hz, which is fast enough to detect not only tsunami generation, but also transient acoustic waves. Figure 2b clearly exhibits the static pressure shift $\Delta P \approx$ 4KPa. This shift in pressure must result from the residual vertical seafloor displacement associated with the fault rupture. Figure 2 also exhibits the high-frequency pressure fluctuation. As we will explore later, this pressure fluctuation is associated with compression and expansion of the water column, namely acoustic waves.

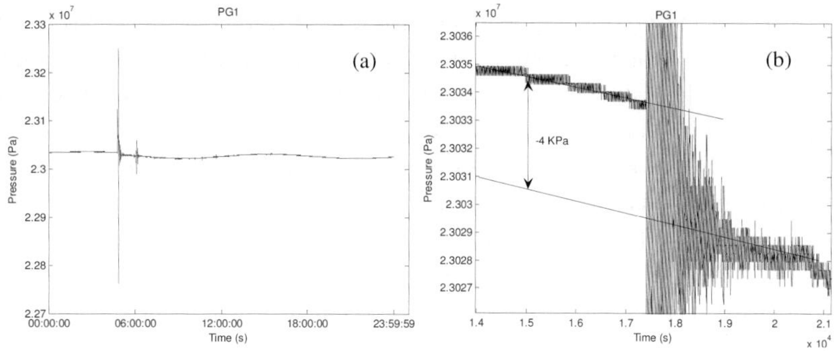

Figure 2. Time histories of the pressure variations at PG1 (a) from 00:00:00 to 23:59:59, (b) in an expanded pressure scale and static pressure shift due to residual bottom uplift.

The role of water compressibility in submarine earthquakes has been discussed in the past. Kajiura [6] pointed out that when the sea bottom is uplifted during a submarine earthquake, the energy is transferred from the solid bottom to the overlying water by generating acoustic waves and then partly converted to gravity waves. Kajiura [6] examined relative importance of acoustic and gravity waves in terms of the generated energy based on the 1-D model in the vertical direction. His 1-D model yields the ratio of the acoustic-wave energy E_1 to the gravity-wave energy E_2:

$$\frac{E_1}{E_2} = \frac{2c}{g\tau}, \tag{1}$$

where c is the acoustic-wave speed in water, g is the gravity acceleration, and τ is the duration of seafloor displacement (assuming that the vertical velocity of the bottom movement is constant for the duration of $[0, \tau]$). Considering the acoustic-wave speed (1500 m/sec) and a short duration of bottom displacement (tens of seconds), Eq. (1) implies that the considerable portion of energy expends in generating acoustic waves instead of gravity waves, which indicates the generation of tsunamis is an inefficient process.

Nosov [7] demonstrated that the significant feature at a tsunami source is the presence of high-frequency water-surface oscillations, which is a consequence of acoustic waves traveling back and forth between the seafloor and the water surface. Nosov [8] examined the ocean-bottom pressure variations recorded during the 2003 Tokachi-Oki earthquake, and suggested the existence of elastic oscillations of water column — the theme of the present paper. In his analysis, the seafloor-uplift speed was estimated 0.1 m/sec rather arbitrarily by his "numerical experience". He also used the final seafloor displacement of 0.4 m at

PG1 that was the preliminary estimate given by JAMSTEC immediately after the earthquake. (Later on, Hirata [9] carefully computed the total seafloor displacement by taking into account the sensor's temperature-effect compensation; Hirata reported that the co-seismic uplift at PG1 was 0.33 m.) Nosov [8] also assumed the following ranges of the physical parameters: 2256 ~ 2578 m for the water depth, 47 ~ 2000 m for the soft sediment thickness, 1740 ~ 2300 m/sec for the P-wave speed in sediments, and 1816 ~ 2053 kg/m^3 for the sediment density. Using these very rough estimates, he showed a trend that inclusion of underlying sediment layer could improve the prediction of the observed acoustic-wave frequency, although his estimated frequency range was wide and no quantitative discussion was made.

In this paper, we follow Nosov's [8] analysis but using more accurate geologic/oceanic information and data, and quantitative analyses of the pressure data shown in Fig. 2 are presented.

2. Numerical Model

The horizontal dimension of seafloor displacement by the 2003 Tokachi-Oki earthquake was in the order of 100 km, which is much greater than the vertical dimension: the water depth (~ 2 km). Hence it is reasonable to assume that the acoustic-wave energy would bounce up and down vertically in the water column, i.e. the 1-D problem. Thus we consider the phenomenon analogous to the water-hammer problem in a closed conduit. Unlike the closed conduit system, there is no pipe-wall boundary to confine the acoustic wave energy laterally. Although there is no friction effect along the lateral boundaries, the acoustic wave does attenuate by dissipation due to the viscous effect within the fluid, lateral energy leakage, and absorption to the atmosphere at the top as well as to the sediment layers at the bottom.

Suppose the distance between the solid bottom and the free water surface is H and the bottom is uplifted with the velocity V_0. The idealized pressure fluctuations are illustrated in Fig. 3. The reflection of acoustic wave at the water surface and the bottom causes a periodic fluctuation of pressure.

The 1-D momentum equation for inviscid fluid can be expressed by

$$\frac{\partial V}{\partial t} + V\frac{\partial V}{\partial z} = -\frac{1}{\rho}\frac{\partial p}{\partial z} - g\,, \tag{2}$$

where V is the fluid velocity, p is the pressure, ρ is the fluid density, z is the vertical coordinate pointing upward from the seafloor, and t is the time. We can write the variables p and ρ in the form

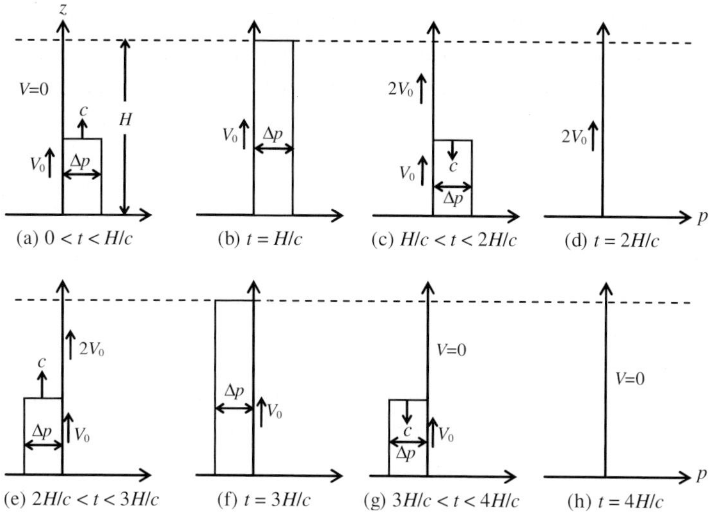

Figure 3. Stages of pressure fluctuations after the bottom uplift (damping effects neglected).

$$p = p_0(z) + p'(z,t) \text{ and } p = p_0(z) + p'(z,t), \tag{3}$$

where p_0 and ρ_0 are the pressure and the density at the equilibrium state, and p' and ρ' are their variations in the acoustic wave. With $\partial p_0 / \partial z = -\rho_0 g$ and $\rho' << \rho_0$. Eq. (2) can be written as:

$$\frac{\partial V}{\partial t} + V\frac{\partial V}{\partial z} + \frac{1}{\rho_0}\frac{\partial p'}{\partial z} = 0. \tag{4}$$

The acoustic wave decays in the field data as seen in Fig. 2, so the momentum equation can be modified, by adding a damping term to Eq. (4), as

$$\frac{\partial V}{\partial t} + V\frac{\partial V}{\partial z} + \frac{1}{\rho_0}\frac{\partial p'}{\partial z} + \alpha V = 0, \tag{5}$$

where α is the damping coefficient. Note that we simply added the damping term because the attenuation can be caused by multiple factors (dissipation, radiation, and absorption) and cannot pinpoint the physics. The damping term yields approximately the exponential decay in velocity V.

The equation of mass conservation can be written as

$$\frac{\partial \rho}{\partial t} + V\frac{\partial \rho}{\partial z} + \rho\frac{\partial V}{\partial z} = 0. \tag{6}$$

Assuming that the propagation of acoustic wave is an adiabatic process, the acoustic wave speed c can be written as

$$c^2 = \frac{dp}{d\rho}. \tag{7}$$

Using Eq. (7), Eq. (6) can be expressed in terms of the pressure p instead of the density ρ:

$$\frac{\partial p}{\partial t} + V\frac{\partial p}{\partial z} + \rho c^2 \frac{\partial V}{\partial z} = 0. \tag{8}$$

With $p = p_0 + p'$, $\frac{\partial p_0}{\partial z} = -\rho_0 g$ and $\rho' << \rho_0$, Eq. (8) can be written as

$$\frac{\partial p'}{\partial t} + V\frac{\partial p'}{\partial z} + \rho_0 c^2 \frac{\partial V}{\partial z} - \rho_0 g V = 0. \tag{9}$$

Equations (5) and (9) provide a pair of partial differential equations for the dependent variables V and p'. These partial differential equations can be transformed to the ordinary differential equations by the method of characteristics. Approximating $c >> V$, the method of characteristics yields the following solutions in the finite difference forms:

$$p_i^{n+1} - p_{i-1}^n + (\rho_0 c)_{i-1}(V_i^{n+1} - V_{i-1}^n) + (\rho_0 c\alpha - \rho_0 g)_{i-1}\frac{(V_{i-1}^n + V_i^{n+1})}{2}\Delta t = 0, \tag{10}$$

$$p_i^{n+1} - p_{i+1}^n - (\rho_0 c)_i(V_i^{n+1} - V_{i+1}^n) - (\rho_0 c\alpha + \rho_0 g)_i\frac{(V_{i+1}^n + V_i^{n+1})}{2}\Delta t = 0, \tag{11}$$

where the subscript i and the superscript n denote the spatial and temporal grid locations, respectively. We take a constant time-step Δt and divide the distance between the bottom and ocean surface into N cells with the size of the i-th cell to be: $(\Delta z)_i = z_{i+1} - z_i = c_i\,(t_{n+1} - t_n) = c_i\,\Delta t$, as shown in Fig. 4. Solving these two equations simultaneously for unknowns V_i^{n+1} and p_i^{n+1} at any grid intersection point yields

$$V_i^{n+1} = \frac{2(p_{i+1}^n - p_{i-1}^n) - (g\Delta t + (2-\alpha\Delta t)c_{i-1})\rho_{0,i-1}V_{i-1}^n + (g\Delta t - (2-\alpha\Delta t)c_i)\rho_{0,i}V_{i+1}^n}{(g\Delta t - (2+\alpha\Delta t)c_{i-1})\rho_{0,i-1} - (g\Delta t + (2+\alpha\Delta t)c_i)\rho_{0,i}}, \tag{12}$$

$$p_i^{n+1} = p_{i-1}^n + \rho_{0,i-1}c_{i-1}(V_{i-1}^n - V_i^{n+1}) + \rho_{0,i-1}(g - c_{i-1}\alpha)\Delta t\frac{(V_{i-1}^n + V_i^{n+1})}{2}. \tag{13}$$

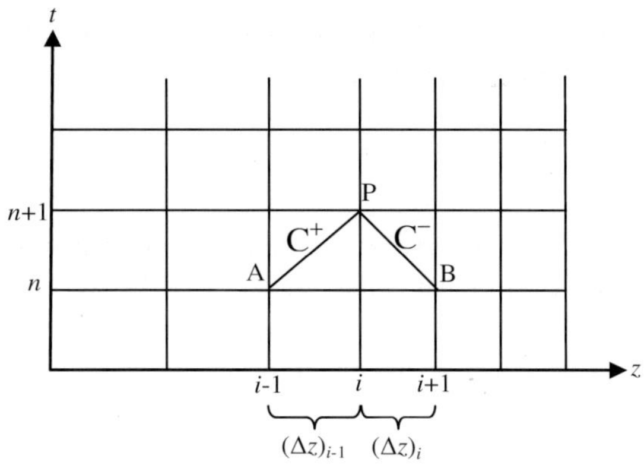

Figure 4. $z-t$ grid for solving characteristic equations.

Initial Conditions:

At time $t = 0$, the fluid is at rest, except the layer adjacent to the bottom, and no acoustic wave is generated.

$$V_1^0 = V_{BT}(t=0), \quad V_i^0 = 0, \quad i = 2,3,...,N+1, \tag{14}$$

$$p_i^0 = 0, \quad i = 1,2,\ldots,N+1, \tag{15}$$

where V_{BT} is the velocity of the bottom.

Boundary Conditions:

The fluid adjacent to the bottom has the same velocity as that of the bottom.

$$V_1^n = V_{BT}(t = t_n), \quad n \ge 0, \tag{16}$$

and the acoustic pressure p at the bottom for each time step is determined by the direct solution of Eq. (11):

$$p_1^{n+1} = p_2^n + \rho_{0,1}c_1\left(V_1^{n+1} - V_2^n\right) + \rho_{0,1}(g + c_1\alpha)\frac{(V_2^n + V_1^{n+1})}{2}\Delta t. \tag{17}$$

At the sea surface, the fluid is always under the atmospheric pressure, so the acoustic pressure vanishes all the time.

$$P_{N+1}^n = 0, \quad n \ge 0. \tag{18}$$

The fluid velocity V at the sea surface for each time step is obtained by Eq. (10):

$$V_{N+1}^{n+1} = \frac{2(p_{N+1}^{n+1} - p_N^n) - (g\Delta t + (2-\alpha\Delta t)c_N)\rho_{0,N}V_N}{(g\Delta t - (2+\alpha\Delta t)c_N)\rho_{0,N}}. \tag{19}$$

In order to model the time history of the realistic seafloor displacement, we used a half cosine curve:

$$d = \frac{\Delta D}{2}\left[1-\cos\left(\frac{\pi}{\tau}t\right)\right] \quad \text{for } 0 \le t \le \tau, \tag{20}$$

where τ is the duration of the seafloor displacement and ΔD is the net seafloor uplift. We use ΔD = 0.33 m that is the estimation of the co-seismic uplift given by Hirata [9]. Neither the duration of bottom displacement τ nor the damping coefficient α has effect on the natural frequencies of the acoustic wave; the natural frequencies are determined by the elastic properties of the medium alone.

3. Analyses of Measured Pressure Data

Prior to the analyses, we first roughly estimate the natural frequencies of acoustic-wave oscillations by the model depicted in Fig. 3:

$$f_k = (2k-1)c_w / 4H_w \quad \text{for } k = 1, 2, 3, \ldots \tag{21}$$

With the water depth $H_w = 2283\,\text{m}$ and the acoustic speed in water $c_w = 1500$ m/sec, the first three solutions are $f_1 = 0.164$ Hz, $f_2 = 0.493$ Hz, and $f_3 = 0.821$ Hz. Note that our numerical solutions by the characteristics method also yield the identical peak frequencies (not shown here for brevity).

To extract acoustic waves from the raw data of PG1 shown in Fig. 2, we applied a 5th order Butterworth band-pass filter with the cutoff frequencies 0.01 and 2 Hz; the filter should capture the acoustic waves while eliminating high-frequency noise. Figure 5 shows the spectrum of the filtered data for the main shock ($250 < t < 1750$ sec): hereinafter, the time origin ($t = 0$) is set at the time of earthquake, i.e. 04:50:06 JST. The spectrum in Fig. 5 indicates that the dominant frequency of the measured pressure fluctuations is 0.131 Hz, which is substantially lower than that of the computed acoustic wave, 0.164 Hz.

Following how Nosov [8] explained the discrepancy in the dominant frequency, we include the underlying soft sediment layers in the acoustic-wave-penetration media, in addition to the water column, to test its importance. Figure 6 shows the estimated acoustic speed structure at PG1, based on the data provided by Tsuru [10]: the acoustic wave speeds were obtained from the multi-channel seismic-profile data taken at the location close to PG1 (approximately 10 km west). We assume that the underlying sediment structure is the same. The density variations in sediments are estimated based on the drilled core samples taken at the location close to PG1, given by Oil and Natural Gas Resources in

Japan [11]. Again, we assume the physical properties of the sediments are the same as those at PG1 and the adjusted density profile is presented in Table 1.

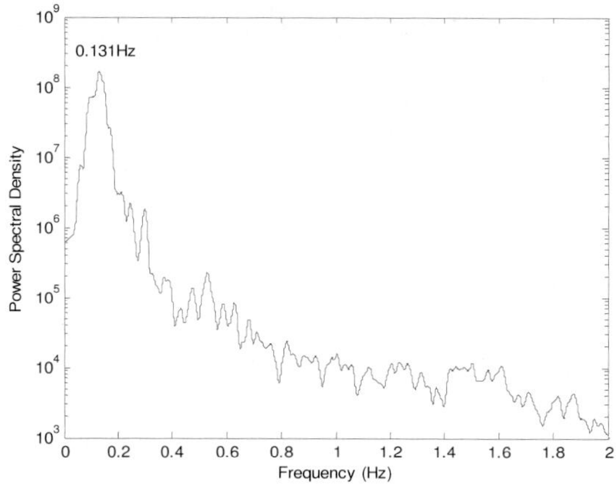

Figure 5. Power Spectrum of the 250–1750 s filtered field data. The dominant frequency is 0.131 Hz.

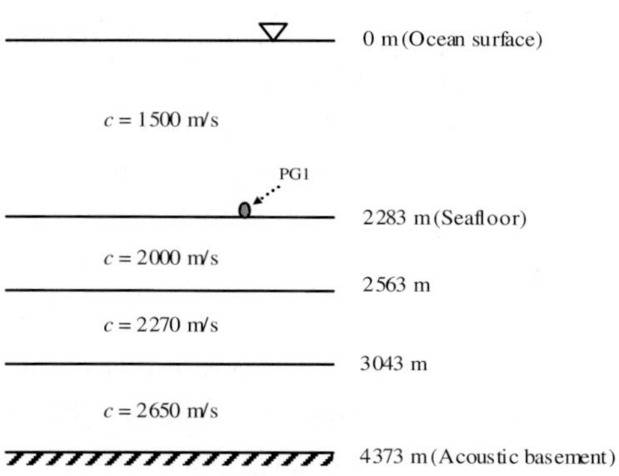

Figure 6. Acoustic speed structure at PG1.

Table 1. Density profile below the sea level.

Depth (m)	Density (kg/m^3)
0–2283	1028
2283–2700	1869–1935
2700–3195.4	1935–2184
3195.4–4373	2184–2381

Our hypothesis is that the sediment layer, if it is soft enough, should play a role in the acoustic wave transmission. As a rough estimate (used by Nosov [8]), the natural frequencies of the coupled elastic oscillations of water column and single sediment layer can be computed analytically by

$$\tan\left(\frac{2\pi f H_s}{c_s}\right)\tan\left(\frac{2\pi f H_w}{c_w}\right)=\frac{\rho_s c_s}{\rho_w c_w}. \qquad (22)$$

The values of the sediment elastic properties are obtained by taking weighted average of the data given in Fig. 6 and Table 1: $c_s = 2476$ m/sec, $\rho_s = 2154$ kg/m^3 (H_s = 2090 m). The dominant frequency computed by Eq. (22) is 0.138 Hz, which is in good agreement with that of the measurements (0.131 Hz). This suggests that the sediment layer must have played a role in the formation of the acoustic wave, which supports Nosov's [8] conclusion.

The sediment properties in Fig. 6 and Table 1, without averaging, are used for our numerical simulation, which also yields the dominant frequency at 0.138 Hz. To simulate more detailed acoustic-wave behavior (e.g. the amplitude and its decay), we must first estimate the duration of the seafloor displacement τ and the damping coefficient α. Comparing the e-folding time for the simulated and measured pressure fluctuations, we found that the damping coefficient α is approximately 0.0264 sec^{-1}. (It can be shown that α is about twice the reciprocal of the e-folding time of the amplitude decay.) In order to estimate the value of τ, we match the maximum peak-to-peak amplitudes of the simulated and measured acoustic waves for the duration of 0–2400 seconds; the result yields τ = 6.8 sec as demonstrated in Fig. 7a. Alternatively, we match their total power by summing the power spectra density (PSD): $E=\sum(PSD)=\sum p^2\Delta t$, which yields τ = 7.9 sec as shown in Fig. 7b. The resulting duration of the seafloor uplift, 6.8 ~ 7.9 seconds, is much longer than Nosov's [8] guess: he used τ = 4 sec. It must be noted that our numerical model uses the half-cosine curve (Eq. (20)) for the seafloor motion, while Nosov [8] assumed the constant uplift speed of 0.1 m/sec.

For detailed analyses of pressure fluctuations, the band-pass-filtered pressure data are re-plotted in Fig. 8 (it is essentially the same as the measured data shown in Fig. 7). The formation of amplitude modulation can be readily detected in Fig. 8. To study the modulation characteristics, we use the complex demodulation method (CDM). The first step of CDM is to estimate the demodulation frequency, which is set at the dominant peak frequency. In order to keep track of the variation of dominant peak frequency in the duration of 2400 seconds, the power spectra for segmented time intervals (300 seconds each) are computed. Since the main shock lasted about 40 seconds, we start the computation from $t = 50$ seconds, eliminating the segment of forced oscillations. The peak frequencies of the spectra and the corresponding time intervals are listed in Table 2, which indicates that the peak frequency is fairly constant varying between 0.120 and 0.133 Hz except the very early and very late stages.

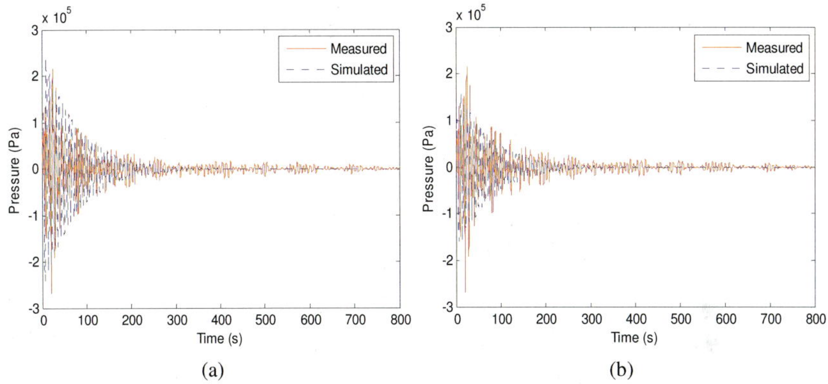

Figure 7. Time histories of the filtered measured (solid) and simulated pressure data (dotted) at PG1 for the duration of 800 seconds with $\alpha = 0.0264$ sec^{-1}, and (a) $\tau = 6.8$ sec or (b) $\tau = 7.9$ sec. The time origin corresponds to the start of the main shock.

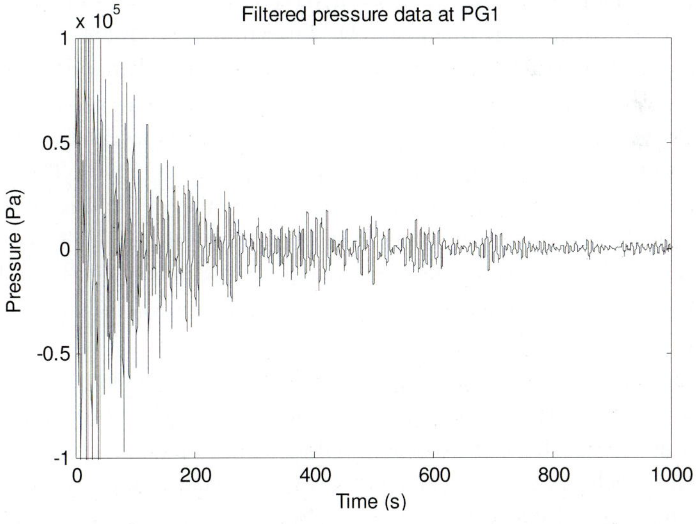

Figure 8. Time history of the filtered pressure data at PG1 for the duration of 1000 seconds.

Table 2. Dominant peak frequencies of the resulting spectra of the filtered pressure data in segmented time intervals.

Time interval (sec)	Peak frequency (Hz)
50–350	0.150
150–450	0.117
250–550	0.130
350–650	0.133
450–750	0.133
550–850	0.127
650–950	0.123
750–1050	0.123
850–1150	0.120
950–1250	0.130
1050–1350	0.133
1150–1450	0.133
1450–1750	0.120
1550–1850	0.113
1950–2250	0.093
2050–2350	0.100

Because the carrier-wave frequency is not exactly constant (Table 2), we apply the variable-frequency complex demodulation method (VFCDM). This method is a modification to the CDM. Instead of a constant demodulation frequency used in the CDM, the VFCDM follows the (slowly varying) evolution of the carrier-wave frequency so as to yield the temporal variations of the amplitude and relative phase of the evolving dominant oscillation [12]: the detailed VFCDM analysis of the present data is referred to Li [13]. The temporal variation of the carrier-wave frequency computed by the VFCDM is presented in Fig. 9, which approximately agrees with the results shown in Table 2. The amplitude of the carrier-wave frequency component is shown with the bold solid line in Fig. 10. The amplitude presents a good approximation for the envelope of the pressure data, and indicates a regular oscillatory pattern.

Figure 11 presents the spectrum of the de-trended modulating signal for the duration of 100–2400 seconds. The dominant peak frequency of the modulation is found at 0.0087 Hz. The existence of dominant frequency component in the amplitude envelopes means that there exist side-band frequency components around the carrier-frequency component. Note that our 1-D numerical model could not simulate such amplitude modulation in spite of the inclusion of multiple sediment layers. We conjecture that this modulating behavior must be related to the wave propagation medium: the interference of scattering acoustic waves from the neighboring areas may be a possibility.

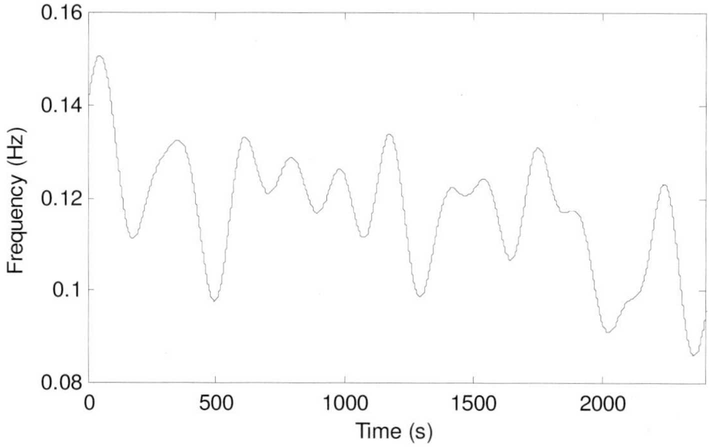

Figure 9. Time-varying dominant frequency of the filtered pressure fluctuations at PG1.

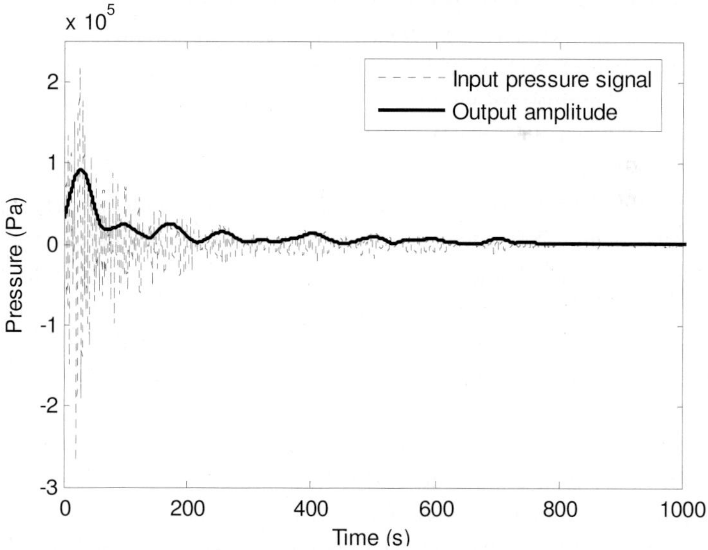

Figure 10. Time series of the filtered pressure data at PG1 (dotted) and the output amplitude for the carrier-wave frequency component (bold solid) of the duration of 1000 seconds.

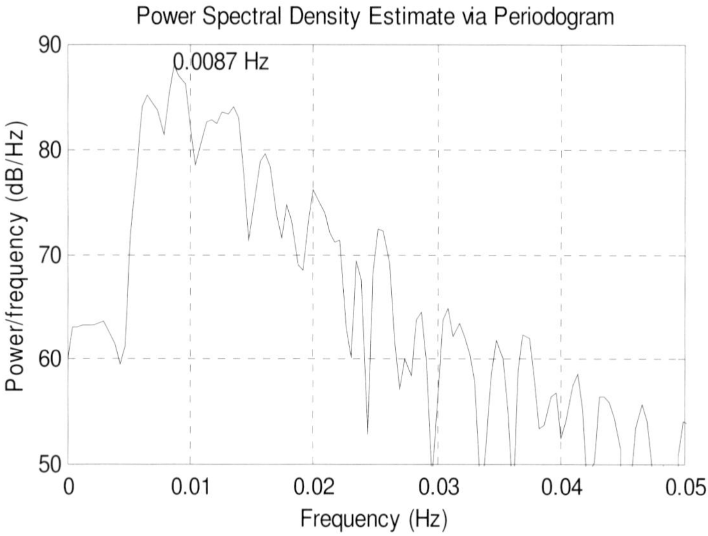

Figure 11. Power spectrum of the de-trended amplitude signal of 100–2400 seconds. The dominant peak frequency is 0.0087 Hz.

Temporal evolution of the amplitude spectra is computed with the standard CDM, and the results are shown in Fig. 12. (We divide the frequency range of [0, 0.2 Hz] into 20 sections ($\Delta f = 0.01$Hz) and apply the CDM to the frequency at the center of each section with a 0.005 Hz low-pass cutoff frequency.) Here the amplitude is normalized with the maximum value in the frequency range at a given instant so that the maximum amplitude is constant over the time, and plotted in the logarithm scale since it varies several orders of magnitude over the frequency. Figure 12 clearly shows the gradual decrease in the carrier frequency of the data over 2200 seconds. While the actual amplitude decays, the (normalized) spectral characteristics of the acoustic waves appear to persist. Another observation that can be made from Fig. 12 is the rapid decay in energy of the low frequency components, say, less than 0.04 Hz; the energy decays quickly in the first 500 seconds or so. The careful observation of Fig. 12 indicates that the decay process is oscillatory with the period of approximately 100 seconds. This observed decay behavior at low frequency must be related to the gravity waves (tsunamis) generated by the earthquake. Note that progressive gravity waves with frequency less than 0.04 Hz can penetrate the water column of 2283 m where PG1 is located. The observed modulation with the period of 100 seconds must be caused by the tsunamis.

The band-pass-filtered pressure data including the aftershocks are plotted in Fig. 13: recall that the main shock occurred at 4:50:06 JST and the largest aftershock followed at 06:08 local time. It is seen that the amplitude of pressure fluctuations attenuates after every shock.

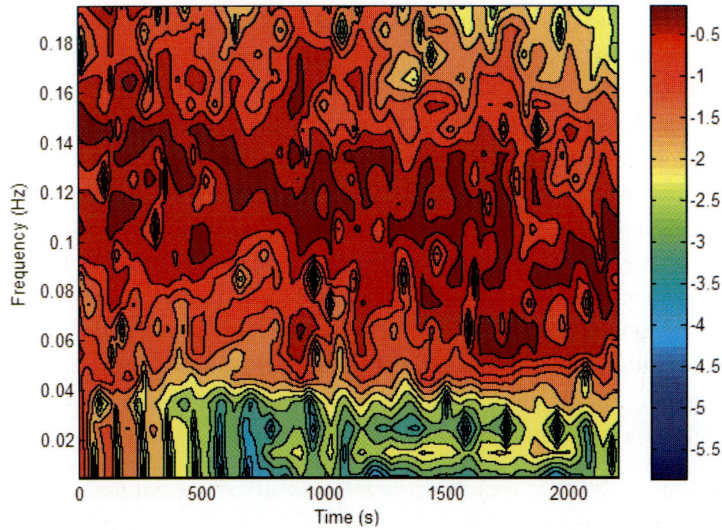

Figure 12. Temporal evolution of the amplitude spectra (in logarithm scale) for the filtered pressure data at PG1 of 0–2200 seconds.

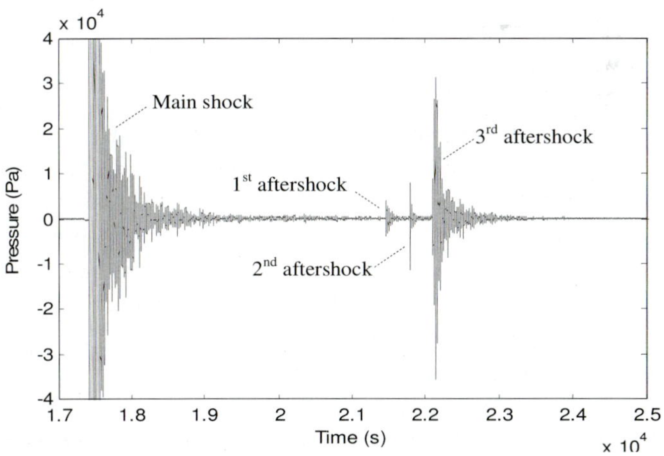

Figure 13. Time series of filtered pressure data at PG1 for the duration of 8000 seconds starting at 04:43:20 JST.

To examine the attenuation rates of the main shock and the subsequent aftershocks, we first calculate the peak-to-peak amplitude of the acoustic wave, and then plot the envelope curves of the peak-to-peak amplitudes as shown in Fig. 14. For the main shock and the 3rd aftershock, it can be seen that the attenuation is faster in the first 150 seconds and slows down afterward. Using linear regression to the semi-log plots for the first 150 seconds, straight lines for the best fit are drawn in Fig. 14. Then the e-folding time of attenuation is calculated for each event and presented in Table 3. For the acoustic waves generated at the same position, we expect that the attenuation characteristic be the same for all the shocks. Table 3 indicates that this is indeed the case except the 2nd aftershock: attenuation of the 2nd aftershock is much quicker than the other three events. The different decay rate of the 2nd aftershock must be due to a different source mechanism: such as the rupture of a very small area (i.e. quicker lateral radiation losses) or the seafloor displacement occurred somewhere away from the location of PG1.

The e-folding time for the main shock is 75.4 seconds, which is equivalent to the damping coefficient $\alpha \approx 0.0265$ sec^{-1}. Recall that we use $\alpha = 0.0264$ sec^{-1} in our numerical simulations.

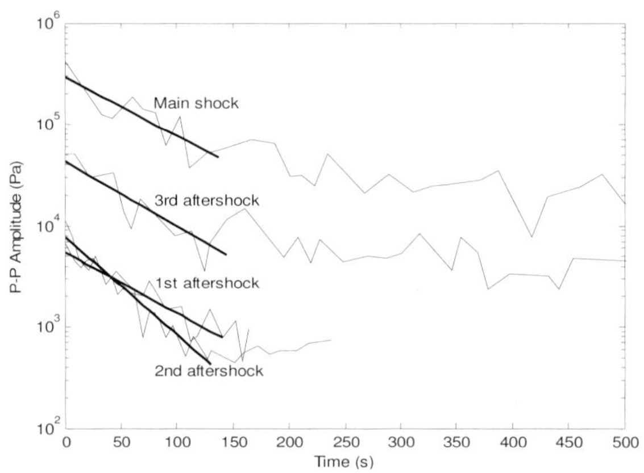

Figure 14. Envelopes of the peak-to-peak amplitudes (semi-log plots) for the main shock and the following three aftershocks (thin solid lines). The thick straight lines are the best-fitting curves of the envelopes for the first 150 seconds using linear regression. The time origin corresponds to the time of the highest peak for each envelope.

Table 3. E-folding time for the main shock and the following three aftershocks.

Shock	Main shock	1st aftershock	2nd aftershock	3rd aftershock
e-folding time (sec)	75.4	72.0	45.3	67.9

4. Conclusions

During the 2003 Tokachi-Oki earthquake, the ocean-bottom pressures were captured by the cabled seafloor observatory located in the source region. The data at relatively high sampling rate (10 Hz) enable us to explore the characteristics of acoustic waves induced by the co-seismic seafloor displacement. Based on the field-data analyses together with the numerical simulations, we found:

- The co-seismic seafloor displacement generated the acoustic waves bouncing up-and-down between the hard bottom and the sea surface. The dominant frequency of the fluctuation indicates that soft sediment layers must have played a role in the formation of the acoustic waves.
- The modulation of the pressure fluctuations was found to be quite regular; it suggests that local variations in bathymetry and sediment layers may have caused the interference of the acoustic waves from different origins in the proximity of the measurement site.
- Comparing the simulated and measured acoustic waves, the damping coefficient α and the duration of the seafloor displacement τ are estimated to be $\alpha = 0.0264$ sec^{-1} and $\tau = 6.8 \sim 7.9$ sec, respectively.
- The amplitude of measured pressure fluctuations attenuates quickly after every shock. The attenuation rate of the 2nd aftershock is much faster than the other events, which suggests that the 2nd aftershock must result from the rupture of a very small area (i.e. quicker lateral radiation losses) or the seafloor displacement occurred somewhere away from the location of PG1.

Considering the current developments of ocean-bottom observatory systems in the United States and elsewhere, the present analyses demonstrate how the data from a future observatory can be used when and if an earthquake occurs right at the location of the observatory.

Acknowledgment

This work was supported by the US National Science Foundation (CMS-0245206).

References

1. Y. Yamanaka, and M. Kikuchi, *Earth Planets Space*, **55**, e21 (2003).
2. J. Ueno, S. Takada, and Y. Kuwata, *Proceedings of the Third Taiwan-Japan Workshop on Lifeline Performance and Disaster Mitigation* (2004).

3. Y. Tanioka, Y. Nishimura, K. Hirakawa, F. Imamura, I. Abe, Y. Abe, K. Shindou, H. Matsutomi, T. Takahashi, K. Imai, K. Harada, Y. Namegawa, Y. Hasegawa, Y. Hayashi, F. Nanayama, T. Kamataki, Y. Kawata, Y. Fukasawa, S. Koshimura, Y. Hada, Y. Azumai, K. Hirata, A. Kamikawa, A. Yoshikawa, T. Shiga, M. Kobayashi, and S. Masaka, *Earth Planets Space*, **56(3)**, 359 (2004).
4. K. Hirata, Y. Tanioka, K. Satake, S. Yamaki, and E. L. Geist, *Earth Planets Space*, **56(3)**, 367 (2004).
5. K. Hirata, T. Baba, H. Sugioka, H. Matsumoto, E. Araki, T. Watanabe, H. Mikada, K. Mitsuzawa, R. Iwase, R. Otsuka, S. Morita, and K. Suyehiro, *2003 AGU Fall Meeting*, **84(46)**, Abstract S52L–04 (2003).
6. K. Kajiura, *Bulletin of Earthquake Research Institute*, **48**, 835 (1970).
7. M. A. Nosov, *Physics and Chemistry of the Earth*, pt B, **24(5)**, 437 (1999).
8. M. A. Nosov, *22nd IUGG International Tsunami Symposium*, Chania, Greece, June 27-29th, 2005.
9. K. Hirata, Personal Communication (2004).
10. T. Tsuru, Personal Communication (2004).
11. Oil and Natural Gas Resources in Japan, *Association of Continental Shelf Oil Development, Natural Gas Mining Society* (1992).
12. H. Gasquet, and A. J. Wootton, *Review of Scientific Instruments*, **68(1)**, pt 2, 1111 (1997).
13. W. Li, Master's Thesis (2006).

TSUNAMI HYDRODYNAMIC MODELING: STANDARDS AND GUIDELINES

COSTAS SYNOLAKIS

Coastal Engineering and Natural Hazards Laboratory, Technical University of Crete
Chanea, 73100, Greece
Viterbi School of Engineering, University of Southern California
Los Angeles, California, 90089-2531, USA

UTKU KÂNOĞLU

Department of Engineering Sciences, Middle East Technical University
Ankara, 06531, Turkey

We review standards and guidelines (S + G) for inundation models that are used to evaluate hazards from tectonic- and landslide-triggered tsunamis. Standards and guidelines have evolved in the past fifteen years through the contributions of Professor Phil Liu in numerical, analytical, and laboratory studies. These S + G have been developed over the past two years to assist in the assessment of computational models used in the production of inundation maps in the US and elsewhere. When unvalidated models are used for making inundation projections, they may result in excessive estimates — one celebrated case assessing the impact of a 1700-type tsunami from the Cascadian Subduction Zone, the authors projected four times the inundation heights that had been observed in the sediment record for the 1700 event; panic resulted among the local emergency management who did not know whether to rely on the new projections or on the inundation maps which had been obtained with validated models. The consequences of low projections are equally ghastly. To limit these occurrences, S + G are urgently needed. The standards refer to benchmark tests for numerical model validation, while the guidelines refer to procedures to ensure that the adoption of any particular set of predictions for the impact of future tsunamis.

1. Tsunami Hydrodynamics

The evolution of tsunamis from their source region to their target is one of the quintessential problems in coastal engineering (Liu *et al.*, 1991). In all but the simplest one dimensional problem with idealized initial conditions, this evolution needs be calculated numerically to obtain estimates of tsunami currents, forces, and runup on coastal structures or inundation of coastlines.

The unavailability of instrumental recordings of tsunamis in the open ocean — until recently (Titov *et al.*, 2005; Bernard *et al.*, 2006) — resulted in tsunami science and engineering evolving differently than studies in other extreme natural hazards. While the basic equations for analysis have been known for decades, the existing "grand" synthesis had to await the development of sophisticated modeling tools, the large-scale laboratory experiments in the

1980s, the field survey results of the 1990s, and the tsunameter recordings of 2003 and since. Free-field tsunameter recordings are now available online from the National Oceanic and Atmospheric Administration (NOAA) of the US (http://www.ndbc.noaa.gov/dart.shtml). Deep-ocean Assessment and Reporting of Tsunamis (DART) system provides time histories of the free surface elevation of the passing tsunami in the open ocean. The field results in the 1990s served as crude proxies to free-field tsunami recordings — field surveys typically provide measurements of the maximum onland penetration of the tsunami or of the maximum flow depth at select locations. Nonetheless, the field measurements of the elevation of the maximum penetration of the tsunami compared with the initial shoreline (a height often referred to as runup), they have allowed for the validation of numerical procedures. For a more detailed history of tsunami hydrodynamics, refer to Synolakis and Bernard (2006).

The need for standards for numerical codes whose predictions are used for structural design or for hazard assessment is obvious. By some accounts, more than ten new tsunami codes have been claimed in the aftermath of the 2004 Boxing Day tsunami. Given the extensive development effort even for one dimensional codes, such claims need to be carefully examined. For example, the code TsunAWI being developed in Germany is now undergoing rigorous testing, with intermediate results presented to assess its veracity. On the other end, some numerical predictions for the impact of past tsunamis in the US kept being presented in international meetings in 2005, even though they differed by factors up to 4 from established paleotsunami measurements. These "new" predictions were based largely on untested models, or extrapolation of procedures developed for processes of different physical scales. In some cases, the authors use higher order models, at considerable computational costs, without any understanding of grid resolution issues that are the ultimate limiting step of any numerical exercise. Some of the new predictions attracted press attention leading to additional efforts from hazard mitigation professionals to explain to the public the differences between peer-reviewed models/predictions and the alternatives.

The process of establishing standards has been the outcome of three landmark scientific meetings that have significantly contributed to our understanding of tsunami hydrodynamics. All three have been supported by the National Science Foundation (NSF) of the US, i.e., the 1990 Catalina Island, California (Liu *et al.*, 1991); 1995 Friday Harbor, Seattle, Washington (Yeh *et al.*, 1996); and 2004 Catalina Island, California (Liu *et al.*, 2008) workshops on Long-Wave Runup Models. In these workshops, benchmark problems were proposed and modelers were asked to present comparisons of their model predictions with analytical solutions and/or laboratory data and/or field data. These workshops and their resultant publications established the basis for the S + G for tsunami inundation models, as they have evolved to date. Here, we will provide a short summary of the S + G for tsunami hydrodynamic modeling. For more detail discussion, refer to Synolakis *et al.* (2007).

2. Benchmarking for Tsunami Inundation Models

Numerical codes used for current tsunami inundation mapping in the US are first validated with analytical solutions and laboratory measurements and then verified with field data. While, in principle, there is no absolute certainty that a numerical code that has performed well in all existing benchmark tests will also produce realistic inundation prediction with any particular source motion, validated codes reduce the level of uncertainty to hopefully that in the geophysical initial conditions. In what follows, steps for validating and verifying 2 + 1 codes are discussed in four categories: basic considerations, analytical benchmarking, laboratory benchmarking, and field data benchmarking. Additional steps need to be considered for forecast models.

2.1. *Basic Considerations*

A most basic step in making sure that a numerical model works for predicting evolution adequately is ensuring that the model conserves mass and checking convergence of the numerical code to a certain asymptotic limit, presumably the actual solution of the equations solved. While the conservation of mass equation is one of the equations of motion that are solved in any numerical procedure, cumulative numerical approximations can sometimes produce results that violate mass conservation. This is particularly the case when friction factors are used, or smoothing to stabilize computations for breaking waves, or when large grid sizes are inadvertent due to inadequate bathymetry resolution. The numerical predictions should be shown to converge to a certain value with further reductions in step sizes not producing any change of the results. The optimal locations to check convergence are the extreme runup and rundown.

2.2. *Analytical Benchmarking*

Any numerical code that will be used in tsunami modeling should perform well in wide range of parameters and should solve the governing equations with consistent accuracy in geophysical scales. This is assured through analytical benchmarking. Exact solutions of the shallow-water wave (SW) equations are useful for validating the complex numerical models which are used for final design and which often involve ad-hoc assumptions, particularly during inundation computations when grid points are introduced during the runup process on what was dry land initially. Even though analytical solutions involve approximations, they are exact and approximations involved are quantifiable. Comparisons of numerical predictions with analytical solutions can identify systematic errors, as when using friction factors or dissipative terms to augment the idealized equations of motion.

Over two decades, analytic solutions of the linear and nonlinear SW equations of motion have been validated themselves with laboratory data. Both linear and nonlinear versions of the SW equations have demonstrated a remarkable and surprising capability to model complex evolution phenomena, and in particular the maximum runup/inundation. Certain numerical models have been validated with analytical solutions, laboratory data and verified with field data, thus setting a golden standard. No modeling of any real tsunami should ever be undertaken with any code, before comparing its predictions with the benchmark solutions described here.

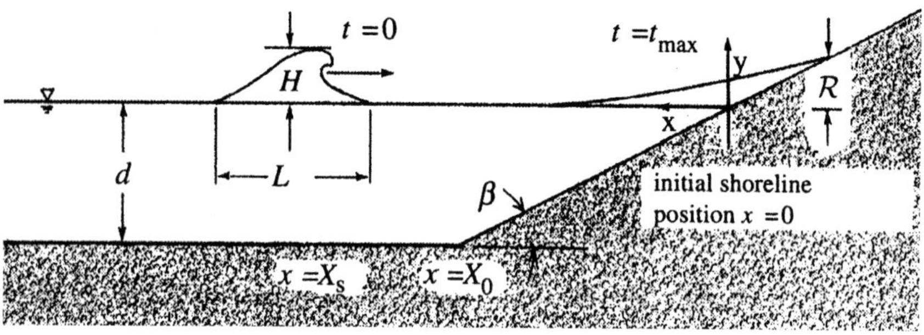

Figure 1. Definition sketch for canonical bathymetry.

2.2.1. *Solitary Wave Runup over Canonical Bathymetry*

The analytical work of Synolakis (1986) was a significant advance in exploring the evolution and runup characteristics of nonbreaking solitary waves for canonical problem, i.e., propagation of long-wave first over a constant depth then sloping beach (Fig. 1). Synolakis (1986, 1987) solved a combined initial value problem and was able to obtain a solution to the linear SW equation considering a solitary wave with offshore height H, propagating over constant-depth d and then climbing up a beach of angle β, with profile $\eta(x,t=0)/d=(H/d)\,\mathrm{sech}^2[\gamma(x-x_1)]$; $\gamma=\sqrt{(3/4)(H/d^3)}$. Synolakis (1986, 1987) used direct contour integration, summed up the resulting series asymptotically and derived an expression for maximum runup with exact coefficient and exponent, i.e., $R/d=2.831\cot\beta\,(H/d)^{5/4}$ which is known as the runup law (Fig. 2). Most of the later numerical solutions have used the runup law for validation.

Hence, Synolakis (1986, 1987) was able to mathematically justify the empirical relationship of Hall and Watts (1953); Hall and Watts (1953) had presented the first laboratory study of a single long wave approaching a sloping beach. They introduced what is now known as the canonical problem of one-dimensional long-wave theory. They used empirically generated water waves which resembled the classical solitary wave shape. Using dimensional analysis,

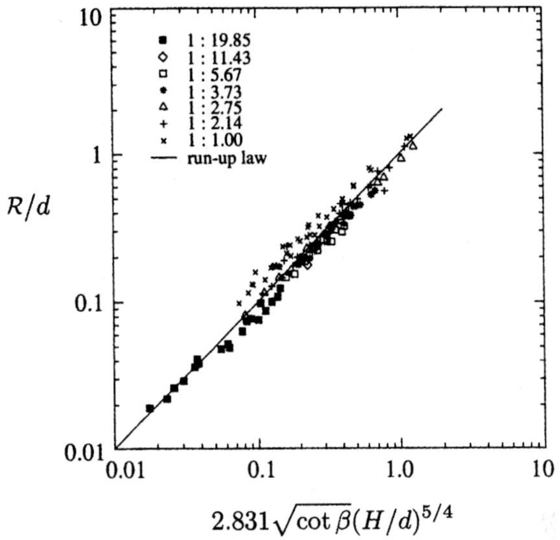

Figure 2. Laboratory data for maximum runup of nonbreaking waves climbing up different beach slopes: ■ 1:19.85 (Synolakis, 1986); ◇ 1:11.43, □ 1:5.67, ∗ 1:3.73, + 1:2.14, × 1:1.00 (Hall and Watts, 1953); △ 1:2.75 (Pedersen and Gjevik, 1983). Solid line represents the runup law of Synolakis (1986, 1987).

$\alpha(\beta)$ and $\lambda(\beta)$ being empirical coefficients dependent on the beach slope β, they concluded that $R/d = \alpha(\beta)(H/d)^{\lambda(\beta)}$.

In terms of the solution of the nonlinear SW equations, the major analytical advance was presented by Carrier and Greenspan (1958). Carrier and Greenspan (1958) introduced spatial and temporal like variables, (σ, λ), and were able to reduce the nonlinear SW equations into a single second-order partial differential equation. However, there were two basic difficulties in Carrier and Greenspan (1958)'s solution. The first lies in the derivation of an equivalent initial condition over the transform (σ, λ)-space for a given initial wave in the physical (x, t)-space. The second difficulty is the conversion of the solution to the physical (x, t)-space once obtained in the transformed (σ, λ)-space.

Synolakis (1986, 1987) resolved these difficulties considering the boundary value problem solution of the nonlinear SW equations for Boussinesq solitary wave. To achieve this, Synolakis (1986, 1987) used the Keller and Keller (1964) formalism to derive a boundary condition at the toe of the sloping beach from the linear problem. Then he linearized the Carrier and Greenspan (1958) transformation at the transition point, arguing that far off the beach nonlinear effects were expected to be small and proceeded to derive a new solution to the nonlinear problem for general offshore initial conditions and calculated the evolution of a solitary wave with nonlinear theory, through the entire runup and rundown process. Unlike Carrier and Greenspan (1958), Synolakis (1986, 1987) was able to back calculate the evolution of the wave into the physical (x, t)-space, and successfully compared the analytical predictions with laboratory measurements. It is important to note that, during runup, individual monochromatic waves reflect with phase-shifts that are slope dependent. Early SW formulations did not account for reflection. While their predictions for the Carrier and Greenspan (1958) sinusoids were correct, they exhibited significant errors when modeling solitary waves or N-waves. Hence the analytical solution of Synolakis (1986, 1987) should be used for code validation. Comparison of results with the predictions for a spectrum of frequencies is far more useful and realistic than a comparison for a single frequency.

2.2.2. *N-wave Runup over Canonical Bathymetry*

Tsunamis sometimes arrive to the target coastlines as a leading-depression N-wave (LDN), i.e., elevation wave with a trough in front (Fig. 3) as oppose to leading-elevation N-wave (LEN), i.e., crest arriving first with a trough behind it. One major observation during the field survey of 1992 Nicaraguan tsunami were the reports of the incoming tsunami causing the shoreline to recede first. However, before 1992, LDNs were believed to be hydrodynamically unstable; the crest was supposed to quickly overtake the trough. On 9 October 1995 Manzanillo, Mexico tsunami also arrived at the target coastline with an LDN and International Tsunami Survey Team was able to acquire two series of photographs from eyewitnesses noting LDN (Borrero *et al.*, 1997). This is believed to be the first documented observation of a leading-depression N-wave causing shoreline recession.

By noting that the first arrival in Nicaragua was waves that caused the shoreline to recede before advancing, Tadepalli and Synolakis (1994) proposed a model for the leading wave of tsunamis with an N-wave shape, i.e., $\eta(x,t=0)/d = \varepsilon(H/d)(x-x_2)\,\mathrm{sech}^2[\gamma(x-x_1)]$, where x_1 and x_2 define the

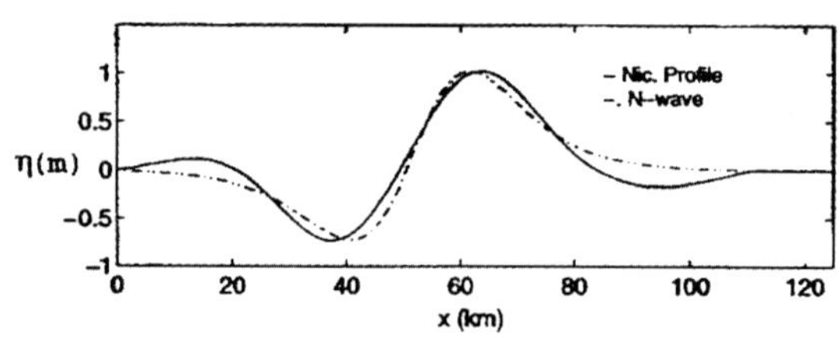

Figure 3. Leading-depression N-wave at the toe of the beach for the 1992 Nicaragua event (Tadepalli and Synolakis, 1996).

locations of the crest and the trough; $\gamma = \sqrt{(3/4)(H/d^3)}$; and ε is a scaling parameter to allow for direct comparison with solitary waves. Tadepalli and Synolakis (1994) used the methodology of Synolakis (1987) to evolve these N-waves over the canonical geometry and found that N-waves with the trough first run-up higher than N-waves with the crest arriving first (Fig. 4). Since N-wave theory does provide a conceptual framework for analysis and for explaining certain field observations it can be used as analytical benchmark.

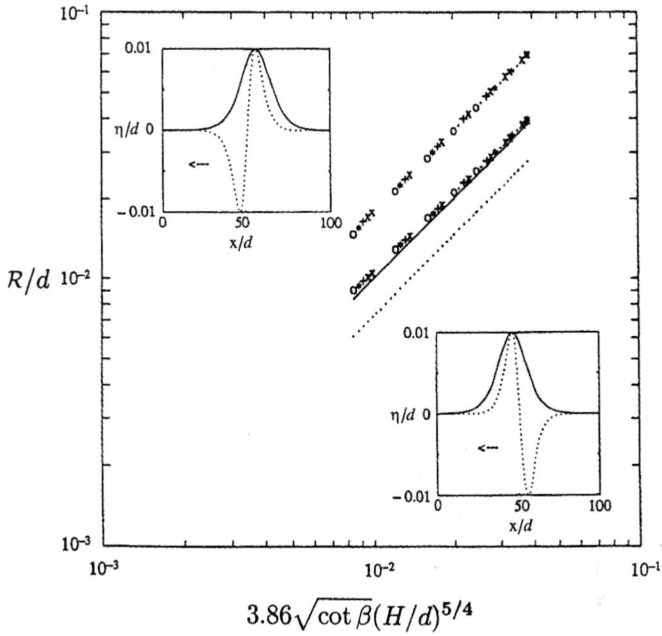

Figure 4. Maximum runup of isosceles — the same heights for depression and elevation — N-waves and solitary wave. Top and lower set of points are results for the maximum runup of leading-depression and -elevation isosceles N-waves respectively. Dotted line represents the maximum runup of solitary waves. Lower and upper insets compare a solitary wave profile to an isosceles leading-depression and -elevation N-waves respectively. After Tadepalli and Synolakis (1994).

2.2.3. *Composite Beach, Revere Beach*

Kânoğlu and Synolakis (1998) extended Synolakis (1986)'s approach to the wave propagation over piecewise linear topographies. Kânoğlu and Synolakis (1998) were able to present matrix formulation which can apply different piecewise linear topographies. Kânoğlu and Synolakis (1998) were then able to evaluate evolution and maximum runup of solitary wave over three piecewise linear slopes of 1:53, 1:150, and 1:13 (seaward to shoreward) fronted with a vertical wall at the shoreline (Fig. 5). This is the geometry of Revere Beach, Massachusetts for which experimental data was available. Kânoğlu and Synolakis (1998) analytically derived a simple formulation, $R/d = 2\,(h_w/d)^{-1/4}\,(H/d)$ which shows that the maximum runup is driven by the water depth h_w at the shoreline (Fig. 6). Since most topographies of engineering interest can be approximated by piecewise linear segments analytical result of Kânoğlu and Synolakis (1998) is an essential benchmark case for code validation, as it helps identify the grid resolution needed near barriers.

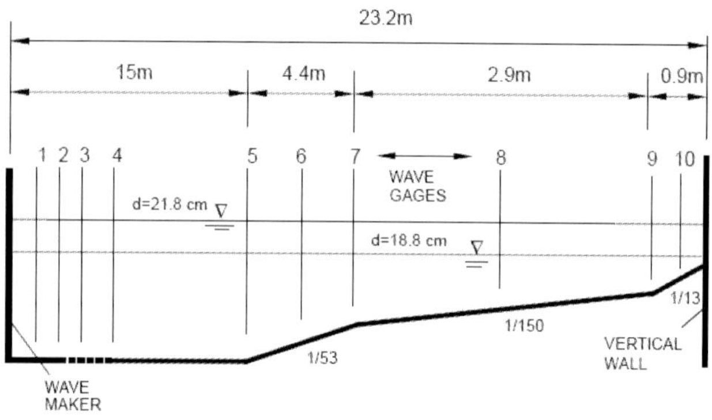

Figure 5. Definition sketch for Revere Beach. Not to scale. After Kânoğlu and Synolakis (1998).

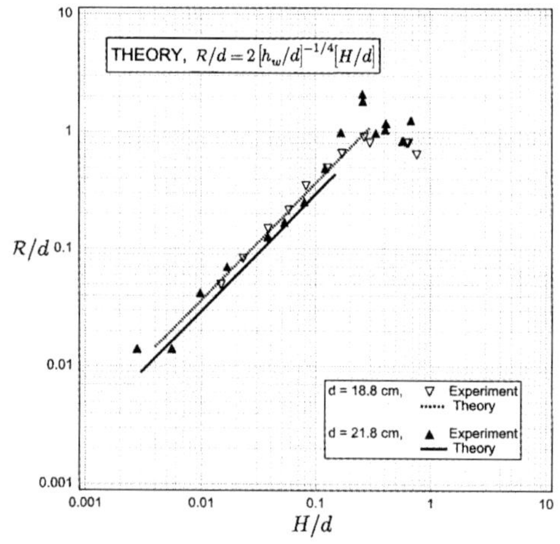

Figure 6. Comparison of maximum runup values for the linear analytical solution and laboratory results for two different depths, i.e., $d = 18.8$ cm and $d = 21.8$ cm. After Kânoğlu and Synolakis (1998).

2.2.4. *Initial Value Problem Solution over a Sloping Beach*

Carrier *et al.* (2003) revisited the 1 + 1 initial value problem (IVP) solution of the nonlinear SW equations for a uniformly sloping beach with a new hodograph transformation. However, the Carrier *et al.* (2003) solution includes integrals with singularities, issues addressed shortly thereafter by Kânoğlu (2004). Later, Kânoğlu and Synolakis (2006) were able to present a more general solution to the IVP of the nonlinear SW equations, i.e., initial wave with velocity. Unlike Carrier *et al.* (2003), Kânoğlu and Synolakis (2006) were able to solve the case with initial velocity without any difficulty for larger initial wave heights. The analytical solution of Kânoğlu (2004), without initial velocity, and Kânoğlu and Synolakis (2006), with initial velocity, provide good benchmark cases for numerical estimates for the shoreline position and velocity. One example is provided in Fig. 7.

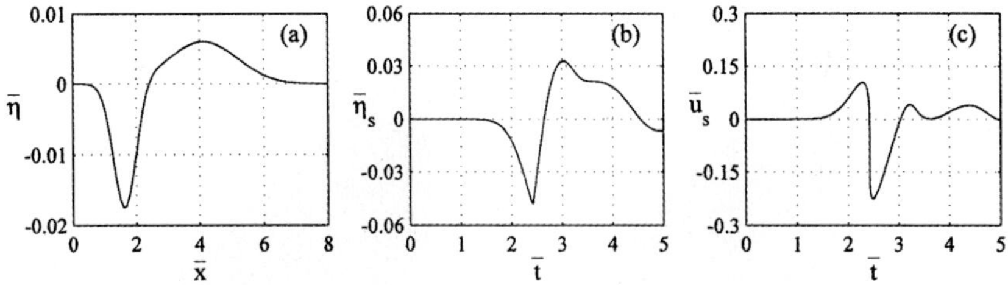

Figure 7. (a) The leading-depression initial waveform $\eta(x,t=0)/d=(H_1/d)\exp(-c_1(x/l-x_1/l)^2)-(H_2/d)\exp(-c_2(x/l-x_2/l)^2)$ presented by Carrier *et al.* (2003) with $H_1/d=0.006$, $c_1=0.4444$, $x_1/l=4.1209$, $H_2/d=0.018$, $c_2=4.0$, and $x_2/l=1.6384$, (b) shoreline wave height, and (c) shoreline velocity. For details refer to Kânoğlu (2004). Here l is the reference length and $d=l\tan\beta$. Note that nondimensional quantities $\bar{\eta}$, $\bar{\eta}_s$, $\bar{u}_s$, $\bar{x}$, and $\bar{t}$ are defined as η/d, η_s/d, $u_s/\sqrt{gd}$, x/l, and $t\sqrt{g\tan\beta/l}$ respectively. Here g is the gravitational acceleration.

2.2.5. *Subaerial Landslide on Simple Beach*

Landslide wave generation remains a frontier for numerical modeling, particularly for subaerial slides. Specifically, subaerial landslides involve not only the calculation of rapid change of the seafloor, but also the impact of the generated wave on the shoreline. Therefore numerical codes that will be used to model landslide generated tsunamis need to be tested against analytical solution given in Liu *et al.* (2003). Liu *et al.* (2003) considered tsunami generation by a translating — Gaussian shaped — slide on a uniformly sloping beach. Once in motion, the slide moves at constant acceleration. Liu *et al.* (2003) solved a forced linear SW equation for the translating Gaussian shape. An example is presented in Fig. 8. Given the complexity of calculating the shoreline evolution over a deforming coastline, landslide codes need to evolve from one-dimensional to two-dimensional computations. The Liu *et al.* (2003) solution is a very useful test that helps identify computational assumptions, sometimes inadvertent at the moving shoreline boundary.

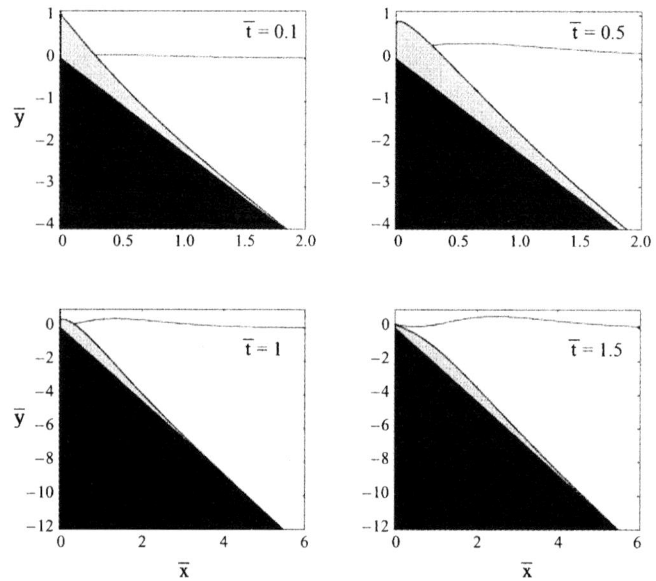

Figure 8. Spatial snapshots of the analytical solution at four different times for a beach slope, $\beta = 5°$, and landslide aspect ratio — the ratio of the maximum vertical thickness δ of the sliding mass to its horizontal length L — $\mu = \delta / L = 0.05$. The slide mass is indicated by the light shaded area, the solid beach slope by the black region, and η / δ by the solid line. Refer to Liu *et al.* (2003) for details. Note that nondimensional quantities $\bar{x}$, $\bar{y}$, and $\bar{t}$ are defined as x / L, y / δ, and $t / (\sqrt{\delta / g} \,/\, \mu)$ respectively. Here g is the gravitational acceleration.

2.3. *Laboratory Benchmarking*

Long before the availability of numerical codes, physical models at small scale have been used to visualize wave phenomena in the laboratory, and then predictions were scaled to the prototype. Even today, when designing harbors, laboratory experiments — scale model tests — are used to confirm different flow details and validate the numerical model used in the final design and analysis.

For the purpose of validating inundation models, scale differences are believed not to be important. Certain numerical codes developed in the 1990s produce predictions in excellent agreement with measurements from small scale laboratory experiments; the same have been also shown to model geophysical scale tsunamis well. For example, a numerical code that quantitatively models the inundation in a 1 m-deep model adequately is also expected to model the inundation of a 1 km-deep geophysical geometry. Scale models, in general, do not have similar bottom-friction characteristics as real ocean floors or sandy beaches, but this has turned out not to be a severe limitation. Tsunamis are such long waves, that bottom friction tends to be less important than the inertia of the motion. Friction may be important in cases of extreme inundation, as observed in Banda Aceh with 3 km penetration distances during the 2004 Boxing Day tsunami. However, it has been observed that even with numerical codes that use friction factors, the predictions are not sensitive to first order to the choice of friction factors, within of course reasonable limits.

Given the small number of laboratory measurements in three dimensional wave basins, 1 + 1 versions of 2 + 1 models should first be tested with 1 + 1 laboratory models in addition to the tests with 1 + 1 analytical models.

2.3.1. *Solitary Wave Experiments on Canonical Bathymetry*

Synolakis (1986, 1987) performed solitary wave experiments on the canonical bathymetry (Fig. 1). Synolakis (1986, 1987) used the California Institute of Technology, Pasadena, California wave tank with water at the constant depth region varying from 6.25 cm to 38.32 cm and then a 1:19.85 sloping beach. Solitary waves are uniquely defined by their maximum height H to depth d ratio and the constant offshore depth d. H/d ranged from 0.021 to 0.626. Breaking occurred when $H/d > 0.045$, for this particular beach. The maximum runup analytical predictions presented in section 2.2.1 were validated with this same set of experiments. This set of laboratory data has been used extensively for numerical code validation, in particular, the data sets for the $H/d = 0.0185$ nonbreaking and $H/d = 0.3$ breaking solitary waves are the most useful — notice that such a comparison would help validate results over one order of magnitude variation in the wave height.

2.3.2. *Solitary Wave Experiment on Composite Revere Beach*

1 + 1 models that perform well with the solitary wave experiments on canonical bathymetry must still be tested with the composite beach geometry of Revere Beach, Massachusetts, for which both analytical and laboratory data exist, with solitary waves as inputs (Kânoğlu and Synolakis, 1998). A physical model of the beach was constructed at the Coastal Engineering Laboratory of the US Army Corps of Engineers, Vicksburg, Mississippi facility, earlier known as Coastal Engineering Research Center (Fig. 5). The laboratory tests were performed at two water depths, i.e., 18.8 cm and 21.8 cm. In the experiments, solitary waves

of different heights were generated, water surface elevations were measured with 10 wave gages, and the maximum runup of solitary waves striking a vertical wall was measured visually (Fig. 6). The results were used as benchmark data to validate 1 + 1 numerical codes in the 1995 Friday Harbor, Seattle, Washington workshop, as discussed in Yeh *et al.* (1996) and by Kânoğlu (1998), Kânoğlu and Synolakis (1998) in validating their analytical formulation. This additional test will ensure that the code is stable enough for large waves which are near the breaking limit offshore, beyond the ranges where the analytical results described earlier are applicable.

2.3.3. *A Solitary Wave Attacking a Conical Island*

The Flores 1992 earthquake triggered a large tsunami and is mostly remembered by the catastrophe in Babi Island. Babi is a volcanic island, of conical shape with a diameter at the shoreline of about 2 km. While the tsunami attacked from the north, most of the inundation was in the south, normally protected from wind waves. The two fishing villages in south side of Babi were completely inundated and runup values ranged up to 7 m (Fig. 9). Professor Philip Liu was one of the scientists who visited Babi Island in the immediate aftermath of the tsunami and helped describe the unusual phenomenon. Fortuitously, and in response to the Catalina 1990 workshop, large-scale 2 + 1 laboratory experiments had been under way at a 27 m-wide, 30 m-long, and 60 cm-deep wave basin of what is now known as the Coastal Engineering Laboratory of the US Army Corps of Engineers, Vicksburg, Mississippi. A laboratory model of a 7.2 m base-diameter conical island was constructed with a slope angle of 14° (Fig. 10). The experiments demonstrated that once the wave hits the side of the island across from the generator, the crest splits into two waves (Fig. 9). These two waves move with the crest perpendicular to the shoreline, propagate around the island and collide behind it, in a spectacular demonstration of constructive interference (Yeh *et al.*, 1994). A series of laboratory tests were undertaken with different wave crest lengths and wave heights of solitary wave around the conical island.

The circular island data measurements were then used to refine shoreline 2 + 1 algorithms under development. Preliminary modeling results were published by Yeh *et al.* (1994), and more comprehensive analyses by Liu *et al.* (1995), Kânoğlu (1998), Kânoğlu and Synolakis (1998), and Titov and Synolakis (1995, 1998), where the code MOST used by NOAA for real-time forecasting was most comprehensively validated. The experimental data formed the basis of one of the four benchmark problems used in a NSF 1995 Friday Harbor, Washington workshop for inter-model comparison and code validation. 2 + 1 dimensional calculations should be tested with the measurements from the conical island experiment. It is important to ensure that the numerical procedure models two wave fronts that split in front of the island, colliding behind it and remaining stable.

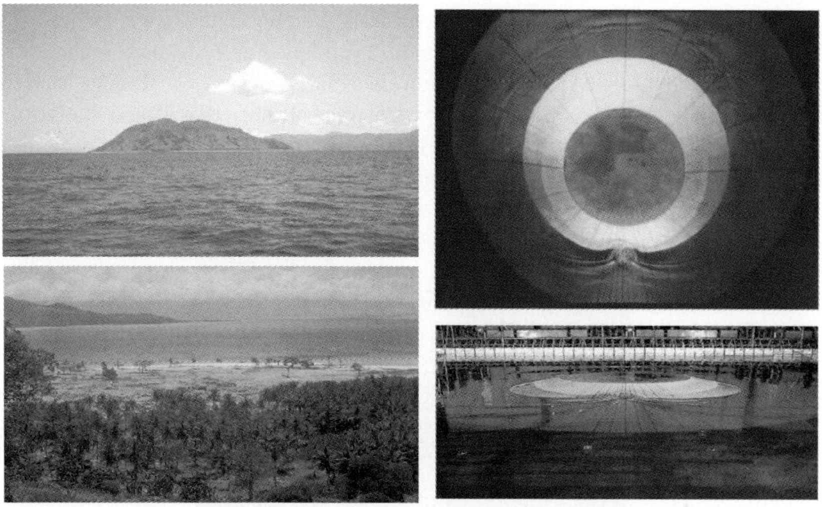

Figure 9. (top left) Babi Island and (bottom left) catastrophe on the back side of it. After Yeh *et al.* (1994). (top right) Top view and (bottom right) side view of the laboratory manifestation of a solitary wave attacking a conical island of slope 1:2 to model the catastrophe in Babi Island in 1992. After Kânoğlu and Synolakis (1998).

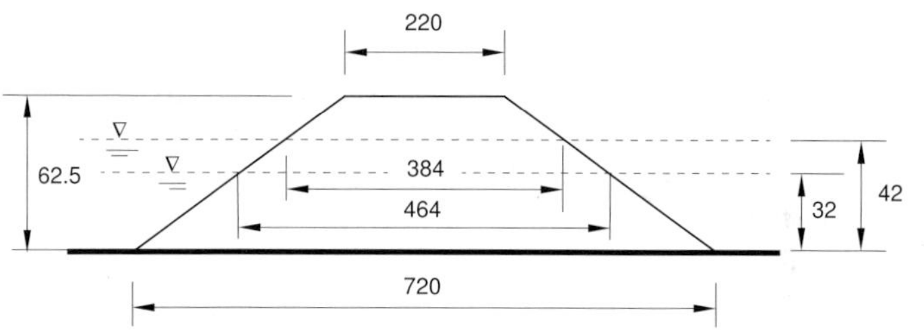

Figure 10. Definition sketch for conical island. All dimesions are in cm. Not to scale.

2.3.4. *Okushiri Island Experiment*

The 12 July 1993 Hokkaido-Nansei-Oki tsunami hit northern Japan and devastated the island of Okushiri. Extreme runup of just over 30 m was observed at Monai Valley. A laboratory model closely representing the actual bathymetry of Monai Valley, Okushiri Island, Japan at a scale of 1/400 was constructed at the Central Research Institute for Electric Power Industry (CRIEPI) in Abiko, Japan (Fig. 11). 2 + 1 numerical computations should be tested with the laboratory model of Monai Valley. Because of excellent bathymetry data before and after the event, it was possible to identify an appropriate initial condition, and the experimental basin closely resembles the actual bathymetry. Hence, this data set is another benchmark test for model validation and was used in the 2004 Catalina Island, Los Angeles, California NSF Long-Wave Runup Models Workshop. The initial condition is a LDN (Takahashi, 1996), and the entire simulation allows the identification of how well a numerical code performs in a rapid sequence of withdrawal and runup.

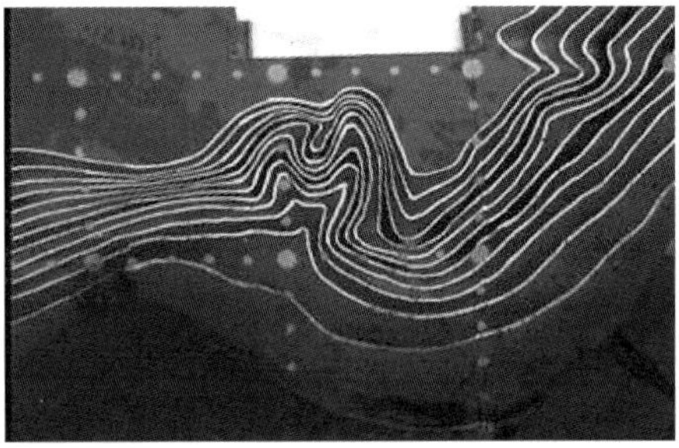

Figure 11. Bathymetric profile for experimental setup for Monai Valley, Okushiri Island, Japan experiment.

2.3.5. *Tsunami Generation due to Three-dimensional Landslide*

The event that shook tsunami science, on a scale analogous to the 2004 megatsunami, was the 17 July 1998 Papua New Guinea (PNG) tsunami. Despite the relatively small size of the parent earthquake, this tsunami resulted in over 2100 fatalities. The source mechanism of the PNG tsunami had remained controversial — field evidence was suggestive of a landslide trigger for the main tsunami, but initially it was not sufficient to overcome arguments as to its timing, not to mention the unusual fault geometry that had been claimed to have produced a larger than expected tsunami. The controversy was settled by Synolakis *et al.* (2002). The identification of a submarine landslide as the source of the 1998 tsunami resulted in renewed sensitivity for the hazards created by underwater landslides.

In an effort to better understand the generation and runup of waves from submarine and subaerial slides, Raichlen and Synolakis (2003) conducted large scale experiments in a 104 m-long, 3.7 m-deep, and 4.6 m-wide wave channel with a plane slope (1:2) located at one end of the tank. They first used a 91 cm-long, 61 cm-wide with a 46 cm-high vertical face wedge block (Fig. 12), and then semi-spherical and rectangular boxes of equal volume. By varying the weight of the blocks, they were able to vary the initial acceleration, while the initial position varied from totally subaerial to totally submerged. The experiments revealed a three-dimensional depression forming over the wedge as motion initiates. The depression forms the leading portion of the LDN propagating towards shore and running up the slope, while the LEN propagates offshore. For the aspect ratio of their boxes, they observed that wave generation became rapidly inefficient as the submergence of the blocks exceeded one block height. These experiments have since been used as another benchmark test for model validation for landslide models (Liu *et al.*, 2008).

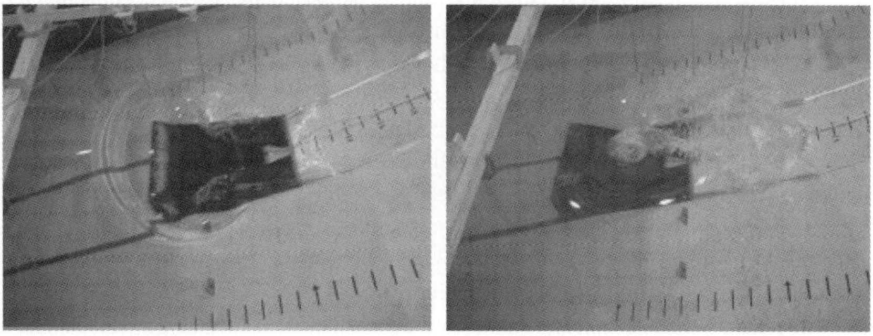

Figure 12. A moving wedge as a model of submarine landslide. (left) The initial wave from a submerged slide, immediately after motion started; note the LDN. (right) The wave about 1 sec later, as the runup forms per Raichlen and Synolakis (2003).

2.4. *Field Data Benchmarking*

Verification of a model in a real-world setting is an important part of model validation especially for operational models that attempt real time tsunami forecasts. No analytical or laboratory data comparisons (or any limited number of tests, for that matter) can assure robust model performance in the operational environment. Test comparisons with real-world data provide an additional important step in the validation of a model to perform well during operation implementation.

2.4.1. *Okushiri Island*

For validation, 2 + 1 computations need to be compared with the field measurements from the 12 July 1993 Hokkaido-Nansei-Oki tsunami around Okushiri Island, Japan. This was the worst local tsunami-relate death toll in fifty years in Japan, with estimated 10–18 m/sec overland flow velocities and 30 m runup. This event was surveyed extensively and produced densely distributed field measurements. Having high resolution bathymetric and reasonable ground deformation data, field measurements from this event have been as one of the benchmark problems during the 1995 Friday Harbor workshop (Yeh *et al.*, 1996). While it is likely that measurements from Banda Aceh for the 2004 event will become the ultimate test, until the time the data set is presented in an intercomparison workshop, the Okushiri data remains a very valuable benchmark.

2.4.2. *Rat Islands Tsunami*

The 17 November 2003 Rat Islands tsunami provided the most comprehensive test for forecast methodology — a real time forecast was successfully performed. The 15 November 2006 Kuril Islands, the 13 January 2007 Kuril Islands, and the 01 April 2007 Solomon Islands forecasts — the only ones since the expanded network of tsunameter buoys in the Pacific Ocean — were equally successful in the sense that an initial warning was cancelled on the basis of the forecast. The far field impact of these events has been evaluated qualitatively correctly, with the exception of the harbor oscillations observed in Crescent City following the 2006 event — inside ports, computations at much higher resolution are required. While more tests needed to ensure repeatability of the forecast technology, these first three tests indicate that the methodology for tsunami forecast as described in Titov *et al.* (2005) works well.

For operational codes, testing should employ the tsunameter signal of the 17 November 2003 Rat Islands tsunami to scale an estimate of the source based on the model's implementation of the generation of the initial surface based on the seismic fault solution. The forecast should then provide an estimate of the inundation in Hilo, Hawaii. This is the most difficult but most realistic test for any operational model, for it involves a forecast (now hindcast) and has to be done much faster than real time. Here, at least first 4 waves have to be simulated, with accuracy of at least 75% in the amplitude and period estimates with arrival time error of not more than 3 minutes. Maximum amplitude has to be reproduced with at least 90% accuracy.

3. Scientific and Operational Evaluation

We suggest one further step for the assessment of impact metrics or predictions of tsunami impact, the formal scientific evaluation of the model used. Here, there are no quantifiable standards, instead, the existing practice suggests the adoption of guidelines.

Model validation and verification is a continuing process. Any model used for inundation mapping or operational forecasts needs to be presented in peer-reviewed scientific journals with impact factors greater than 0.5. The publications need to include comparisons of the model predictions with all the above benchmarks. Any model used for the evaluation of tsunami impact or inundation mapping or operational forecast needs to go through the same formal evaluation process as the models used for the assessment of safety of critical structures. This process includes the solicitation of additional reviews of the model's veracity by experts, or the requirement that additional testing be performed. This process helps set the standard for best available practice at any given time, and it will hopefully eliminate the liability to code developing institutions, states, engineers, and geophysicists who collaborate on the development of inundation maps.

While the evaluation process may identify models that are realistic and computationally correct, some of them may not be sufficiently versatile for inundation mapping or operational applications, as is often the case with university-based research codes. An additional approval process needs to be established by governments to assess operational factors of the model, such as special implementation hardware/software issues, ease of use, computation time, etc. If a model is not approved at this phase, it is recommended that the approval body recommend what additional steps or improvement will be helpful for eventual approval.

4. Conclusions

Tsunamis used to be thought of as more extreme hazards than earthquakes, hurricanes, and tornadoes. They occur less frequently, and, except possibly in Japan, historic records are unsystematic. Smaller tsunamis have highly localized impact, and tsunamis in the past centuries probably were under-reported and unrecognized, as the recent once-a-year, on average, worldwide incidence suggests. Measurements from NOAA's tsunameters since 2003 have shown that tsunamis occur in the Pacific a few times per year.

Overall, substantial progress has been achieved in 50 years of tsunami science in assessing tsunami hazards, largely through the contributions of Professor Phil Liu in analytical, numerical, laboratory and field studies. Tsunami inundation forecasts integrating seismic and tsunameter recordings in real time and leading to cancellation of warnings have been attempted successfully (Titov *et al.*, 2005). Based on a single reading off a NOAA tsunameter in the North Pacific for the 17 November 2003 tsunami, Titov *et al.* (2005) scaled their precomputed scenario event that most closely matched the parent earthquake to evaluate the leading wave height off Hilo, Hawaii. They then proceeded with a real-time NSW computation that resulted in the first tsunami forecast and a warning was cancelled. Their comparison of the forecast to the measured wave is the new golden standard for operational models.

In this brief paper, standards for validating tsunami hydrodynamic codes have been proposed, and guidelines for the evaluation of numerical models. Uncertainties arise in the specification of the initial condition, the calculation of the evolution of the geophysical process and finally assessing the repeat interval of the phenomenon.

References

1. E. N. Bernard, H. O. Mofjeld, V. Titov, C. E. Synolakis and F. I. González, *Phil. Trans. R. Soc. A* **364**, 1989 (2006).
2. J. Borerro, M. Ortiz, V. V. Titov and C. E. Synolakis, *EOS, Transactions American Geophysical Union* **78(8)**, 85 (1997).
3. G. F. Carrier and H. P. Greenspan, *J. Fluid Mech.* **17**, 97 (1958).
4. G. F. Carrier, T. T. Wu and H. Yeh, *J. Fluid Mech.* **475**, 79 (2003).
5. J. V. Hall and J. W. Watts, *Laboratory investigation of the vertical rise of solitary waves on impermeable slopes*, Tech. Memo. 33, Beach Erosion Board, USACE (1953), pp. 14.
6. U. Kânoğlu, The runup of long waves around piecewise linear bathymetries, PhD thesis, University of Southern California, (CA, USA, 1998), pp. 273.
7. U. Kânoğlu and C. E. Synolakis, *J. Fluid Mech.* **374**, 1 (1998).
8. U. Kânoğlu, *J. Fluid Mech.* **513**, 363 (2004).
9. U. Kânoğlu and C. Synolakis, *Phys. Rev. Lett.* **97**, 148501 (2006).
10. J. B. Keller and H. B. Keller, *Water wave run-up on a beach*, ONR Research Report NONR-3828(00), Dept. of the Navy (Wash. DC., 1964), pp. 40.
11. P. L.-F. Liu, C. E. Synolakis and H. H. Yeh, *J. Fluid Mech.* **229**, 675 (1991).
12. P. L.-F. Liu, Y.-S. Cho, M. J. Briggs, U. Kânoğlu and C. E. Synolakis, *J. Fluid Mech.* **320**, 259 (1995).
13. P. L.-F. Liu, P. Lynett and C. E. Synolakis, *J. Fluid Mech.* **478**, 101 (2003).
14. P. L.-F. Liu, T.-R. Wu, F. Raichlen, C. E. Synolakis and J. Borrero, *J. Fluid Mech.* **536**, 107 (2005).
15. P. L.-F. Liu, H. Yeh and C. Synolakis (eds.), *Advanced Numerical Models for Simulating Tsunami Waves and Runup* (Advances in Coastal and Ocean Engineering, Vol. 10, World Scientific, 2008).
16. G. Pedersen and B. Gjevik, *J. Fluid Mech.* **135**, 283 (1983).
17. F. Raichlen and C. E. Synolakis, Runup from three dimensional sliding mass, in *Proc. Long Waves Symposium*, eds. M. Briggs and C. Koutitas (Thessaloniki, Greece, 2003), pp. 247–256.
18. C. E. Synolakis, The runup of long waves, PhD thesis, California Institute of Technology (CA, USA, 1986), pp. 228.
19. C. E. Synolakis, *J. Fluid Mech.* **185**, 523 (1987).
20. C. E. Synolakis and E. N. Bernard, *Phil. Trans. R. Soc. A* **364**, 2231 (2006).
21. C. E. Synolakis, E. Bernard, V. Titov, U. Kânoğlu and F. Gonzalez, *Standards and Guidelines for Inundation Models*, NOAA report (2007).
22. C. E. Synolakis, J. P. Bardet, J. Borrero, H. Davies, E. Okal, E. Silver, J. Sweet and D. Tappin, *Proc. of the Royal Society of London A* **458**, 763 (2002).
23. S. Tadepalli and C. E. Synolakis, *Proc. R. Soc. Lond. A* **445**, 99 (1994).

24. S. Tadepalli and C. E. Synolakis, *Phys. Rev. Lett.* **77**, 2141 (1996).
25. T. Takahashi, Benchmark problem 4; the 1993 Okushiri tsunami — Data, conditions and phenomena. In *Long-Wave Runup Models* (World Scientific, 1996), pp. 384–403.
26. V. V. Titov and C. E. Synolakis, *J. Waterway Port Ocean Coast. Eng.* **121(6)**, 308 (1995).
27. V. V. Titov and C. E. Synolakis, *J. Waterway Port Ocean Coast. Eng.* **124**, 157 (1998).
28. V. V. Titov, F. I. Gonzalez, E. M. Bernard, M. C. Eble, H. O. Mofjeld, J. C. Newman and A. J. Venturato, *Nat. Hazards* **35(1)**, 45 (2005).
29. H. Yeh, P. L.-F. Liu, M. Briggs and C. E. Synolakis, *Nature* **372**, 6503 (1994).
30. H. Yeh, P. L.-F. Liu and C. E. Synolakis (eds.), *Long-Wave Runup Models* (World Scientific Publishing, Singapore, 1996), pp. 403.

PREDICTING RUN-UP OF BREAKING AND NONBREAKING LONG WAVES BY APPLYING THE CORNELL COMCOT MODEL

HONGQIANG ZHOU and MICHELLE H. TENG

Department of Civil and Environmental Engineering
University of Hawaii at Manoa
Honolulu, HI 96822, USA

PENGZHI LIN

Department of Civil Engineering, National University of Singapore
117576, Singapore

EDISON GICA

NOAA Center for Tsunami Research, PMEL
Seattle, WA 98115, USA

KELIE FENG

M & E Pacific Inc.
Honolulu, HI 96813, USA

The present numerical and experimental study examines the performance of the Cornell COMCOT tsunami model developed under the direction of Professor Philip L.-F. Liu. The 2-D depth-averaged model was initially developed for nonbreaking wave run-up as it did not include an empirical term approximating the energy dissipation due to wave breaking. In this study, numerical simulations based on the 2-D model of both breaking and nonbreaking wave run-up were performed. The results were then compared with the experimental results (Synolakis [1], Feng [2]) and also with the results based on a more advanced depth-resolving RANS model for breaking waves developed by Lin *et al.* [3,4], with good agreement. An analysis of the numerical scheme used by COMCOT revealed that the truncation error has the similar effect as the empirical diffusion term typically used for simulating energy dissipation caused by wave breaking for depth-averaged models, thus eliminating the need to include this additional term in the COMCOT model. A verification of the COMCOT model against a historical tsunami event is also presented. The results show that the COMCOT model is a simple, yet effective model for predicting both breaking and nonbreaking wave run-up for practical applications.

1. Introduction

Hurricane-induced storm surges and earthquake-generated tsunamis are typical examples of ocean long waves. These long waves can cause severe property damage and loss of human lives when they run up onto coastal land causing flooding of the coastal community. The December 26, 2004 Indian Ocean Tsunami demonstrates the destructive power of a tsunami and the urgent need to develop efficient computer models for predicting the maximum run-up and inundation limit so that proper tsunami warnings can be issued in a timely manner.

Many of the existing run-up models are based on depth-integrated shallow water long wave equations [5,6,7]. In recent years, more advanced depth-resolving RANS models based on the Reynolds Averaged Navier-Stokes equations have also been developed [3,4]. In general, 2-D models based on depth-integrated equations require less CPU time and are more efficient. However, most of them usually cannot predict the turbulent features associated with breaking waves accurately. The RANS model, on the other hand, may use more CPU time but it can simulate the detailed turbulent features and the dynamic development of breaking waves, enabling us to understand the fundamental mechanism behind breaking waves more thoroughly.

This paper is a further extension of an earlier brief conference paper by Zhou *et al.* [8], and focuses its discussions on the 2-D depth-averaged COMCOT tsunami model developed by Liu *et al.* [9,10]. This model is simple, efficient and suitable for predicting large-scale (e.g., across an ocean) long wave propagation and run-up. However, since the equations did not include a term approximating the effect of breaking waves, it remains to be studied whether the model is valid for predicting breaking wave run-up. Our study reported here attempts to answer this question through numerical and experimental investigations. In reality, both breaking and nonbreaking wave run-up can occur, therefore, it would be helpful if a model can predict both while remaining simple at the same time.

2. Theory

This study focuses on the accuracy and efficiency of the Cornell depth-averaged COMCOT model for predicting long wave run-up, and will use solitary waves on 1-D smooth beaches as a case study, as shown in Fig. 1. The effects of bottom friction will not be included. To model long wave run-up, both the depth-averaged shallow water equations (SWE) and the Reynolds Averaged Navier-Stokes Equations (RANS) can be used.

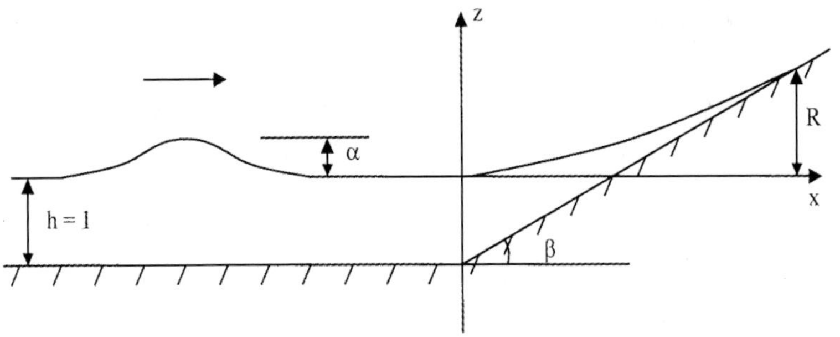

Figure 1. Solitary wave run-up on a smooth plane beach.

2.1 *The SWE-Based COMCOT Numerical Model*

The dimensionless nonlinear depth-averaged shallow water equations (SWE) are given as:

$$\frac{\partial \eta}{\partial t} + \frac{\partial Q}{\partial x} = 0, \tag{1}$$

$$\frac{\partial Q}{\partial t} + \frac{\partial}{\partial x}\left(\frac{Q^2}{H}\right) + \frac{\partial \eta}{\partial x} = 0, \tag{2}$$

where η is the wave elevation, Q the horizontal flux, and $H = h + \eta$ the total depth.

A finite difference scheme was constructed by Liu *et al.* [9,10] to solve the above equations (1)-(2). Specifically, the linear derivative terms are approximated by a leap-frog scheme in time and central difference in space while the nonlinear convective term in the momentum equation is discretized by an upwind scheme. In addition, the flux Q and wave elevation η are evaluated at staggered grid points. The discretized form of equations (1) and (2) can be written as

$$\eta_i^{n+1/2} = \eta_i^{n-1/2} - r_x\left(Q_{i+1/2}^n - Q_{i-1/2}^n\right), \tag{3}$$

$$Q_{i+1/2}^{n+1} = Q_{i+1/2}^n - r_x H_{i+1/2}^{n+1/2}(\eta_{i+1}^{n+1/2} - \eta_i^{n+1/2})$$
$$- r_x\left[\lambda_1 \frac{(Q_{i+3/2}^n)^2}{H_{i+3/2}^n} + \lambda_2 \frac{(Q_{i+1/2}^n)^2}{H_{i+1/2}^n} + \lambda_3 \frac{(Q_{i-1/2}^n)^2}{H_{i-1/2}^n}\right], \tag{4}$$

where $r_x = \Delta t / \Delta x$ and λ_i, $i = 1,2,3$ are defined as

$$\lambda_1 = 0, \quad \lambda_2 = 1, \quad \lambda_3 = -1, \quad \text{if } Q^n_{i+1/2} \geq 0,$$

and

$$\lambda_1 = 1, \quad \lambda_2 = -1, \quad \lambda_3 = 0, \quad \text{if } Q^n_{i+1/2} < 0.$$

This scheme is explicit for both η and Q, and accurate to the first order in time and space. The stability criterion is $\Delta t / \Delta x < 1$. The detailed run-up scheme for tracking the moving wave front onto dry land can be found in Liu *et al.* [9, 10].

We would like to mention that the shallow water equations listed here are a set of traditional long wave equations. They have been used by many different research groups in modeling long wave propagation and run-up. For these models, although the governing equations can be the same, their performance may be different depending on the specific numerical scheme that each model applies. In this study, we are particularly interested in the numerical scheme constructed by Liu *et al.* [9, 10].

2.2 *The RANS Model*

The Reynolds Averaged Navier-Stokes equations, or RANS, are derived from the Navier-Stokes equations by separating the instantaneous flow properties into the mean terms and the turbulent fluctuations. RANS can be written as follows:

$$\frac{\partial \langle u_i \rangle}{\partial x_i} = 0, \tag{5}$$

$$\frac{\partial \langle u_i \rangle}{\partial t} + \langle u_j \rangle \frac{\partial \langle u_i \rangle}{\partial x_j} = -\frac{1}{\rho}\frac{\partial \langle p \rangle}{\partial x_i} + g_i + \frac{1}{\rho}\frac{\partial \langle \tau_{ij} \rangle}{\partial x_j} - \frac{\partial \langle u'_i u'_j \rangle}{\partial x_j}, \tag{6}$$

where "$\langle\ \rangle$" denotes the mean quantities and the prime represents the turbulent fluctuations, u_i, x_i, g_i are the velocities, coordinates and gravitational accelerations in different directions, and τ^m_{ij} are the viscous stresses. A turbulence closure model is required to solve the terms of Reynolds stress $R_{ij} = -\langle \rho \rangle \langle u'_i u'_j \rangle$. In Lin and Liu [3] and Lin *et al.* [4], the authors adopted a well-established $k - \varepsilon$ model, which reads

$$\frac{\partial k}{\partial t} + \langle u_j \rangle \frac{\partial k}{\partial x_j} = \frac{\partial}{\partial x_j}\left[\left(\frac{\nu_t}{\sigma_k} + \nu\right)\frac{\partial k}{\partial x_j}\right] - \langle u_i' u_j \rangle \frac{\partial \langle u_i' \rangle}{\partial x_j} - \varepsilon, \tag{7}$$

where $\nu = \mu / \rho$ is the kinematic viscosity, ν_t is the eddy viscosity, and σ_k is an empirical coefficient. After k and ε are obtained, the Reynolds stresses can be approximated through the following equations:

$$\langle u_i' u_j' \rangle = \frac{2}{3} k \delta_{ij} - C_d \frac{k^2}{\varepsilon} \left(\frac{\partial \langle u_i \rangle}{\partial x_j} + \frac{\partial \langle u_j \rangle}{\partial x_i} \right)$$

$$- \frac{k^3}{\varepsilon^2} \left[C_1 \left(\frac{\partial \langle u_i \rangle}{\partial x_l} \frac{\partial \langle u_l \rangle}{\partial x_j} + \frac{\partial \langle u_j \rangle}{\partial x_l} \frac{\partial \langle u_l \rangle}{\partial x_i} - \frac{2}{3} \frac{\partial \langle u_l \rangle}{\partial x_k} \frac{\partial \langle u_k \rangle}{\partial x_l} \delta_{ij} \right) \right.$$

$$+ C_2 \left(\frac{\partial \langle u_i \rangle}{\partial x_k} \frac{\partial \langle u_j \rangle}{\partial x_k} - \frac{1}{3} \frac{\partial \langle u_l \rangle}{\partial x_k} \frac{\partial \langle u_l \rangle}{\partial x_k} \delta_{ij} \right)$$

$$\left. + C_3 \left(\frac{\partial \langle u_k \rangle}{\partial x_i} \frac{\partial \langle u_k \rangle}{\partial x_j} - \frac{1}{3} \frac{\partial \langle u_l \rangle}{\partial x_k} \frac{\partial \langle u_l \rangle}{\partial x_k} \delta_{ij} \right) \right] . \quad (8)$$

A numerical model based on the RANS equations was developed by Lin and Liu [3], Lin *et al.* [4] for simulating solitary wave propagation and run-up on sloping beaches. The detailed numerical techniques and the values of different parameters can be found in these two papers.

2.3 *Synolakis' Analytical Solution*

Synolakis [1] obtained an asymptotic solution based on the nonlinear shallow water equations for predicting wave run-up on a smooth plane slope. In his solution, the maximum run-up is predicted by

$$\frac{R}{h} = 2.831 (\cot \beta)^{1/2} \left(\frac{\alpha}{h} \right)^{5/4} , \quad (9)$$

where R denotes the maximum run-up, h is the constant still water depth immediately upstream of the sloping beach, β is the angle of the beach, and α is the initial wave height. Equation (9) is valid for nonbreaking waves with:

$$\frac{\alpha}{h} < 0.479 (\cot \beta)^{-10/9} . \quad (10)$$

In addition to the analytical solution, Synolakis [1] also carried out experiments to measure solitary wave run-up on mild slopes. He studied the run-up of both breaking and nonbreaking waves. His analytical solution as well as his experimental results from the 1987 study has often been used as the benchmark results for validating numerical models.

3. Numerical and Experimental Results

In this study, numerical simulations of solitary wave run-up on smooth plane beaches based on the Cornell COMCOT model (1)-(4) were carried out. The simulated results were compared with the experimental data from Feng [2] and Synolakis [1] and with the results based on the RANS model in order to examine the accuracy and efficiency of the COMCOT model. In Feng's [2] study, wave run-up on both smooth and rough artificial beaches was measured in a wave tank, and the effect of different roughness on reducing the maximum run-up was investigated. Since the present study is focused on examining the prediction for breaking wave run-up, the roughness effect will not be discussed. Only run-up data on smooth plane beaches will be used for validation purposes. The results on the effect of coastal terrain roughness on wave run-up will be reported in a separate publication.

Figure 2 shows the comparative results of solitary wave run-up R versus the initial wave amplitude α on plane beaches of 20° and 5° slopes. The experimental results presented in Figure 2 were obtained by Feng [2]. All the results presented in this paper are in dimensionless form with the length scaled by the unperturbed water depth h.

Results from Figure 2 show that Lin's RANS model gives excellent prediction for solitary wave run-up for both breaking and nonbreaking waves. The COMCOT model based on the shallow water equations gives equally excellent prediction for the nonbreaking run-up on the 20° slope. For wave run-up on the milder slope of 5° where breaking wave run-up occurred, the prediction by the COMCOT model is not as perfect as the RANS model, however, the agreement between the COMCOT prediction and the experimental results is reasonably good, as also evidenced by the quantitative comparison presented in Table 1.

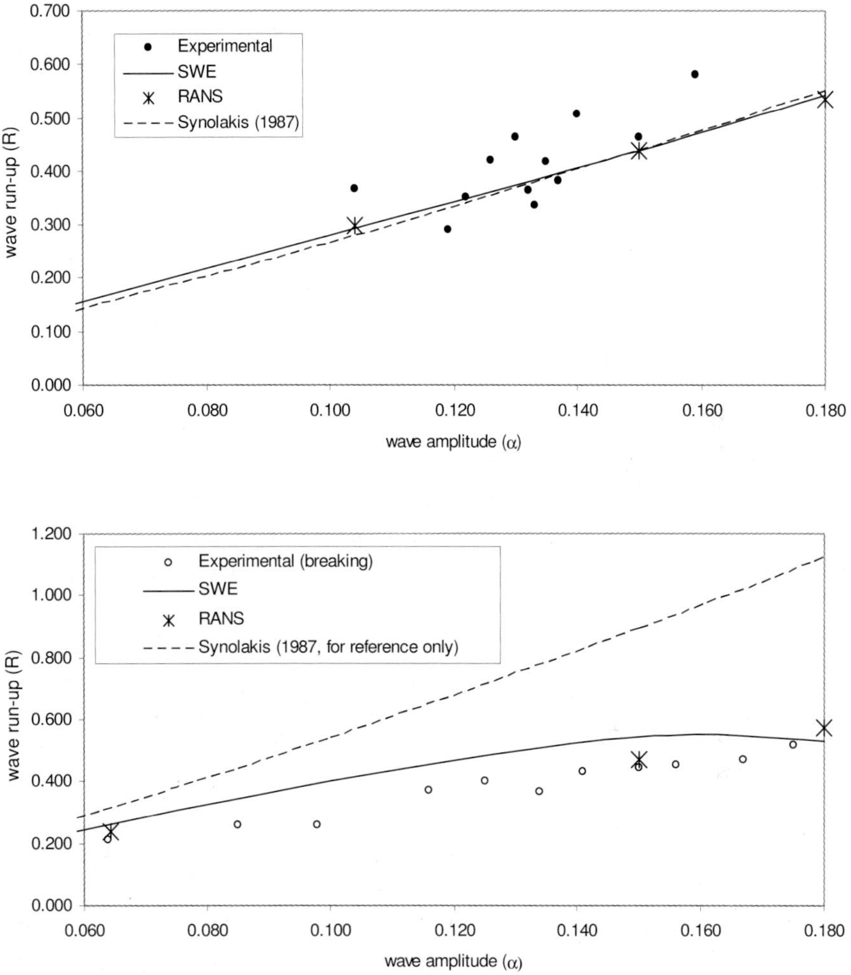

Figure 2. Comparison between the COMCOT prediction of solitary wave run-up and the experimental results (Feng [2]) and also the predicted results based on the RANS model; top: 20° slope, nonbreaking wave run-up; bottom: 5° slope, breaking wave run-up. The dashed line represents Synolakis' [1] analytical solution (9).

Table 1. Comparison of the run-up values based on the COMCOT model, the RANS model and the experimental data.

β	5°		20°	
	breaking		nonbreaking	
α	0.150	0.202	0.150	0.192
Exp.	0.444	0.549	0.464	0.566
RANS	0.469	0.582	0.438	0.573
COMCOT	0.545	0.616	0.439	0.586
Diff.	16%	6%	0.2%	2%

In the above table, the percentage difference in the last row is the relative difference between the RANS and the COMCOT results.

Further validation of the COMCOT model was done by comparing the COMCOT results for wave run-up on a 2.88° slope with the experimental data obtained by Synolakis [1]. The comparative results are presented in Figure 3. Since the RANS model was validated against the Synolakis experiments in an earlier publication and showed excellent agreement, no further comparison with the RANS results was presented in Figure 3 in this study. The results in Figure 3 show that the COMCOT model can predict the run-up for both breaking and nonbreaking waves quite satisfactorily.

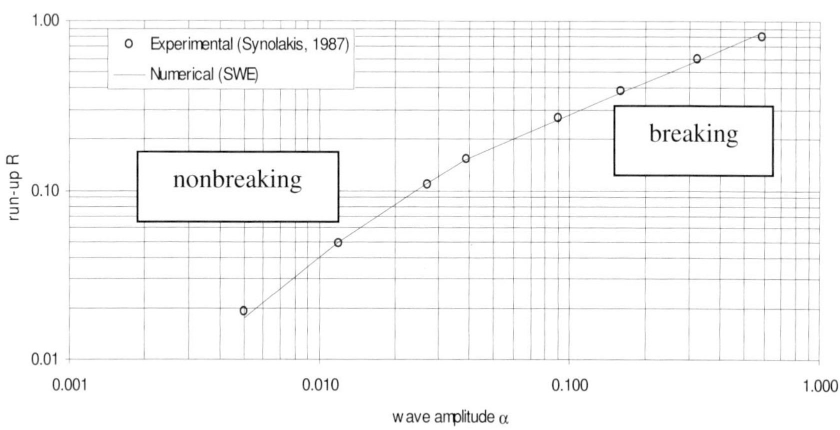

Figure 3. Comparison between the simulated run-up with Synolakis' experimental results [1] for both breaking and nonbreaking waves on a 2.88° slope. The simulation is based on the COMCOT model.

Figure 4 shows the detailed wave profile during run-up on the 2.88° slope simulated by the COMCOT model. We note that before reaching the sloping beach, the wave travels over a constant depth for a short distance. Over this distance, we used the linearized version of the SWE to simulate the wave propagation. The nonlinear effect was included over the sloping beach. All run-up values based on the COMCOT model presented in this paper were obtained using the same simulation steps.

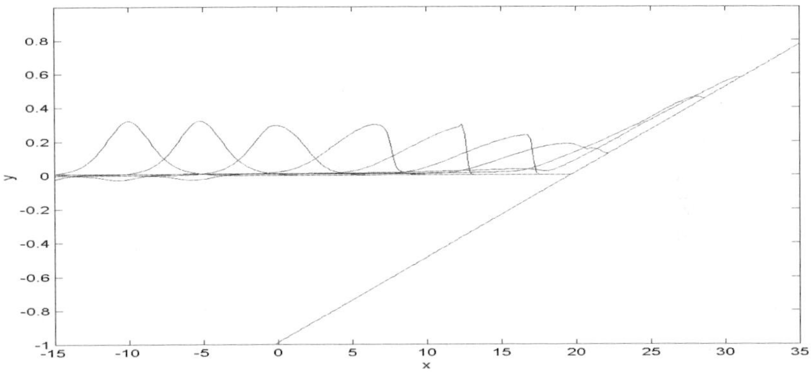

Figure 4. Simulation of wave shoaling and run-up on a 2.88° slope by the COMCOT model. From $x = -15$ to $x = 0$, the model was linearized. The nonlinear effect started at $x = 0$.

In the numerical simulation, the spatial increment Δx and time step Δt were selected by satisfying the stability criterion and by performing convergence tests. If Δx and Δt are too large, the run-up prediction will not be accurate. In our convergence test, we simulated the run-up by using different Δx and Δt values, usually starting from large values and then lowering to smaller values. We selected the optimal values which ensure that the run-up result would not change significantly with further decreased values of Δx and Δt from the optimal values. The convergence test is important and necessary in numerical simulations because incorrect Δx and Δt values may lead to inaccurate predictions. In the present study, the values for Δx and Δt range from 0.05 to 0.2 and 0.025 to 0.1, respectively. In general, for run-up on very mild slopes, smaller Δx and Δt should be used.

Besides validation against experimental data and the RANS results, we also carried out simulations of an actual tsunami event, i.e., the 1960 Chilean tsunami, that caused severe damage in the Pacific Basin. The COMCOT model was applied to simulate the entire tsunami process from its generation by the 1960 earthquake in Chile, the propagation across the Pacific Ocean and its run-up and inundation in Hawaii. The specific location we choose to study is Kahana Bay on Oahu Island in Hawaii as shown in Figure 5.

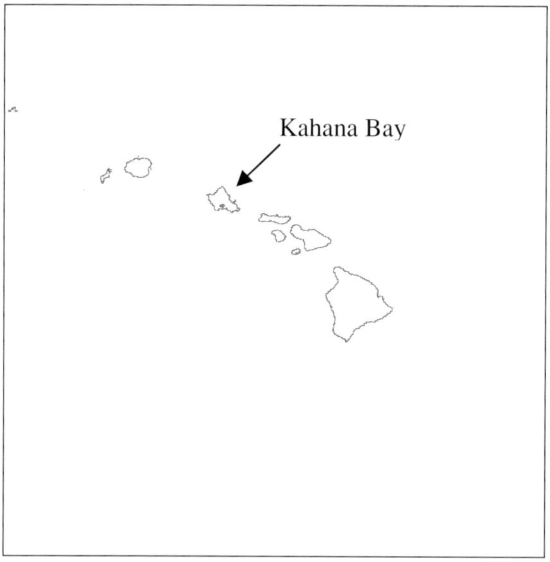

Figure 5. Location of Kahana Bay in the Hawai'ian Islands.

The simulated tsunami inundation in Kahana Bay is presented in Figure 6. Due to the different slopes of topography at different locations in the bay, the run-up height along the inundation boundary is different. Based on the numerical results, the run-up height ranges from 2.5 ft to 8.7 ft (0.7 m to 2.65 m) in the bay. There are no field records of the detailed inundation boundary in the bay. However, the maximum run-up height in the bay was recorded as 8 ft (2.4 m) which is consistent with the numerical simulation result.

In the simulation, a three-layer grid system was used with the open Pacific Ocean covered with coarser grids and the coastal area near Kahana Bay covered with fine grids. The detailed input parameters for the 1960 Chilean earthquake, the grid size and the time step as well as more validation cases can be found in Gica [11].

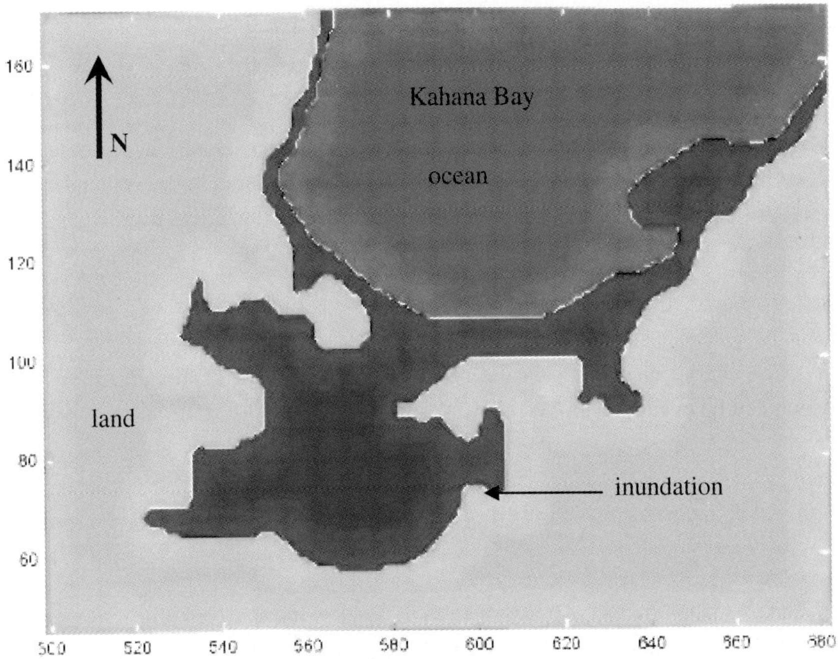

Figure 6. Simulated inundation due to the 1960 Chilean tsunami at Kahana Bay.

4. Discussions

So far, the numerical results show that the Cornell COMCOT tsunami model can predict the maximum run-up of both breaking and nonbreaking waves reasonably well. In the run-up study by Heitner and Housner [5], an empirical term was added in the momentum equation in the shallow water model to approximate the energy dissipation associated with turbulent wave breaking. Specifically, an artificial viscous force F was introduced, which can be estimated by the following formula:

$$F = l^2 H \left(\frac{\partial u}{\partial x} \right)^2 \mathrm{He}\left(-\frac{\partial u}{\partial x} \right), \tag{11}$$

where $\mathrm{He}(\ )$ is the Heavyside function defined as

$$He\left(-\frac{\partial u}{\partial x}\right)=\begin{cases}1, \text{for} -\partial u/\partial x \geq 0\\ 0, \text{for} -\partial u/\partial x < 0\end{cases} \tag{12}$$

where l is the shock length. Since equation (11) is an empirical formula and wave breaking is a complex phenomenon to model accurately, there is no exact value for l. In Heitner and Housner [5], the following range was used:

$$l^2 \approx 0.001 \sim 0.1 \,. \tag{13}$$

The energy dissipation in the momentum equation is expressed as $\partial F / \partial x$ which can be evaluated numerically as follows

$$\left(\frac{dF}{dx}\right)_{i+1/2}^{n+1/2}=\frac{F_{i+1}^{n+1/2}-F_i^{n+1/2}}{\Delta x}\,. \tag{14}$$

Mathematically, $\partial F / \partial x$ leads to the second order derivative of u in x which is usually a presentation of the diffusive effect. Similar empirical terms have also been used by other researchers (e.g., Zelt [12]) for approximating the dissipation associated with wave breaking in shallow water long wave modeling.

The Cornell COMCOT model is based on the shallow water equations. It did not include the empirical term for modeling the turbulent dissipation due to wave breaking, yet our simulation results showed that the COMCOT model can predict the maximum run-up of breaking wave satisfactorily. We analyzed the numerical scheme for the momentum equation in the COMCOT model by using Taylor expansion, and obtained the result for the truncation error as

$$E_{i+1/2}^{n+1/2}=-\frac{\Delta x}{2}\frac{\partial^2}{\partial x^2}\left(\frac{Q^2}{H}\right)_{i+1/2}^{n+1/2}-\frac{\Delta t}{2}\frac{\partial^2}{\partial x \partial t}\left(\frac{Q^2}{H}\right)_{i+1/2}^{n+1/2}, \quad \text{for } Q_{i+1/2}^n > 0\,, \tag{15}$$

$$E_{i+1/2}^{n+1/2}=\frac{\Delta x}{2}\frac{\partial^2}{\partial x^2}\left(\frac{Q^2}{H}\right)_{i+1/2}^{n+1/2}-\frac{\Delta t}{2}\frac{\partial^2}{\partial x \partial t}\left(\frac{Q^2}{H}\right)_{i+1/2}^{n+1/2}, \quad \text{for } Q_{i+1/2}^n < 0\,. \tag{16}$$

These equations show that the truncation error happens to represent a diffusive effect. To examine the effect of the numerical truncation error and compare with the empirical energy dissipation, a numerical test is performed. In this test, a solitary wave with amplitude $\alpha = 0.3$ travels over a constant water depth from the left to the right as shown in Figure 7. Since the shallow water equations applied for this test are not linearized, i.e., they are nonlinear and nondispersive, the front of the wave will steepen and eventually break. The simulation shows that there is a decrease in the wave amplitude after the wave "breaks". In the simulation, the original solitary wave starts at $x = 10.0$. After it travels for about 10 dimensionless time units, the wave reaches the location of about $x = 21.5$. The magnitude of the truncation error at $x = 21.5$ is evaluated as a function of time and compared with the empirical energy dissipation at the same location. The time and spatial intervals are chosen to be $\Delta t = 0.025$ and $\Delta x = 0.05$, respectively. Shock length of $l = 0.05$ and 0.1 is tested for the empirical dissipation estimate. The time-series of the numerical and empirical energy dissipations is plotted together in Figure 8. From this figure, we can see that the numerical dissipation shows the same trend as the empirical dissipation. If we use $l = 0.05$, the magnitude of the two also matches each other well. However, if we use $l = 0.1$, then the magnitude based on the empirical term and the truncation error is different. If the empirical parameter l can be determined more accurately through experimental measurement of breaking waves and the associated turbulence features, it can help to make the empirical prediction as well as the comparison between the empirical prediction and the numerical dissipation more certain. However, regardless of the uncertain in the value of l, the results show that the numerical dissipation has the similar qualitative effect as the empirical prediction. We should mention that in Figure 8, the negative values of energy dissipation were set to zero based on (11)-(12).

Although the results presented so far show that the COMCOT model can predict the maximum run-up reasonably well, we do not imply that the COMCOT model based on the depth-averaged shallow water equations can substitute for the more advanced RANS model for turbulent modeling or for predicting the detailed flow features of a breaking wave. To illustrate this point, the propagation and shoaling of a nonlinear long wave over a plane slope is simulated by using both the COMCOT model and the RANS model. The comparative results are shown in Figure 9. This figure shows that the COMCOT model cannot predict the detailed dynamic development of a breaking wave as accurately as the RANS model can. If we are interested not only in the final run-up height, but also in the detailed turbulent features of the breaking wave, then the RANS model should be used.

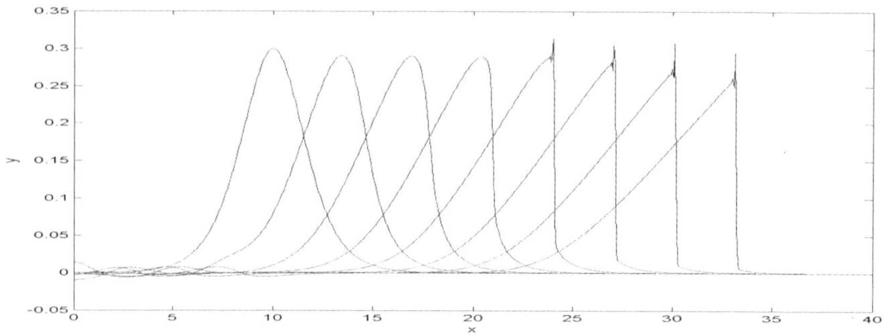

Figure 7. Simulated wave propagation over constant depth by the nonlinear and nondispersive COMCOT model.

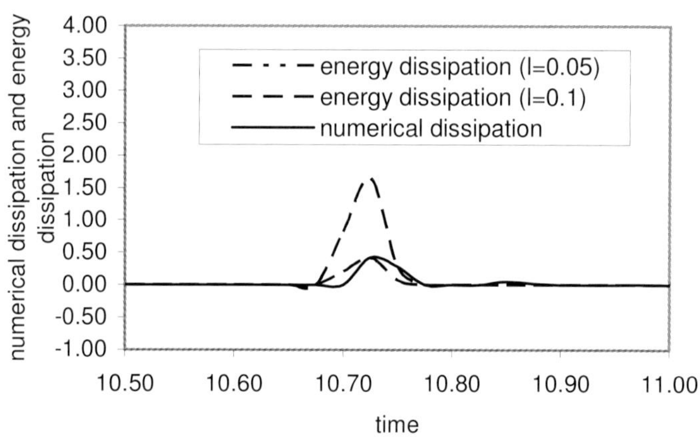

Figure 8. Comparison between the numerical truncation error and the empirical estimate for the energy dissipation due to wave breaking.

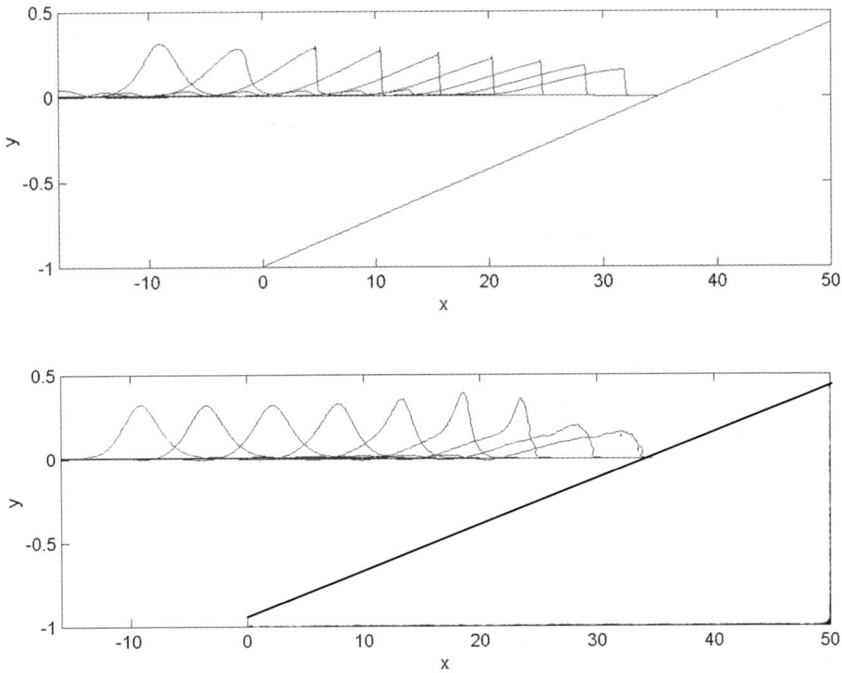

Figure 9. Comparison of the simulated wave shoaling between the COMCOT model (top, without linearization over the constant depth) and the RANS model (bottom).

In practical applications, predicting the maximum inundation limit and run-up height is often an important objective. Also important is how fast the model can obtain the simulation result especially when real-time forecasting is required. In these cases, the COMCOT model can serve as an efficient and practical model for tsunami simulation in engineering applications.

In other situations, we desire to obtain the detailed 3-D flow and wave features that can help us to understand better the fundamental mechanism of breaking wave and turbulence. In addition, the detailed wave-structure interaction can be also of critical importance in engineering design of coastal structures. In these cases, the RANS model would be much more appropriate than the SWE models for the simulations.

5. Conclusion

Based on the comparison between the numerical and experimental results as well as the field data presented in this paper, we conclude that the Cornell COMCOT model can predict both the breaking and nonbreaking wave run-up reasonably well. It was found that the numerical truncation error of the scheme happens to represent a diffusive effect, very similar to the empirical term that researchers have used to approximate the energy dissipation associated with wave breaking, thus eliminating the need to include such a term in the shallow water equations. Based on our experiences applying the COMCOT model for simulating tsunamis in the Pacific Basin in the past several years, the COMCOT model has been found to be a very stable and effective model in simulating tsunami generation, propagation and run-up. In practical applications, the required computational time is an important concern especially in real-time forecasting. So far, the COMCOT model has proved to be an efficient and reliable model for predicting the inundation limit and run-up height.

Acknowledgments

The study was partially funded by the NOAA Sea Grant at the University of Hawaii. The authors are very grateful to Professor Philip L-.F. Liu for his guidance, help and support for our tsunami research.

References

1. C. E. Synolakis, *J. Fluid Mech.*, **185**, 523-545 (1987).
2. K. Feng, Experimental study of the effect of coastal terrain roughness on coastal long wave run-up, *MS thesis*, Department of Civil and Environmental Engineering, University of Hawaii at Manoa, Honolulu, Hawaii (2001).
3. P. Lin and P. L.-F. Liu, *J. Fluid Mech.*, **359**, 239-264 (1998).
4. P. Lin, K. A. Chang and P. L.-F. Liu, *J. Waterway, Port, Coast. Ocean Eng.*, **125**, 247-255 (1999).
5. K. L. Heitner and G. W. Housner, *J. Waterway, Port, Coast. Ocean Eng.*, **96 (WW3)**, 701-719 (1970).
6. F. Imamura, N. Shuto and C. Goto, *Proceedings of the 6th Congress of Asian and Pacific Regional Division*, IAHR, Japan, 265-272 (1988).

7. V. V. Titov and C. E. Synolakis, *J. Waterway, Port, Coast. Ocean Eng.*, **124** (**4**), 157-171 (1998).
8. H. Zhou, M. H. Teng and P. Lin, *CD-ROM Proceedings of the Joint ASME/ASCE/SES Conference on Mechanics and Materials (not copyrighted)*, Baton Rouge, Louisiana, paper No. 88-383 (2005).
9. P. L.-F. Liu, Y. S. Cho, M. J. Briggs, U. Kanoglu and C. E. Synolakis, *J. Fluid Mech.*, **302**, 259-285 (1995).
10. P. L.-F. Liu, S. B. Woo and Y. S. Cho, *Technical Report*, School of Civil and Environmental Engineering, Cornell University, Ithaca, New York (1998).
11. E. Gica, Risk analysis of coastal flooding due to distant tsunamis, *Ph.D. thesis*, Department of Civil and Environmental Engineering, University of Hawaii at Manoa, Honolulu, Hawaii (2005).
12. J. A. Zelt, *Coastal Engineering*, **15**, 205-246 (1991).

BOUNDARY LAYER EFFECTS ON THE PROPAGATION OF WEAKLY NONLINEAR LONG WAVES

G. SIMARRO

Universidad de Castilla–La Mancha
Ciudad Real, 13071, Spain
E-mail: Gonzalo.Simarro@uclm.es

A. ORFILA

Institut Mediterrani d'Estudis Avançats (CSIC–UIB)
Esporles, 07190, Spain
E-mail: a.orfila@uib.es

This paper presents a set of Boussinesq-type equations where the bottom effects are introduced through the analysis of the boundary layer, either laminar or turbulent. The equations are written for the velocity at an arbitrary elevation and they are derived in a nondimensional form tracking the order of the different errors, so that the range of conditions under which the present approach is valid can be discussed. The equations are written for the Fourier components assuming time periodicity, and the parabolic approximation is used to convert the elliptic equations into parabolic ones, which can be solved in an efficient way to obtain the velocity at the arbitrary elevation, the free surface elevation and the bed shear stress.

1. Introduction

The understanding of water wave propagation from deep to shallow water is a crucial aspect for many kinds of coastal issues. Coastal morphodynamics, sediment transport processes, wave breaking currents or the design of coastal structures are but a few examples of processes demanding accurate knowledge of wave conditions in relatively shallow waters.

As water waves propagate from the region where they were generated to the coast, both wave amplitude and wavelength are modified due to refraction, diffraction and loss of energy. The accurate description and modeling of these processes has received increasing attention in the last three decades. An overview of the major advances in the wave phase modeling can be found in Liu and Losada.[1]

Early efforts to explain the wave transformation were based on the geometrical optic theory so as to include wave refraction. This so-called "ray theory", however, ignored both nonlinear and diffraction effects. The "mild slope" equation, a modification of the linear wave theory, appears as an improvement to the former since it includes wave diffraction effects. The mild slope equation assumes that the water depth varies *slightly* in a wave length. Further, the equation can be converted into a parabolic type by assuming that the wave amplitude is primarily a function of the water depth. This "parabolic approximation" can be seen as a modification of the ray theory where wave energy is allowed to be diffused across the rays.

In addition, near the coast the wave amplitude increases, so that the linear theory is no longer valid. The Boussinesq equations for uneven bottoms as derived by Mei and LeMehaute[2] and Peregrine[3] are meant to meet the requirements of weakly nonlinear and weakly dispersive waves. The equations can be used to model nonlinear wave transformation due to refraction and diffraction. Moreover, in order to extend the range of applicability of the Boussinesq equations to deeper waters, Nwogu[4] introduced the equations written for an arbitrary depth z_α so that the dispersion behaviour was improved maintaining the order of the equations.

Furthermore, as the water depth decreases, boundary layer effects at the bottom become important. Moreover, bottom shear stress is responsible for the sediment transport and morphological changes. In order to include the effects of the bottom friction in the wave models, a bed shear stress term is usually added to the momentum equation. The bed shear stress is usually modelled as

$$\boldsymbol{\tau} = C_f \rho \, \mathbf{u_b} \|\mathbf{u_b}\|, \tag{1}$$

where C_f is the corresponding friction coefficient, ρ is the water density and $\mathbf{u_b}$ is the near bottom velocity. A frictional model such as (1) is appropriate if the primary concern is the amount of energy dissipated, as long as C_f is conveniently evaluated. However, the above model fails to represent the phase of the bottom stress relative to the bottom velocity. It is well known[5] that for the laminar boundary layer the phase lag between the bottom shear stress and the bed velocity is $\pi/4$, being smaller in the turbulent case.

Recently, Liu and Orfila[6] and Orfila *et al.*[7] have introduced analytical solutions of the boundary layer in the Boussinesq-type equations for the flat bottom case, for both laminar and turbulent boundary layers respectively. In their approach, the boundary layer effects are introduced into the Boussinesq equations through the kinematic bottom boundary condition.

The present paper describes a long-wave weakly nonlinear propagation model for uneven bottoms with the bottom boundary layer effects included.

Following Nwogu,[4] the Boussinesq-type equations are presented for the velocity at an arbitrary elevation. Furthermore we shall only focus on periodic waves in absence of mean currents.

The paper is structured as follows. The governing equations for the fluid motion are presented in Section 2 both in the dimensional and nondimensional form. The boundary layer analysis is then presented in Section 3 both for the laminar and turbulent cases. Assuming periodic motion, solution for the rotational component of the velocity is obtained. The influence of the boundary layer on the core is presented in Section 4. The Boussinesq types of equations are derived in Section 5, and they are written in terms of the Fourier components for the periodic case. Section 5 concludes the paper with the analysis of the wave propagation in a 1D case.

2. Governing equations

Throughout the paper, dimensional variables will be denoted with primes, the dimensionless ones being unprimed. The continuity equation reads

$$\boldsymbol{\nabla}' \cdot \mathbf{u}' = 0, \qquad -h' \leqslant z' \leqslant \eta', \tag{2}$$

where $z' = -h'(x', y', t')$ and $z' = \eta'(x', y', t')$ represent respectively the bottom and the free surface elevation. At those boundaries, no flux conditions are

$$\frac{\partial h'}{\partial t'} + u'_x \frac{\partial h'}{\partial x'} + u'_y \frac{\partial h'}{\partial y'} + u'_z = 0, \qquad z' = -h', \tag{3}$$

$$\frac{\partial \eta'}{\partial t'} + u'_x \frac{\partial \eta'}{\partial x'} + u'_y \frac{\partial \eta'}{\partial y'} - u'_z = 0, \qquad z' = \eta'. \tag{4}$$

Assuming constant atmospheric pressure and the gravity being the only volumetric force, the Stokes form of the Navier-Stokes equation at the free surface is

$$\frac{\partial \mathbf{u}'}{\partial t'} + \frac{1}{2}\boldsymbol{\nabla}' \|\mathbf{u}'\|^2 + g'\boldsymbol{\nabla}' z' = \mathbf{u}' \times \boldsymbol{\omega}' - \nu' \boldsymbol{\nabla}' \times \boldsymbol{\omega}', \qquad z' = \eta', \tag{5}$$

where $\boldsymbol{\omega}' \equiv \boldsymbol{\nabla}' \times \mathbf{u}'$ is the vorticity vector and ν' is the molecular kinematic viscosity. Using the Helmholtz decomposition

$$\mathbf{u}' = \mathbf{u}'_\mathbf{i} + \mathbf{u}'_\mathbf{r} = \boldsymbol{\nabla}'\Phi' + \mathbf{u}'_\mathbf{r},$$

where $\mathbf{u}'_\mathbf{i} \equiv \boldsymbol{\nabla}'\Phi'$ and $\mathbf{u}'_\mathbf{r}$ are the irrotational and rotational components of the velocity respectively. Both the irrotational and the rotational components of the velocity satisfy the continuity equation, so that $\nabla'^2\Phi' = 0$ and $\boldsymbol{\nabla}' \cdot \mathbf{u}'_\mathbf{r} = 0$. We anticipate that the rotational velocity is small except in the near bed region. As it is usual, hereinafter we will denote $\mathbf{u}'$ and w' as the horizontal and vertical components of the velocity and $\boldsymbol{\nabla}' = (\partial/\partial x', \partial/\partial y')$ as horizontal gradient. We remark, for future use, that both the irrotational and rotational components of the velocity can be further decomposed into steady (current) and unsteady (oscillatory) components.

2.1. *Dimensionless equations for the potential*

Let $k_0'^{-1}$ and h'_0 be a characteristic horizontal and vertical length scales of the wave motion and a'_0 the characteristic wave amplitude. The dimensionless parameters

$$\mu \equiv k'_0 h'_0, \qquad \text{and} \qquad \epsilon \equiv \frac{a'_0}{h'_0},$$

represent the dispersive and the nonlinear behaviour of the wave respectively. Besides, we define the dimensionless lengths

$$\{x, y\} \equiv k'_0 \{x', y'\}, \qquad \{z, h\} \equiv \frac{1}{h'_0}\{z', h'\}, \qquad \eta \equiv \frac{1}{a'_0}\eta'.$$

Let $\omega_0'^{-1}$ be the characteristic time for the oscillatory component of the motion, so that we can define the dimensionless time

$$t \equiv \omega'_0 t'.$$

Using the above definitions, the continuity equation for the potential is

$$\mu^2\nabla^2\Phi + \frac{\partial^2\Phi}{\partial z^2} = 0, \qquad -h \leqslant z \leqslant \epsilon\eta, \tag{6}$$

and the boundary conditions (3) to (5) are

$$\frac{\mu^2}{\epsilon}\frac{\partial h}{\partial t} + \mu^2\left(\boldsymbol{\nabla}\Phi + \mathbf{u}_\mathbf{r}\right)\cdot\boldsymbol{\nabla} h + \frac{\partial\Phi}{\partial z} + w_r = 0, \tag{7}$$

at $z = -h$ (*i.e.*, at the bottom), and

$$-\mu^2\frac{\partial\eta}{\partial t} - \epsilon\mu^2\boldsymbol{\nabla}\Phi\cdot\boldsymbol{\nabla}\eta + \frac{\partial\Phi}{\partial z} = \mathcal{O}\left(\epsilon_1\right), \tag{8}$$

$$\boldsymbol{\nabla}\frac{\partial\Phi}{\partial t} + \frac{\epsilon}{2}\boldsymbol{\nabla}\left\{\|\boldsymbol{\nabla}\Phi\|^2 + \frac{1}{\mu^2}\left(\partial\Phi/\partial z\right)^2\right\} + \boldsymbol{\nabla}\eta = \mathcal{O}\left(\epsilon_2\right), \tag{9}$$

at $z = \epsilon\eta$ (*i.e.*, at the free surface). Equation (8) is the kinematic free surface boundary condition which establishes that this is a material surface, while equation (9) is the dynamic free surface boundary condition which imposes continuity of stresses along this surface. To obtain the above equations we established $\omega_0' = k_0'\sqrt{g'h_0'}$ and defined the following dimensionless variables

$$\Phi \equiv \frac{\Phi'}{\Phi_0'}, \qquad \mathbf{u_r} \equiv \frac{\mathbf{u_r'}}{u_0'}, \qquad w_r \equiv \frac{w_r'}{w_0'},$$

where

$$\Phi_0' \equiv \frac{\epsilon\omega_0'}{k_0'^2}, \qquad u_0' \equiv k_0'\Phi_0' = \frac{\epsilon\omega_0'}{k_0'}, \qquad w_0' \equiv \frac{\Phi_0'}{h_0'} = \frac{\epsilon\omega_0'}{\mu k_0'}, \tag{10}$$

are, respectively, characteristic values for the potential and for the horizontal and vertical components of the velocity.

Moreover, both ϵ_1 and ϵ_2 in equations (8) and (9) involve the rotational velocity at the free surface, $z = \epsilon\eta$. It will be shown that considering only the bottom boundary layer effects, ϵ_1 and ϵ_2 are negligible, so they will be omitted in the equations hereinafter.

The scaling introduced above uses the characteristic scales of the oscillatory component of the motion. The scales are not representative for the bed variations and therefore, the order of $\beta \equiv \partial h/\partial t$ and $\varpi \equiv \nabla h$ will not be necessarily one. On the contrary, usually we have $\mathcal{O}(\beta, \varpi) \ll \mathcal{O}(1)$.

Furthermore, since u_0' is defined from the oscillatory component of the motion, the dimensionless velocity $\mathbf{u} = \mathbf{u}'/u_0'$ will be order one only as long as $\mathcal{O}(\psi) \leqslant \mathcal{O}(1)$, where ψ is defined as

$$\psi \equiv \frac{u_0^{s\prime}}{u_0'} \quad \left\{ = \frac{u_0^{s\prime}}{\epsilon\sqrt{g'h_0'}} \right\}, \tag{11}$$

where $u_0^{s\prime}$ is the characteristic velocity of the steady (current) horizontal velocity. We anticipate that we will consider $\mathcal{O}(\psi) \leqslant \mathcal{O}(\epsilon)$. Finally, hereafter we consider

$$\mathcal{O}(\beta, \epsilon, \mu, \varpi, \psi) \leqslant \mathcal{O}(1), \tag{12}$$

although some more restrictive conditions will eventually be required.

2.2. *Depth integrated equation*

Using kinematic boundary conditions (7) and (8) and depth integrating the continuity equation (6) we get

$$\nabla \cdot \int_{-h}^{\epsilon\eta} \nabla\Phi\, dz + \frac{1}{\epsilon}\frac{\partial h}{\partial t} + \frac{\partial \eta}{\partial t} + \mathbf{u_r}\cdot\nabla h + \frac{w_r}{\mu^2} = 0, \tag{13}$$

where the rotational velocity $\mathbf{u_r}$ and w_r are evaluated at the bottom.

In the following section we evaluate the rotational velocity as a function of the potential Φ near the bed.

3. Boundary layer analysis

The boundary layer is a thin layer attached to the bottom where the velocity gradients are large and therefore the viscous effects become important. Denoting δ'_0 as the characteristic thickness of the boundary layer ($\delta'_0 \ll h'_0$), we define the dimensionless parameter χ as

$$\chi \equiv \frac{\delta'_0}{h'_0} \; (\ll 1) \, .$$

Following Liu *et al.*[8] we introduce a local orthonormal coordinate system attached to the bottom for the boundary layer analysis (see Figure 1). The coordinates and variables related to this system are denoted with hats, so that $\hat{\mathbf{u}}'_{\mathbf{r}}$ and $\hat{w}'_r$ stand for the parallel and normal to the bottom components of the velocity. The scales for $\hat{\mathbf{u}}'_{\mathbf{r}}$ and $\hat{w}'_r$ are considered the same as those for $\mathbf{u}'_{\mathbf{r}}$ and w'_r respectively.

In order to introduce the Navier-Stokes and boundary conditions in the new coordinate system, it is assumed here that the bed is quasi rigid and

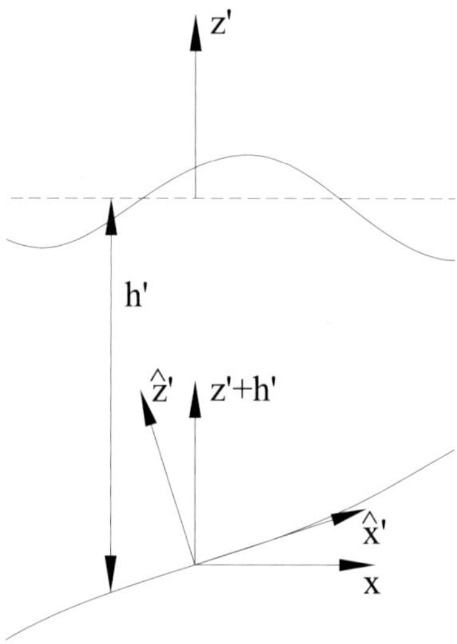

Fig. 1. Boundary layer coordinate system.

quasi plane (*i.e.* $\mathcal{O}(\varpi, \beta) \lll \mathcal{O}(1)$). These requirements will be later on required alternatively. Using the new coordinate system, note that

$$\hat{w}_r = w_r\left(1 + \mathcal{O}\left(\mu^2\varpi^2\right)\right) + \mu^2 \mathbf{u_r}\cdot\nabla h,$$

so that the equation (13) can be rewritten as

$$\nabla\cdot\int_{-h}^{\epsilon\eta} \nabla\Phi\, dz + \frac{1}{\epsilon}\frac{\partial h}{\partial t} + \frac{\partial \eta}{\partial t} + \frac{\hat{w}_r}{\mu^2} = \mathcal{O}\left(w_r\varpi^2\right), \tag{14}$$

with the rotational velocity $\hat{w}_r$ evaluated at the bottom.

Within the boundary layer, we further scale the $\hat{z}'$ with δ_0', *i.e.* $\hat{z} \equiv \hat{z}'/\delta_0'$ so that $\mathcal{O}(\hat{z}) = \mathcal{O}(1)$ within the boundary layer.

3.1. *Laminar boundary layer*

Dimensionless continuity and momentum equations for the rotational component of the velocity in the laminar boundary layer are, respectively

$$\hat{\nabla}\cdot\hat{\mathbf{u}}_\mathbf{r} + \frac{1}{\chi\mu^2}\frac{\partial \hat{w}_r}{\partial \hat{z}} = 0, \tag{15}$$

and

$$\frac{\partial \hat{\mathbf{u}}_\mathbf{r}}{\partial t} + \epsilon\left(\hat{\mathbf{u}}_\mathbf{i}\cdot\hat{\nabla}\right)\hat{\mathbf{u}}_\mathbf{r} + \frac{\epsilon}{\chi\mu^2}\,\hat{w}_i\frac{\partial \hat{\mathbf{u}}_\mathbf{r}}{\partial \hat{z}} + \epsilon\left(\hat{\mathbf{u}}_\mathbf{r}\cdot\hat{\nabla}\right)\hat{\mathbf{u}}_\mathbf{i} + \frac{\epsilon}{\chi\mu^2}\,\hat{w}_r\frac{\partial \hat{\mathbf{u}}_\mathbf{i}}{\partial \hat{z}}$$
$$+ \epsilon\left(\hat{\mathbf{u}}_\mathbf{r}\cdot\hat{\nabla}\right)\hat{\mathbf{u}}_\mathbf{r} + \frac{\epsilon}{\chi\mu^2}\,\hat{w}_r\frac{\partial \hat{\mathbf{u}}_\mathbf{r}}{\partial \hat{z}} = \alpha^2\hat{\nabla}^2\hat{\mathbf{u}}_\mathbf{r} + \frac{\alpha^2}{\chi^2\mu^2}\frac{\partial^2 \hat{\mathbf{u}}_\mathbf{r}}{\partial \hat{z}^2}, \tag{16}$$

where $\hat{\mathbf{u}}_\mathbf{i}$ and $\hat{w}_i$ are the components of the irrotational velocity and

$$\alpha \equiv \sqrt{\frac{\nu' k_0'^2}{\omega_0'}}\ (\ll 1), \tag{17}$$

is a dimensionless parameter accounting for the viscosity.

Far from the boundary layer the velocity gradient diminishes and viscous effects become negligible so that the rotational velocity tends to zero, *i.e.*

$$\hat{\mathbf{u}}_\mathbf{r} = \mathbf{0}, \qquad \text{for } \hat{z} \to \infty, \tag{18}$$

$$\hat{w}_r = 0, \qquad \text{for } \hat{z} \to \infty. \tag{19}$$

At the bottom, the total velocity equals the bed velocity, *i.e.*

$$\nabla\Phi + \mathbf{u_r} = \mathbf{u_b}, \qquad \text{at } \hat{z} = 0,$$

$$\frac{\partial \Phi}{\partial z} + w_r = w_b, \qquad \text{at } \hat{z} = 0,$$

where $\mathbf{u_b}$ and w_b stand for the bed velocity. Recalling expression (7), now

$$\frac{\mu^2}{\epsilon}\frac{\partial h}{\partial t} + \mu^2 \mathbf{u_b}\cdot\boldsymbol{\nabla} h + w_b = 0,$$

and we will assume $\mathcal{O}(\mathbf{u_b}) \leqslant \mathcal{O}(\beta\epsilon^{-1}\varpi^{-1})$ and $\mathcal{O}(w_b) \leqslant \mathcal{O}(\beta\epsilon^{-1}\mu^2)$.

From the above no-slip boundary conditions, within the boundary layer $\mathcal{O}(\mathbf{u_r}) = \mathcal{O}(1)$. Moreover, it is anticipated that, also in the boundary layer, $\mathcal{O}(\hat{\mathbf{u}}_\mathbf{r}) = \mathcal{O}(1)$ and $\mathcal{O}(\hat{w}_r) = \mathcal{O}(\chi\mu^2)$.

The above no-slip boundary conditions can thus be written in terms of $\hat{\mathbf{u}}_\mathbf{r}$ and $\hat{w}_r$ as

$$\boldsymbol{\nabla}\Phi + \hat{\mathbf{u}}_\mathbf{r} = \mathcal{O}\left(\beta\epsilon^{-1}\varpi^{-1}, \mu^2\varpi^2, \chi\mu^2\varpi\right), \qquad \text{at } \hat{z} = 0, \tag{20}$$

$$\frac{\partial\Phi}{\partial z} + \hat{w}_r - \mu^2\hat{\mathbf{u}}_\mathbf{r}\cdot\boldsymbol{\nabla} h = \mathcal{O}\left(\beta\epsilon^{-1}\mu^2, \chi\mu^4\varpi^2\right), \qquad \text{at } \hat{z} = 0. \tag{21}$$

Equations (15) to (21) allow to solve the rotational velocity for a given potential field. Before we present the solution, let us make a brief discussion about of the importance of the stationary component of the solution.

The rotational velocity is the addition of steady (current) and unsteady (oscillatory) components, *i.e.*

$$\hat{\mathbf{u}}_\mathbf{r} = \hat{\mathbf{u}}_\mathbf{r}^s + \hat{\mathbf{u}}_\mathbf{r}^u, \qquad \hat{w}_r = \hat{w}_r^s + \hat{w}_r^u,$$

where the supra-indices "s" and "u" stand for "steady" and "unsteady" respectively. As mentioned, the same decomposition holds for the irrotational velocity. We obtain the equations for both the steady and unsteady components from the above equations in a similar way that RANS equations are usually obtained. Let us focus only on the boundary condition (20): from this condition, $\mathcal{O}(\hat{\mathbf{u}}_\mathbf{r}^s) = \mathcal{O}(\mathbf{u}_\mathbf{i}^s)$ at the bottom. Anticipating that the latter will be considered $\mathcal{O}(\epsilon)$, we get $\mathcal{O}(\hat{\mathbf{u}}_\mathbf{r}^s) = \mathcal{O}(\epsilon)$ and, from the continuity equation (15), $\mathcal{O}(\hat{w}_r^s) = \mathcal{O}(\epsilon\chi\mu^2)$ within the boundary layer.

3.1.1. *Linearized problem*

Since the steady components $\hat{\mathbf{u}}_\mathbf{r}^s$ and $\hat{w}_r^s$ of the rotational velocity are small, the rotational velocity is composed by the unsteady component to the leading order. Moreover, within the boundary layer it is considered that viscous effects balance the local acceleration. Recalling momentum equation (16), this requires[6]

$$\frac{\alpha^2}{\chi^2\mu^2} = 1, \qquad \text{and hence} \qquad \delta_0' = \sqrt{\frac{\nu'}{\omega_0'}}, \tag{22}$$

is to be the characteristic thickness of the boundary layer.

Anticipating the potential expansion that will be presented in expression (60), within the boundary layer $\mathcal{O}(\hat{z}) = \mathcal{O}(1)$, $\mathcal{O}(z+h) = \mathcal{O}(\chi)$ and the momentum equation (16) can be written as

$$\frac{\partial \hat{\mathbf{u}}_{\mathbf{r}}}{\partial t} = \frac{\partial^2 \hat{\mathbf{u}}_{\mathbf{r}}}{\partial \hat{z}^2} + \mathcal{O}(\epsilon_m), \tag{23}$$

where it can be shown that the error ϵ_m is proportional to the rotational velocity and satisfies $\mathcal{O}(\epsilon_m) < \mathcal{O}(1)$ if

$$\mathcal{O}(\epsilon) < \mathcal{O}(1), \qquad \mathcal{O}(\varpi) < \mathcal{O}(\chi\epsilon^{-1}), \qquad \mathcal{O}(\chi^2) < \mathcal{O}(\mu^{-2}). \tag{24}$$

The above conditions must hold for the equation (23) to be meaningful, and they will be assumed hereafter. Note that, since parameter χ will easily be very small, $\mathcal{O}(\varpi) < \mathcal{O}(\chi\epsilon^{-1})$ can yield a strong condition on bed slope.

3.1.2. *Rotational velocity and bed shear stress*

The velocities $\hat{\mathbf{u}}_{\mathbf{r}}$ and $\hat{w}_r$ are found by solving equations (15) and (23) with boundary conditions (18) to (21). To the leading order, the solution reads[6]

$$\hat{\mathbf{u}}_{\mathbf{r}}(\hat{z}, t) = -\frac{\hat{z}}{\sqrt{4\pi}} \int_0^t \frac{\boldsymbol{\nabla}\Phi(z=-h,\xi)}{(t-\xi)^{3/2}} \exp\left(-\frac{\hat{z}^2}{4(t-\xi)}\right) d\xi, \tag{25}$$

$$\hat{w}_r(\hat{z}, t) = -\frac{\chi\mu^2}{\sqrt{\pi}} \int_0^t \frac{\nabla^2\Phi(z=-h,\xi)}{\sqrt{t-\xi}} \exp\left(-\frac{\hat{z}^2}{4(t-\xi)}\right) d\xi, \tag{26}$$

so that, at the bottom ($\hat{z}=0$),

$$\hat{\mathbf{u}}_{\mathbf{r}}(0) = -\boldsymbol{\nabla}\Phi(z=-h,t), \tag{27}$$

$$\hat{w}_r(0) = -\frac{\chi\mu^2}{\sqrt{\pi}} \int_0^t \frac{\nabla^2\Phi(z=-h,\xi)}{\sqrt{t-\xi}} d\xi. \tag{28}$$

From (25) and (26), and being $\mathcal{O}(\hat{z}) = \mathcal{O}(\chi^{-1}) \ggg 1$ at the free surface, the rotational velocity at the free surface is negligible and so they are errors ϵ_1 and ϵ_2 introduced in expressions (8) and (9), as anticipated.

The above solution for the rotational velocity $\hat{\mathbf{u}}_{\mathbf{r}}$ allows to compute the bed shear stress. To the leading order

$$\hat{\boldsymbol{\tau}} = \left.\frac{\partial \hat{\mathbf{u}}_{\mathbf{r}}}{\partial \hat{z}}\right|_{\hat{z}=0}, \tag{29}$$

where $\hat{\boldsymbol{\tau}}$ is the dimensionless stress tangent to the bed, defined as

$$\hat{\boldsymbol{\tau}} \equiv \frac{\hat{\boldsymbol{\tau}}'}{\chi\epsilon\mu\rho' g' h_0'}, \tag{30}$$

where $\hat{\boldsymbol{\tau}}'$ is the dimensional stress. From the solution (25) we get[8]

$$\hat{\boldsymbol{\tau}} = \frac{1}{\sqrt{\pi}} \left[\frac{\boldsymbol{\nabla}\Phi\,(z = -h, 0)}{\sqrt{t}} + \int_0^t \frac{\partial\,(\boldsymbol{\nabla}\Phi\,(z = -h, \xi))\,/\partial\xi}{\sqrt{t-\xi}}\, d\xi \right].$$

Note that the second term in the bed shear stress is a convolution integral of the near bed local acceleration and the first one vanishes for $t \to \infty$.

3.2. *Turbulent boundary layer*

The flow within the boundary layer is rarely laminar. In the turbulent case, the above approach can be used only if we consider constant eddy viscosity. To analyze the turbulent boundary layer case in a more accurate way, we follow Kajiura[9] and Grant and Madsen[10] approaches, considering that the eddy viscosity ν_t' is time independent and that can be expressed as,

$$\nu_t' = \kappa u_{*c}' \hat{z}',$$

where u_{*c}' is a characteristic frictional velocity. We define the characteristic eddy viscosity ν_0' as the eddy viscosity at $\hat{z}' = \delta_0'$, *i.e.* $\nu_0' = \kappa u_{*c}' \delta_0'$. Hence, the above expression can be written as,

$$\nu_t' = \nu_0' \hat{z}, \tag{31}$$

and the condition (22) reads

$$\delta_0' = \sqrt{\frac{\nu_0'}{\omega_0'}}, \qquad i.e. \qquad \delta_0' = \frac{\kappa u_{*c}'}{\omega_0'}. \tag{32}$$

Recalling the equation (31), the linearized momentum equation for the rotational component is now, already in dimensionless form,

$$\frac{\partial \hat{\mathbf{u}}_{\mathbf{r}}}{\partial t} = \frac{\partial}{\partial \hat{z}} \left(\hat{z} \frac{\partial \hat{\mathbf{u}}_{\mathbf{r}}}{\partial \hat{z}} \right), \tag{33}$$

while the continuity equation (15) remains valid. Of note, we will not track the errors in this section, since their behaviour is basically the same as the one presented in Section 3.1.

Kinematic boundary conditions far from the boundary layer, expressed in the equations (18) and (19), remain valid for the turbulent case. However, no-slip conditions (20) and (21) will now apply at

$$\hat{z} = \hat{z}_0 \equiv \frac{\hat{z}_0'}{\delta_0'}, \tag{34}$$

where $\hat{z}'_0$ is the dimensional elevation over the bed at which the total velocity cancels. Following the open channel flow theory $\hat{z}'_0$ is approximated by[11]

$$\hat{z}'_0 \approx \frac{k'_s}{30} + \frac{\nu'}{9.2\, u'_{*c}}, \tag{35}$$

so that, in the rough case, $\hat{z}'_0 = k'_s/30$.

In the above equations, both the viscosity ν'_t and the elevation $\hat{z}'_0$ depend on the characteristic frictional velocity u'_{*c}, so far undefined. In order to close the problem, u'_{*c} has to be related to the bed shear stress $\boldsymbol{\tau}'$, which is

$$\hat{\boldsymbol{\tau}}' = \rho' \nu'_t \frac{\partial \hat{\mathbf{u}}'_{\mathbf{r}}}{\partial \hat{z}'}\bigg|_{\hat{z}'=\hat{z}'_0}. \tag{36}$$

Following Orfila *et al.*[7] we consider u'_{*c} so that

$$\rho' u'^2_{*c} = \left\langle \left| \hat{\boldsymbol{\tau}}' \right| \right\rangle, \tag{37}$$

in which $\langle \cdot \rangle$ stands for the time average. Since the bottom stress $\hat{\boldsymbol{\tau}}'$ depends on u'_{*c}, an iterative procedure will be introduced below. Other expressions alternative to (37) are studied, for instance, by Simarro *et al.*[12]

3.2.1. *Periodic waves*

To solve the rotational velocity analytically from equation (33), we assume herein that the movement is $2\pi/\omega'_0$ periodic in time. Following Appendix A and recalling expressions (15) and (33), continuity and momentum equations for the different Fourier components are respectively

$$\boldsymbol{\nabla}\cdot\hat{\mathbf{u}}_{\mathbf{r},n} + \frac{1}{\chi\mu^2}\frac{\partial \hat{w}_{r,n}}{\partial \hat{z}} = 0, \tag{38}$$

$$in\hat{\mathbf{u}}_{\mathbf{r},n} = \frac{\partial}{\partial \hat{z}}\left(\hat{z}\frac{\partial \hat{\mathbf{u}}_{\mathbf{r},n}}{\partial \hat{z}}\right). \tag{39}$$

Boundary conditions (18) and (19) read, for the different components

$$\hat{\mathbf{u}}_{\mathbf{r},n} = \mathbf{0}, \qquad \text{for } \hat{z} \to \infty, \tag{40}$$

$$\hat{w}_{r,n} = 0, \qquad \text{for } \hat{z} \to \infty, \tag{41}$$

while at the bed, recalling (20) and (21),

$$\boldsymbol{\nabla}\Phi_n + \hat{\mathbf{u}}_{\mathbf{r},n} = 0, \qquad \text{at } \hat{z} = \hat{z}_0, \tag{42}$$

$$\frac{\partial \Phi_n}{\partial z} + \hat{w}_{r,n} - \mu^2 \boldsymbol{\nabla} h\cdot\hat{\mathbf{u}}_{\mathbf{r},n} = 0, \qquad \text{at } \hat{z} = \hat{z}_0. \tag{43}$$

For the above mentioned reasons, we restrict hereinafter to the unsteady component of the movement, *i.e.*, to $n \neq 0$. The extension of the present boundary layer analysis to the wave-current system is presented in Simarro *et al.*[12] For $n \neq 0$, the solution of equations (38) to (43) is

$$\hat{\mathbf{u}}_{\mathbf{r},n}(\hat{z}) = -\boldsymbol{\nabla}\Phi_n(z=-h)\,\frac{\mathrm{K}_0\left(2\sqrt{in\hat{z}}\right)}{\mathrm{K}_0\left(2\sqrt{in\hat{z}_0}\right)}, \tag{44}$$

$$\hat{w}_{r,n}(\hat{z}) = -\chi\mu^2\nabla^2\Phi_n(z=-h)\,\frac{1-i}{\sqrt{2n}}\sqrt{\hat{z}}\,\frac{\mathrm{K}_1\left(2\sqrt{in\hat{z}}\right)}{\mathrm{K}_0\left(2\sqrt{in\hat{z}_0}\right)}, \tag{45}$$

where K_i stand for Kelvin functions. At the bottom ($\hat{z} = \hat{z}_0$)

$$\hat{\mathbf{u}}_{\mathbf{r},n}(\hat{z}_0) = -\boldsymbol{\nabla}\Phi_n(z=-h), \tag{46}$$

$$\hat{w}_{r,n}(\hat{z}_0) = -\chi\mu^2\nabla^2\Phi_n(z=-h)\,\frac{1-i}{\sqrt{2n}}\sqrt{\hat{z}_0}\,\frac{\mathrm{K}_1\left(2\sqrt{in\hat{z}_0}\right)}{\mathrm{K}_0\left(2\sqrt{in\hat{z}_0}\right)}. \tag{47}$$

Recalling expressions (36) and (44), the different Fourier components of the bed shear stress are

$$\hat{\boldsymbol{\tau}}_n = \hat{z}_0\frac{\partial\hat{\mathbf{u}}_{\mathbf{r},n}}{\partial\hat{z}}\bigg|_{\hat{z}=\hat{z}_0} = \boldsymbol{\nabla}\Phi_n(z=-h)\sqrt{in\hat{z}_0}\,\frac{\mathrm{K}_1\left(2\sqrt{in\hat{z}_0}\right)}{\mathrm{K}_0\left(2\sqrt{in\hat{z}_0}\right)}. \tag{48}$$

3.3. *Summary of the boundary layer analysis*

Summarizing the above results for future use, the no-slip condition (21) is

$$\frac{\partial\Phi}{\partial z} + \hat{w}_r(\xi_0) - \mu^2\boldsymbol{\nabla}h\cdot\hat{\mathbf{u}}_{\mathbf{r}}(\xi_0) = \mathcal{O}\left(\beta\epsilon^{-1}\mu^2, \chi\mu^4\varpi^2\right), \tag{49}$$

where ξ_0 is given by

$$\xi_0 = \begin{cases} 0, & \text{laminar (constant viscosity) BL;} \\ \hat{z}_0, & \text{turbulent (linear viscosity) BL.} \end{cases} \tag{50}$$

The solutions for $\hat{\mathbf{u}}_{\mathbf{r}}(\xi_0)$ and $\hat{w}_r(\xi_0)$ have already been presented both for the laminar — equations (27) and (28) — and the turbulent case — equations (46) and (47). In the latter case, the solution has been presented in terms of the Fourier components, assuming the flow is $2\pi/\omega_0'$ periodic in time.

Since the velocity potential will be solved in the frequency domain, the solutions for the laminar case can readily be written in Fourier components. The general solution (both for laminar and turbulent cases) can be written in a compact form as

$$\hat{\mathbf{u}}_{\mathbf{r},n}(\xi_0) = -\boldsymbol{\nabla}\Phi_n(z=-h), \tag{51}$$

$$\hat{w}_{r,n}(\xi_0) = -\chi\mu^2\nabla^2\Phi_n(z=-h)\,\frac{1-i}{\sqrt{2n}}\,\vartheta_n, \tag{52}$$

where ξ_0 is given in (50) and

$$\vartheta_n = \begin{cases} 1, & \text{laminar BL;} \\ \sqrt{\hat{z}_0}\,\dfrac{\mathrm{K}_1\left(2\sqrt{in\hat{z}_0}\right)}{\mathrm{K}_0\left(2\sqrt{in\hat{z}_0}\right)}, & \text{turbulent BL.} \end{cases} \tag{53}$$

Similarly, the solution for the different bed stress components is

$$\hat{\boldsymbol{\tau}}_n = \boldsymbol{\nabla}\Phi_n\left(z=-h\right)\sqrt{in}\,\vartheta_n. \tag{54}$$

Finally, in the turbulent case, the value of the $\hat{z}_0$ is still to be determined. The equations (32), (35) and (37) to be solved are, in dimensionless form

$$\chi = \frac{\epsilon\kappa^2 u_{*c}^2}{\mu}, \tag{55}$$

$$\hat{z}_0 \approx \frac{\varepsilon}{30\chi} + \frac{1}{9.2u_{*c}\sqrt{\chi^3\epsilon\mu}\,\mathrm{Re}}, \tag{56}$$

$$u_{*c} = \sqrt{\left\langle\left|\sum\nolimits_{n>0}\Re\left\{\hat{\boldsymbol{\tau}}_n\left(\hat{z}_0\right)\exp\left(int\right)\right\}\right|\right\rangle}, \tag{57}$$

where

$$u_{*c} \equiv \frac{u'_{*c}}{\sqrt{\chi\epsilon\mu g' h'_0}}, \qquad \varepsilon \equiv \frac{k'_s}{h'_0}, \qquad \mathrm{Re} \equiv \frac{h'_0\sqrt{g'h'_0}}{\nu'}, \tag{58}$$

are, the dimensionless characteristic frictional velocity and the corresponding relative roughness and Reynolds number respectively.

The proposed iterative procedure to solve $\hat{z}_0$ is the following:[7] an initial value for $\hat{z}_0$ is guessed, so that u_{*c} and χ are found through equations (55) and (57). The value of $\hat{z}_0$ is then updated through expression (56) and the process continued until a convergence criteria is satisfied.

4. Core region analysis

The equations for the potential have been presented in Section 2. From the analysis of the boundary layer, however, the no flux boundary condition (7) has been completed with no-slip equations (49). In order to solve the potential we consider the usual perturbation approach where the potential is expressed as a power series in the vertical coordinate[13,14]

$$\Phi\left(x,y,z,t\right) = \sum_{n=0}^{\infty}\left(z+h\right)^n\phi_n\left(x,y,t\right). \tag{59}$$

Using the continuity equation (6) and the boundary condition (21), and following the usual recursive approach, we get

$$\Phi = \phi_0 - \mu^2 \frac{(z+h)^2}{2} \nabla^2 \phi_0 + \mathcal{O}\left(\mu^4, \chi\mu^2, \mu^2\varpi, \beta\epsilon^{-1}\mu^2\right). \qquad (60)$$

The above potential automatically satisfies equations (6) and (49). The remaining equations for the potential are the kinematic equation (14) and the dynamic equation (9), *i.e.*

$$\boldsymbol{\nabla}\cdot \int_{-h}^{\epsilon\eta} \boldsymbol{\nabla}\Phi \, dz + \frac{1}{\epsilon}\frac{\partial h}{\partial t} + \frac{\partial \eta}{\partial t} + \frac{\hat{w}_r}{\mu^2} = \mathcal{O}\left(\chi\mu^2\varpi^2\right), \qquad (61)$$

$$\boldsymbol{\nabla}\frac{\partial \Phi}{\partial t} + \frac{\epsilon}{2}\boldsymbol{\nabla}\left\{ \|\boldsymbol{\nabla}\Phi\|^2 + \frac{1}{\mu^2}\left(\partial\Phi/\partial z\right)^2 \right\} + \boldsymbol{\nabla}\eta = 0. \qquad (62)$$

Of note, boundary layer effects are introduced in the kinematic equation (61) through rotational velocity $\hat{w}_r$ at the bottom.

5. Boussinesq equations

Substitution of the potential Φ as expressed in equation (60) into the kinematic and dynamic expressions (61) and (62) allows to solve for the potential ϕ_0 and the free surface η. However, following Nwogu[4] the equations will be presented for the velocity at $z = z_\alpha$.

For the sake of simplicity, it will be assumed henceforth

$$\mathcal{O}(\epsilon) \leqslant \mathcal{O}\left(\mu^2\right), \qquad (63)$$

which is the Boussinesq condition. From (60),

$$\boldsymbol{\nabla}\Phi = \boldsymbol{\nabla}\phi_0 - \mu^2 \frac{(z+h)^2}{2} \boldsymbol{\nabla}\nabla^2 \phi_0 + \mathcal{O}\left(\mu^4, \chi\mu^2, \mu^2\varpi, \beta\epsilon^{-1}\mu^2\right), \qquad (64)$$

$$\frac{\partial \Phi}{\partial z} = -\mu^2 (z+h) \nabla^2 \phi_0 + \mathcal{O}\left(\mu^4, \chi\mu^2, \mu^2\varpi, \beta\epsilon^{-1}\mu^2\right), \qquad (65)$$

and hence the velocity $\mathbf{u}_\alpha$ at $z = z_\alpha$ is

$$\mathbf{u}_\alpha = \boldsymbol{\nabla}\phi_0 - \mu^2 \frac{(z_\alpha+h)^2}{2} \boldsymbol{\nabla}\nabla^2 \phi_0 + \mathcal{O}\left(\mu^4, \chi\mu^2, \mu^2\varpi, \beta\epsilon^{-1}\mu^2\right),$$

so that equation (64) can be rewritten as

$$\boldsymbol{\nabla}\Phi = \mathbf{u}_\alpha + \mu^2 \frac{z_\alpha(z_\alpha+2h) - z(z+2h)}{2} \boldsymbol{\nabla}\left(\boldsymbol{\nabla}\cdot\mathbf{u}_\alpha\right) + \mathcal{O}\left(\mu^4, \beta\epsilon^{-1}\mu^2, \chi\mu^2, \mu^2\varpi\right). \qquad (66)$$

Finally, assuming that z_α behaves as h, *i.e.*

$$\mathcal{O}(\boldsymbol{\nabla} z_\alpha) = \mathcal{O}(\varpi) \qquad \text{and} \qquad \mathcal{O}(\partial z_\alpha/\partial t) = \mathcal{O}(\beta),$$

the equations for $\mathbf{u}_\alpha$ can be obtained by introducing the above results into kinematic and dynamic equations (61) and (62). We get

$$\boldsymbol{\nabla}\cdot((\epsilon\eta + h)\,\mathbf{u}_\alpha) + \left(c_\alpha + \frac{1}{3}\right)\mu^2 h^3 \nabla^2 \boldsymbol{\nabla}\cdot\mathbf{u}_\alpha + \frac{\partial\eta}{\partial t} + \frac{\hat{w}_r(\xi_0)}{\mu^2} = \mathcal{O}\left(\mu^4, \chi\mu^2, \mu^2\varpi, \beta\epsilon^{-1}\right), \quad (67)$$

and

$$\frac{\partial \mathbf{u}_\alpha}{\partial t} + c_\alpha \mu^2 h^2 \boldsymbol{\nabla}\boldsymbol{\nabla}\cdot\frac{\partial \mathbf{u}_\alpha}{\partial t} + \frac{\epsilon}{2}\boldsymbol{\nabla}(\mathbf{u}_\alpha\cdot\mathbf{u}_\alpha) + \boldsymbol{\nabla}\eta = \mathcal{O}\left(\mu^4, \chi\mu^2, \mu^2\varpi, \beta\epsilon^{-1}\mu^2\right), \quad (68)$$

where c_α is defined as

$$c_\alpha \equiv \frac{z_\alpha(z_\alpha + 2h)}{2h^2}.$$

We remark that, according to Nwogu,[4] $c_\alpha = -0.39$ optimizes dispersion properties when neglecting viscous and nonlinear effects. Furthermore, setting $c_\alpha = -1/3$, the equations for the depth averaged velocity are recovered.[7] For the sake of simplicity, hereinafter it will be assumed also

$$\mathcal{O}(\beta) \leqslant \mathcal{O}(\epsilon\mu^4), \qquad \mathcal{O}(\chi) \leqslant \mathcal{O}(\mu^2), \qquad \mathcal{O}(\varpi) \leqslant \mathcal{O}(\mu^2),$$

so that the errors in expressions (67) and (68) are $\mathcal{O}(\mu^4)$.

5.1. *Periodic waves*

Taking into account the results in Appendix A, the equations (67) and (68) read, for the different Fourier components $\mathbf{u}_{\alpha,n}$ and η_n

$$\frac{\epsilon}{2}\sum_s \boldsymbol{\nabla}\cdot(\eta_s \mathbf{u}_{\alpha,n-s}) + h\boldsymbol{\nabla}\cdot\mathbf{u}_{\alpha,n} + \mathbf{u}_{\alpha,n}\cdot\boldsymbol{\nabla}h + \left(c_\alpha + \frac{1}{3}\right)\mu^2 h^3 \nabla^2\boldsymbol{\nabla}\cdot\mathbf{u}_{\alpha,n} + in\eta_n + \frac{\hat{w}_{r,n}(\xi_0)}{\mu^2} = \mathcal{O}\left(\mu^4\right), \quad (69)$$

and

$$in\mathbf{u}_{\alpha,n} + inc_\alpha\mu^2 h^2 \boldsymbol{\nabla}\boldsymbol{\nabla}\cdot\mathbf{u}_{\alpha,n} + \frac{\epsilon}{4}\sum_s \boldsymbol{\nabla}(\mathbf{u}_{\alpha,s}\cdot\mathbf{u}_{\alpha,n-s}) + \boldsymbol{\nabla}\eta_n = \mathcal{O}\left(\mu^4\right). \quad (70)$$

The above equations show how the different harmonics interact due to the nonlinearities order ϵ. Assuming that there are no other sources for the steady component $\mathbf{u}_{\alpha,0}$ than these interactions, by setting $n = 0$ in (69) we get $\mathcal{O}(\mathbf{u}_{\alpha,0}) = \mathcal{O}(\epsilon)$, as we have been anticipating.

Following Liu *et al.*,[16] combining the above equations we get

$$\nabla \cdot (\varphi_{1,n} \nabla \eta_n) + \varphi_{2,n} n^2 \eta_n = \frac{in\epsilon}{2} \sum_{\substack{s\neq 0 \\ s\neq n}} \nabla \cdot (\eta_s \mathbf{u}_{\alpha,n-s}) - \frac{\epsilon h}{4} \sum_{\substack{s\neq 0 \\ s\neq n}} \nabla^2 (\mathbf{u}_{\alpha,s} \cdot \mathbf{u}_{\alpha,n-s}) + \mathcal{O}(\mu^4), \quad (71)$$

where

$$\varphi_{1,n} \equiv h - \frac{n^2\mu^2h^2}{3}, \qquad \varphi_{2,n} \equiv 1 + \frac{\chi}{h} \frac{1-i}{\sqrt{2n}} \vartheta_n. \quad (72)$$

The RHS of expression (71) can also be written in terms of the free surface elevation to get[16]

$$\begin{aligned} &\nabla \cdot (\varphi_{1,n} \nabla \eta_n) + \varphi_{2,n} n^2 \eta_n - \nabla \eta \cdot \nabla h \\ &= -\frac{\epsilon}{2} \sum_{\substack{s\neq 0 \\ s\neq n}} \frac{n+s}{n-s} \nabla \eta_s \cdot \nabla \eta_{n-s} + \frac{\epsilon}{2h} \sum_{\substack{s\neq 0 \\ s\neq n}} \left(n^2 - s^2\right) \eta_s \eta_{n-s} \\ &\quad - \epsilon h \sum_{\substack{s\neq 0 \\ s\neq n}} \frac{1}{s(n-s)} \left[\frac{\partial^2 \eta_s}{\partial x^2} \frac{\partial^2 \eta_{n-s}}{\partial y^2} - \frac{\partial^2 \eta_s}{\partial x \partial y} \frac{\partial^2 \eta_{n-s}}{\partial x \partial y} \right] + \mathcal{O}(\mu^4). \quad (73) \end{aligned}$$

Note that the equation (73) does not include c_α and, hence, the above mentioned dispersion improvements do not hold for this approach.

5.2. *Periodic wave propagation in a wave tank*

We consider now that the movement is one dimensional, so that the equation (73) reduces to

$$\frac{d}{dx}\left(\varphi_{1,n} \frac{d\eta_n}{dx}\right) + \varphi_{2,n} n^2 \eta_n = -\frac{\epsilon}{2} \sum_{\substack{s\neq 0 \\ s\neq n}} \frac{n+s}{n-s} \frac{\partial \eta_s}{\partial x} \frac{\partial \eta_{n-s}}{\partial x} + \frac{\epsilon}{2h} \sum_{\substack{s\neq 0 \\ s\neq n}} \left(n^2 - s^2\right) \eta_s \eta_{n-s} + \mathcal{O}(\mu^4). \quad (74)$$

Moreover, we write η_n as

$$\eta_n = A_n \exp(-inx), \tag{75}$$

where the amplitudes A_n are assumed to vary slowly, so that

$$\frac{dA_n}{dx} \approx \mathcal{O}(\mu^2), \qquad \frac{d^2 A_n}{dx^2} \approx \mathcal{O}(\mu^4).$$

After some manipulation, the expression (74) reads, for the amplitudes

$$-2inh\frac{dA_n}{dx} + \left(\varphi_n n^2 - in\frac{dh}{dx}\right) A_n \\ = \frac{\epsilon}{2}\sum_{\substack{s\neq 0\\ s\neq n}} (n+s)\, sA_s A_{n-s} + \frac{\epsilon}{2h}\sum_{\substack{s\neq 0\\ s\neq n}} \left(n^2 - s^2\right) A_s A_{n-s} + \mathcal{O}(\mu^4), \tag{76}$$

with $\varphi_n \equiv \varphi_{2,n} - \varphi_{1,n}$. The equation (76) is parabolic and allows to solve for the free surface elevation η_n through the amplitudes A_n.

5.2.1. *The velocity components $u_{\alpha,n}$*

Expression (70) can now be used to solve for the velocity components given the free surface elevation. In the one dimensional case it reads

$$inu_{\alpha,n} + inc_\alpha \mu^2 h^2 \frac{d^2 u_{\alpha,n}}{dx^2} \\ + \frac{\epsilon}{4}\sum_{\substack{s\neq 0\\ s\neq n}} \left(\frac{du_{\alpha,s}}{dx} u_{\alpha,n-s} + u_{\alpha,s}\frac{du_{\alpha,n-s}}{dx}\right) + \frac{d\eta_n}{dx} = \mathcal{O}(\mu^4). \tag{77}$$

Similarly, the different velocity components can be written as,

$$u_{\alpha,n} = B_n \exp(-inx), \tag{78}$$

and further

$$\frac{dB_n}{dx} \approx \mathcal{O}(\mu^2), \qquad \frac{d^2 B_n}{dx^2} \approx \mathcal{O}(\mu^4),$$

so that equation (77) reads

$$\left(1 - c_\alpha n^2 \mu^2 h^2\right) B_n - \frac{\epsilon}{4}\sum_{\substack{s\neq 0\\ s\neq n}} B_s B_{n-s} = \frac{i}{n}\frac{dA_n}{dx} + A_n + \mathcal{O}(\mu^4), \tag{79}$$

where A_n have been already obtained. Expression (79) can be solved using a perturbation approach to handle with the nonlinear term. Considering

$$B_n = B_n^0 + \epsilon B_n^1 + \mathcal{O}(\epsilon^2), \tag{80}$$

the equations for B_n^0 and B_n^1 are

$$\left(1 - c_\alpha n^2 \mu^2 h^2\right) B_n^0 = \frac{i}{n}\frac{dA_n}{dx} + A_n + \mathcal{O}\left(\mu^4\right),$$
$$\left(1 - c_\alpha n^2 \mu^2 h^2\right) B_n^1 = \frac{1}{4} \sum_{\substack{s\neq 0\\ s\neq n}} B_s^0 B_{n-s}^0 + \mathcal{O}\left(\mu^4\right),$$

which can be solved explicitly.

5.2.2. *The bed shear stress*

Recalling the above expression (54), we need $\partial \Phi_n / \partial x\,(z = -h)$ to compute the bed shear stress. Combining expressions (66) and (78)

$$\frac{d\Phi_n}{dx}(z = -h) = \left[1 - \left(c_\alpha + \frac{1}{2}\right) n^2 \mu^2 h^2\right] B_n \exp\left(-inx\right) + \mathcal{O}\left(\mu^4\right),$$

so that the Fourier components of the bed shear stress are, from (54),

$$\tau_n = C_n \exp\left(-inx\right), \tag{81}$$

with

$$C_n = \left[1 - \left(c_\alpha + \frac{1}{2}\right) n^2 \mu^2 h^2\right] \sqrt{in}\, \vartheta_n B_n + \mathcal{O}\left(\mu^4\right), \tag{82}$$

the corresponding amplitude for the bed shear stress.

6. Concluding remarks

We presented a Boussinesq-type model for periodic wave propagation when the boundary layer effects are important. The boundary layer can be either laminar or turbulent. In the latter case, the eddy viscosity was considered time independent so as to get an analytical solution. The boundary layer effects on the core region appear as the normal to the bottom component of the rotational velocity. The presented approach can be solved efficiently decomposing the wave field into Fourier harmonics.

Acknowledgments

This document is the summary of the results of nearly four years of work with Professor Liu. The authors would like to thank him for his guidance and friendship. The authors would also like to thank Professors Miguel Ángel Losada and Íñigo Losada for their support.

Appendix A. Fourier analysis

Assuming $2\pi/\omega_0'$ time periodicity, any variable $\xi'(x', t')$ can be written as

$$\xi'(x', t') = \frac{1}{2}\sum_{n=-\infty}^{\infty} \xi_n'(x')\exp(in\omega_0' t'),$$

where the Fourier coefficients $\xi_n'(x')$ are given by

$$\xi_n'(x') = \frac{\omega_0'}{\pi}\int_{-\pi/\omega_0'}^{\pi/\omega_0'} \xi'(x', t')\exp(-in\omega_0' t')\,dt'.$$

In dimensionless form, the above expansion is

$$\xi(x, t) = \frac{1}{2}\sum_{n=-\infty}^{\infty} \xi_n(x)\exp(int), \tag{A.1}$$

with

$$\xi_n(x) = \frac{1}{\pi}\int_{-\pi}^{\pi} \xi(x, t)\exp(-int)\,dt, \tag{A.2}$$

the dimensionless Fourier coefficients. Further, since $\xi_{-n}(x)$ is the complex conjugate of $\xi_n(x)$, the expression (A.1) is also

$$\xi(x, t) = \frac{1}{2}\sum_{n=-\infty}^{\infty} \xi_n(x)\exp(int) = \frac{\xi_0(x)}{2} + \sum_{n>0}\Re\{\xi_n(x)\exp(int)\}.$$

Using the above Fourier expansion, the spatial derivative is

$$\frac{\partial\xi}{\partial x} = \frac{1}{2}\sum_{n=-\infty}^{\infty}\frac{\partial\xi_n}{\partial x}\exp(int) = \frac{1}{2}\frac{\partial\xi_0}{\partial x} + \sum_{n>0}\Re\left\{\frac{\partial\xi_n}{\partial x}\exp(int)\right\},$$

and

$$\frac{\partial\xi}{\partial t} = \frac{1}{2}\sum_{n\neq 0} in\xi_n\exp(int) = \sum_{n>0}\Re\{in\xi_n\exp(int)\},$$

is the time derivative. The product $\varphi = \xi\psi$ can be written as

$$\varphi = \frac{1}{2}\sum_{n=-\infty}^{\infty}\varphi_n\exp(int) = \frac{\varphi_0}{2} + \sum_{n>0}\Re\{\varphi_n\exp(int)\},$$

with

$$\varphi_n = \frac{1}{2}\sum_{s=-\infty}^{\infty}\xi_s\,\psi_{n-s} = \frac{1}{2}\sum_{s=-\infty}^{\infty}\xi_{n-s}\,\psi_s,$$

where ξ_j and ψ_k are the Fourier coefficients for ξ and ψ.

References

1. P. L.-F. Liu and I. Losada, *Journal of Hydraulic Research* **40**, 229 (2002).
2. C. C. Mei and B. LeMehaute, *Journal of Geophysical Research* **71**, 393 (1966).
3. D. H. Peregrine, *Journal of Fluid Mechanics* **27**, 815 (1967).
4. O. Nwogu, *Journal of Waterway, Port, Coastal and Ocean Engineering* **119(6)**, 618 (1993).
5. P. Nielsen, *Coastal Bottom Boundary Layers and Sediment Transport* (World Scientific, Singapore, 1992).
6. P. L.-F. Liu and A. Orfila, *Journal of Fluid Mechanics* **520**, 83 (2004).
7. A. Orfila, G. Simarro and P. L.-F. Liu, *Coastal Engineering* (2007) (doi: 10.1016/j.coastaleng.2007.05.013).
8. P. L.-F. Liu, G. Simarro, J. Vandever and A. Orfila, *Coastal Engineering* **520(2-3)**, 181 (2006).
9. K. Kajiura, *Bull. Eqrthquake Res. Int.* **42**, 147 (1964).
10. W. D. Grant and O. S. Madsen, *Journal of Geophysical Research* **84(C4)**, 1797 (1979).
11. W. H. Graf and M. S. Altinakar, *Fluvial Hydraulics* (John Wiley and Sons, Ltd., England, 1998).
12. G. Simarro, A. Orfila and P. L.-F. Liu, *Journal of Hydraulic Engineering* (in press).
13. C. C. Mei, *The Applied Dynamics of Ocean Surface Waves* (World Scientific, Singapore, 1989).
14. P. L.-F. Liu, Model equations for wave propagation from deep water to shallow water, in Advances in Coastal and Ocean Engineering, Vol. 1, ed. P. L.-F. Liu (World Scientific, 1994), pp. 125–157.
15. P. E. Meyer-Peter and R. Muller, Formulas for bed load transport, in *Proceedings. 2nd Congress, IAHR.* (Stockholm, Sweden, 1948).
16. P. L.-F. Liu, S. B. Yoon and J. T. Kirby, *Journal of Fluid Mechanics* **153**, 185 (1985).

SUBAQUEOUS FLUID DISCHARGE ESTIMATES FROM SEDIMENTS IN SHALLOW AND DEEP WATER (1 TO 1000 M)

GERARD P. LENNON

Civil and Environmental Engineering Department, Lehigh University, 19 Memorial Drive West, Bethlehem, PA 18015, USA

BOBB CARSON

Earth and Environmental Sciences Department, Lehigh University, Williams Hall Bethlehem, PA, 18015, USA

ELIZABETH SCREATON

Geological Sciences Department, University of Florida, Gainesville Florida, USA

MICHAEL D. WETZEL

Maryland State Highway Administration, 2323 W. Joppa Rd. Lutherville, MD 21093 USA

Two hydrogeological techniques are used to determine the subaqueous fluid discharge rate from sediments with excess pore water pressure into the overlying water column. The first is based on determining diffuse flow (Q) as the product of vertical gradient (i), hydraulic conductivity (K), and area (A) over which the near-vertical discharge occurs. Results are presented from Barnegat Bay, New Jersey, USA in shallow water depths ranging from about 1 m to 10 m, and deep water on the order of 1000 m on an active accretionary prism thrust fault accretionary prism off the coast of Oregon, USA. The portable instrument probe developed by the authors is used to obtain the excess pore pressure approximately 1 m into the sediment, from which the gradient (i) is obtained. The decay rate of an induced pressure pulse matched to adapted slug test curves to determine the hydraulic conductivity (K). The second technique was used on Ocean Drilling Program (ODP) Leg 146, Hole 892 at a water depth of 674 m, on the Oregon accretionary prism. The open interval, just over 100 m below the seafloor, intersects a fault identified by seismic imaging, as contributing to concentrated flows at subaqueous vent sites. A constant-rate flow test using a pump aboard the Alvin submersible and subsequent recovery allowed estimation of hydraulic conductivity (K) and transmissivity (T), the latter a product of K and thickness of the fractured interval. By determining the pressure in excess of hydrostatic over the 10 month period, the flow rate contributed by this fault to the seafloor vent site is estimated. These hydrogeologic tests provided the first in situ estimates of fault-zone transmissivities within a modern, hydrogeologically active accretionary prism fault zone, and indicated that fracture dilation is necessary for this fault zone to act as a significant fluid conduit.

1. Introduction

1.1. *Diffuse Flow in Shallow Water: Barnegat Bay, New Jersey, USA*

Kennish and Lutz [1] provide a physical description and an ecology assessment of Barnegat Bay, New Jersey, USA. One of the critical environmental issues identified is the need to improve the understanding of the distribution and rate of groundwater discharge into the bay and its impact on salinity and nutrient levels.

A variety of techniques are available to estimate the groundwater discharge. Permanent piezometers or mini-piezometers can be installed on land, or with more difficulty, in subaqueous areas. Direct seepage measurements using simple seepage meters can directly measure fluid flow. However, because they create a back-pressure to fluid expulsion, they sample only a portion of the total fluid flux, and must be carefully calibrated for sediments with varying hydraulic conductivities [2, 3]. Although inexpensive, they are inherently unreliable. In addition, in coastal areas where tidal fluctuations can have a significant influence on the pressure field, seepage meters provide only long-term integration of fluid flow, and reveal nothing of short-term fluctuations induced by tidal forcing. In situ methods based on a geochemical signature in the discharging fluid are available [3]. Chemical concentrations in near bottom water are compared to concentrations in the groundwater to determine the flux rate. This technique is more difficult to apply in tidal areas which experience periodic changes in or even reversal of the discharge rate.

The U.S. Geological Survey has performed extensive studies of the groundwater system and drainage basins flowing into Barnegat Bay [4]. The measured shallow groundwater levels from land-based wells in the shallow Kirkwood-Cohansey aquifer system are typically several feet above sea level, driving discharge into the bay. They predicted exchange rates between shallow aquifers and the northwestern part of the bay using a site-specific numerical groundwater model [5]. The major objectives were to refine water budget estimates and to assess effects of water use. The model focused on the Toms River, Metedeconk River, and Kettle Creek Basins, but did include Barnegat Bay and the barrier islands. The model predicted the largest discharge to occur is concentrated close to shore and at the mouths of the rivers and creeks. Although most of the local water supply is satisfied by surface water and deep well pumping, in a couple areas such as Seaside Heights, pumping does appear to induce flow from the bay into the shallow aquifer system. However, because Barnegat Bay was not the major focus of the model, the model cells are relatively large (2000 ft on a side) and the discharge distribution is not highly refined, e.g. most of the discharge occurs within one cell of the shoreline. In addition, because the model was reduced to a single layer under the bay, the measured flow rates in this study cannot be directly compared to the flow through the model's constant head cells. Lennon et al. [6] indicate that a fairly extensive silt and clay layer "cap" inhibits direct discharge to the bay, and that only a small portion of the discharge predicted by USGS model can actually penetrate the layer under reasonable hydraulic gradients. Thus a refinement of their model is necessary to provide a direct comparison to the field data reported here; the USGS model does not include any low permeability cap.

For this study, direct measurements of the in situ hydraulic head and hydraulic conductivity allow calculation of the discharge rates as discussed in Section 3, Results. The distribution of groundwater flux into the bay can have a significant impact on salinity and circulation within the bay and may contribute to the movement of nutrients and other nonpoint source pollutants into the bay.

1.2. *Concentrated Flow in Deep Water: Oregon Accretionary Prism*

The second technique was used on Hole 892B installed as part of the Ocean Drilling Program (ODP) Leg 146 at a depth of 674 m [7, 8]. The site is on an active accretionary prism thrust fault where the open interval intersects a fault identified by seismic imaging 107 to 113 m below the seafloor, as contributing to concentrated flows at subaqueous vent sites. The fault outcrop has chemosynthetic clam communities and diagenetic carbonate and actively discharging concentrated groundwater as observed by Alvin.

2. Materials and Methods

2.1. *Description of Equipment — PISPPI*

The PISPPI-2 (second-generation, Portable, In-Situ Pore Pressure Instrument) system is shown in Fig. 1. The PISPPI-2 consists of

(1) A water-tight pressure case houses an electronics package that includes an electronic data logger and a sensitive, differential pressure transducer
(2) A 0.95 cm O.D., 53 cm long rigid stainless steel probe, fitted with a tapered stainless steel tip and porous stone-covered port
(3) A pump (or floating reservoir) to create a pressure pulse.

Using recent advances in sophisticated electronic data loggers with extremely low power consumption allowed the miniaturization of Bennett's [9] original deep-water design, creating a small, portable, in situ system for shallow water. The transducer output is monitored by a Tattletale 4A computer (Onset Computer Corp.), capable of variable sampling rates (1/10 sec to >1 hr) and storage of 1-2 Mb of data. The logger is housed in a pressure case rated to 3000 m for deep-water deployment.

Pressure is sensed with a miniature, solid state, differential pressure transducer (Validyne DP15) capable of resolving pressure differences to 0.22 cm (0.003 psi or 0.1 inches of water at 4°C), yet able to monitor the peak insertion and pressure-pulse pressure extremes encountered of approximately 80 cm. Higher resolution diaphragms were obtained capable of lower pressure ranges down to 0.044 cm. One side of the transducer is connected to the probe port, while the other monitors ambient hydrostatic pressure just above the sediment-water interface. In this way, the probe port records the pressure in the sediment relative to the water column hydrostatic pressure; this differential pressure yields the hydraulic head. The pressure port is located about 6 cm above the tip, and is protected by a 20 μm sintered metal filter.

The probe has been inserted with a deployment rod from a small boat in depths up to 6 m. Once inserted, the system provides remote display to the operator on the boat for monitoring the test. For long duration tests, the readout monitor can be detached and the operator can move to another station. When the device is retrieved, the data are downloaded from the PISPPI-2 data logger to a laptop computer for analysis at a later time.

The PISPPI-2 requires little maintenance beyond flushing the pressure transducer prior to storage, and charging the battery pack. The software which controls the operation of the instrument is written in BASIC, and may be easily modified by the user. Any MS-DOS or MacIntosh computer can be used to download programs to the PISPPI and to recover data collected by the instrument.

The advantages of the PISPPI relative to previous piezometers include:

(1) miniaturization of all components means that the instrument can by deployed readily from a small boat.
(2) the electronic data-logger can be programmed to collect data rapidly for short-duration deployments or slowly for deployments up to 3 weeks.
(3) the reusable probe obviates the need for permanent or semi-permanent driven pipes and well-points required by normal piezometers. Rapid deployment is enabled using a probe connected to the rest of the system via flexible, hard plastic hydraulic tubing.
(4) the small diameter of the probe produces a relatively small pressure pulse that rapidly decays to background, allowing rapid deployment.

(5) for testing a wide range of moderate to low permeability sediments, a pump is connected to the system (Fig. 1) to introduce a pressure pulse for determining the hydraulic conductivity
(6) for higher permeability sediments, the pump pulse dissipates too rapidly; replacing the pump by a reservoir allows for a longer duration test

The test procedure has five different events as shown in by the sample pressure response shown in Fig. 2 for Station 3 of Transect 2 (see Fig. 3 for location). The PISPPI probe was placed on the sediment to verify the excess pressure head (hydraulic head) is zero (zero test), e.g. verifies the probe measures hydrostatic pressure in the water column. Probe insertion results in a nonzero head spike (probe insertion pulse) followed by an induced head by pumping a small amount of groundwater through the probe using a boat-operated, hand-held pump (pump pulse). After the head stabilized at the positive residual pressure head, the probe was removed to verify h = 0 in the water column.

2.2. *Calculation of Seepage Velocity*

According to Darcy's Law, the seepage velocity is the product of hydraulic gradient and hydraulic conductivity. Assuming vertical flow occurs, the

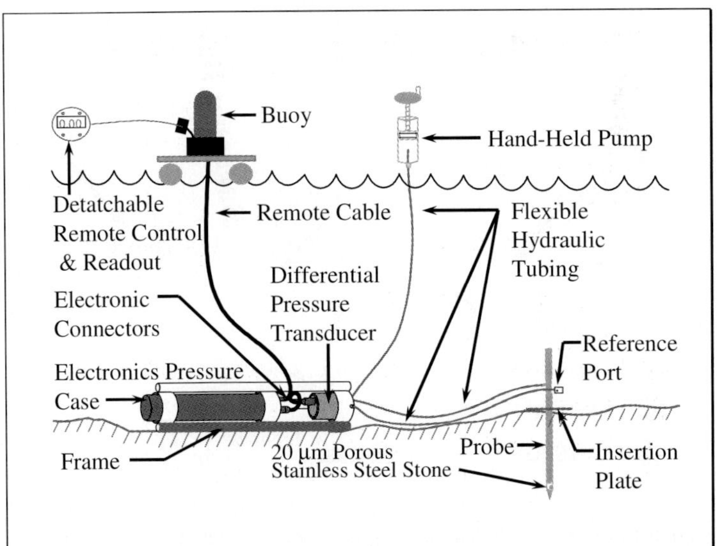

Figure 1. Schematic diagram of deployment of PISPPI with hand-held pump configuration.

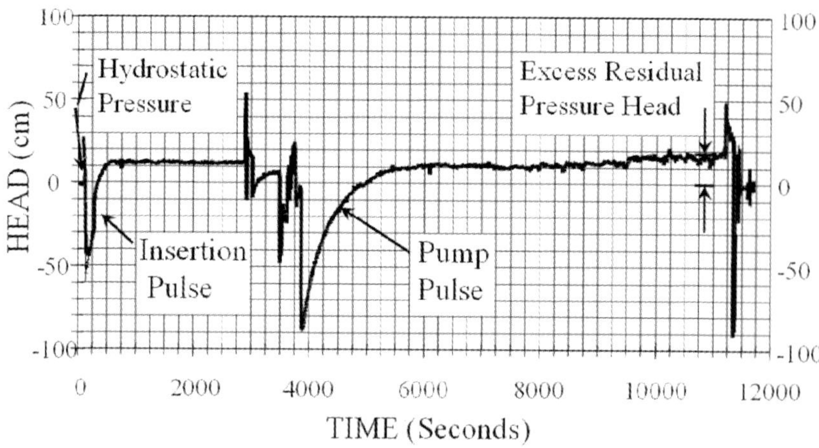

Figure 2. PISPPI raw data for Toms River Transect 2, Station 3. Test lengths vary from 2000 to 15,000 seconds depending on the type of sediment.

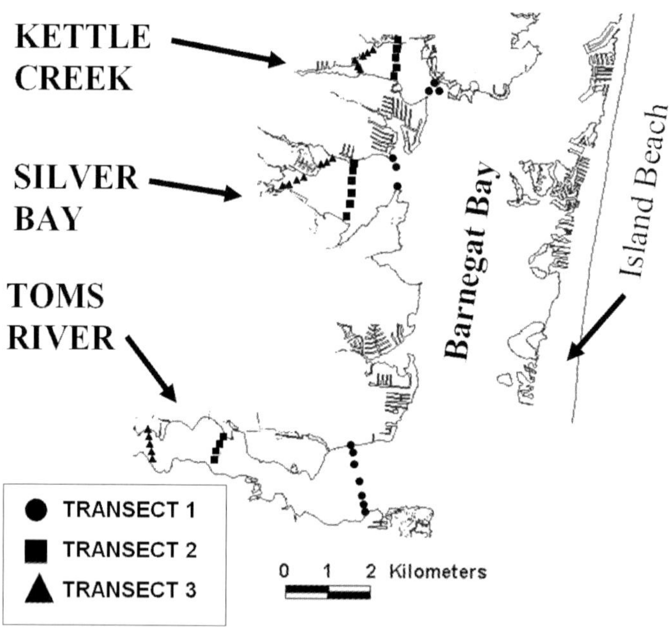

Figure 3. Location of the transects and stations across Kettle Creek, Silver Bay, and Toms River, New Jersey showing three transects in each, with Transect 1 to east, Transect 2 in the middle, and Transect 3 to the west.

hydraulic gradient is the hydraulic head divided by the depth of probe port in the sediment. The hydraulic head is the excess pressure head, e.g. the pressure divided by the unit weight of water; it is the value in excess of hydrostatic.

Instruments such as the ones used here measure differential pressures (pressure relative to the water column), so that the head is easily obtained. Some other instruments measure total pressure which includes hydrostatic pressure. The hydraulic conductivity (K), also known as coefficient of permeability, is related to the permeability (k) and the ratio of the water's viscosity and unit weight.

The hydraulic gradient (i) at each station was calculated by dividing the *positive excess residual pressure head* by the probe length of 53 cm. The hydraulic conductivity (K) is determined by extracting the pump pulse data, normalizing by the maximum head change from *residual*, and setting the zero at the *excess residual pressure head* (see Fig. 2), and matching to type curves [10, 11].

If the hydraulic conductivity is high, the dissipation of the pressure pulse may last for less than a minute making a reliable interpretation of hydraulic conductivity difficult. Because the water must fill the reservoir, the test duration is much longer than the pump pulse test. Using the probe geometry, a standard slug test analysis is performed [11].

2.3. *Shallow Water Measurement Locations*

The original testing of the PISPPI-2 in shallow water was conducted in Ground Sound, New Jersey, USA as reported by Wetzel [12]. Based on these tests, improvements were recommended because of long duration insertion and stress tests with a reservoir system resulted in testing time of hours at come location with low hydraulic conductivity. Although successful vertical seepage velocities were estimated at numerous locations based on hydraulic conductivity and gradient data, an overall discharge volume for Great Sound could was not estimated. Subsequently, the device was modified with the pump system show in Fig. 1.

The Kettle Creek embayment, about 2.4 million square meters, is located in the northwest part of Barnegat Bay (see Fig. 3). Measurements were taken at the three indicated transects. The location of each station was precisely determined using a Global Positioning System (GPS). Positive hydraulic heads were measured at all stations indicating discharge to the embayment as water moves from the recharge areas on land toward the embayment (see Table 1). The sandy Cohansey and Upper Kirkwood formations lie below the area, with a thin, very uniform layer of Holocene age silt and clay overlying the sandy aquifers at the sediment-water interface. Because of its low hydraulic conductivity, the layer significantly inhibits flow and probably encourages water to discharge through concentrated discharge points or displaces discharge eastward into the bay where coarser sediments are exposed at the surface.

A similar analysis was performed on Silver Bay and Toms River estuary (see Fig. 3).

2.4. *Deep Water Measurement — Hole 892B*

Hole 892B installed as part of the Ocean Drilling Program (ODP) Leg 146 at a depth of 674 m [7, 8] off the coast of Oregon. The site is on an active accretionary prism thrust fault where the open interval intersects a fault identified by seismic imaging 107 to 113 m below the seafloor, identified as contributing concentrated flows at subaqueous vent sites. The CORK well head system is shown in Fig. 4, allowing access to the well through the hydraulic port to pump water from the well, which is sealed at the surface. The open-hole interval from the bottom of the casing cement at 93.6 m below sea floor (mbsf) to the bottom of the hole (178.5 mbsf) was tested from the Alvin 10 months after drilling. The site is complicated by tidal fluctuations that must be factored out.

The constant-rate discharge tests were conducted with a positive-displacement pump powered by Alvin to create a stress on the formation above that of the in situ overpressurization. The most successful constant-rate pumping test was conducted on Oct 3, 1993 with a discharge rate of $5.0 \cdot 10^{-6}$ m^3/s for a duration of approximately 2 hours, and then shut in for approximately 2 hours and the recovery monitored and matched to recovery pump curves. The recovery data matched to the modeled recovery curve is shown in Fig. 6, identifying a transmissivity of $3 \cdot 10^{-7}$ m^2/s. Additional details are provided in Screaton et al. [7].

3. Results

3.1. *Shallow Water — Barnegat Bay*

Sediment cores were taken to allow comparison of the calculated hydraulic conductivity to laboratory (permeameter) values [6]. At Kettle Creek Station 4 in Transect 1, the calculated hydraulic conductivity was about twice the laboratory value, and ranged from 20% to 96% of the laboratory value at the other five stations where cores were taken. The accuracy of the hydraulic conductivity tests is generally considered to be an order of magnitude.

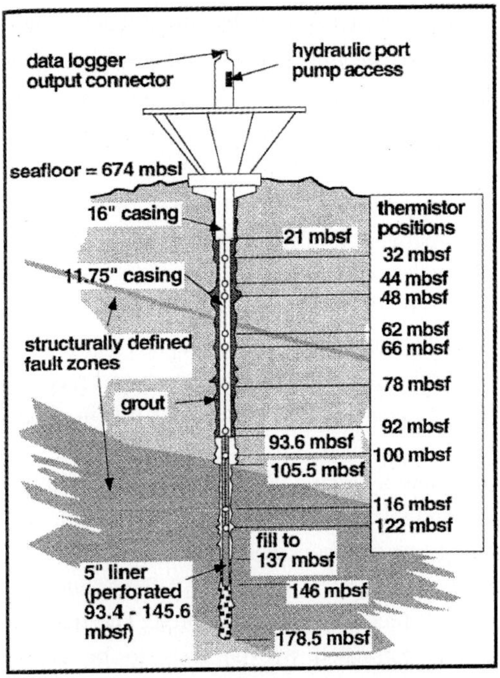

Figure 4. Schematic of Hole 892B sealed borehole and downhole instrumentation, including the CORK (Circulation obviation retrofit kit) well system.

The average flux rates at Transects 1, 2, and 3 are about $1.8{\cdot}10^{-8}$, $1.8{\cdot}10^{-8}$, and $4{\cdot}10^{-9}$ cm/s, respectively. In the embayment at the mouth of Kettle Creek, using a median value of seepage velocity of about $3{\cdot}10^{-10}$ m/s over the approximately $2.4{\cdot}10^{6}$ m^2 of embayment area yields about 0.7 L/s (60 m^3/day) of diffuse discharge flowing up into the embayment.

Table 1 suggests that the gradient is highest at the back of the embayment (Transect 1), and decreases toward the mouth of Kettle Creek (Transect 3). As summarized in Table 1, at each station seepage velocity ($v = Ki = Q/A$) is calculated based on the gradient and hydraulic conductivity. The ground water discharge to Kettle Creek is highest ($> 1{\cdot}10^{-7}$ cm/s) adjacent to the western shoreline and along the southern shoreline ($1{\cdot}10^{-6}$ cm/s to $5{\cdot}10^{-6}$ cm/s) just offshore from the highest topographic point adjacent to the bay. Discharge to the remainder of the bay is less than $1{\cdot}10^{-8}$ cm/s.

Tables 2 and 3 provide similar results for Silver Bay and Toms River estuary sites. Similarly, for the Silver Bay and Toms River estuary the estimated diffuse discharge was estimated to be 0.6 and 1.5 L/s (50 and 130 m^3/day), respectively. These measurements provided an estimate of the diffuse discharge, and did not include areas of concentrated flow, such as localized underwater springs.

3.2. *Deep Water — Hole 892*

Initially slug tests were performed from the deck of the drilling ship, and after letting the pressures recover from disturbances of drilling for 10 months (see Fig. 5), Alvin submersible-based pump tests were conducted [7, 8]. Shut-in pressures varied from 0.0073 to 0.0186 MPa above the pressure at the sediment interface, which was approximately 6.8 MPa and subject to tidal variations. Results are presented for a constant-rate discharge and recovery tests conducted using a positive-displacement pump powered by Alvin in order to induce a stress on the formation.

These tests determine hydraulic conductivity (K) and transmissivity (T), the latter a product of K and thickness of the fractured interval. The most successful constant-rate pumping test was conducted for a flow rate of $5.0 \cdot 10^{-6}$ m^3/s for 7410 s (approximately 2 hours). The hole was then shut in for approximately two hours during which recovery was monitored. The transmissivity (T) of the fractured interval was estimated by the tests to be 2 to $3 \cdot 10^{-7}$ m^2/s. By determining the pressure in excess of hydrostatic over the 10 month period, the flow rate contributed by this fault to the seafloor vent site can be estimated. These hydrogeological tests provided the first in situ estimates of fault-zone transmissivities within a modern, hydrogeologically active accretionary prism fault zone, and indicated that fracture dilation is necessary for this fault zone to act as a significant fluid conduit.

4. Summary

Hydraulic conductivity of the shallow sediments in Kettle Creek sediments showed only slight variations between stations. Positive in situ pressures were found at all PISPPI insertion locations, indicating ground water discharge into the bay. Additional stations would be helpful to better define the variation of seepage velocity. Although consistent data were found in all three study areas, because these values are lower than expected based on prediction flow rates by a numerical groundwater flow model, it is likely that significant additional concentrated discharge occurs in areas where the low permeability surficial layer is absent and from subaqueous springs or seeps. These areas of concentrated flow will be located and measured in the future to obtain a total flow rate, which can then be compared to groundwater flow model predictions.

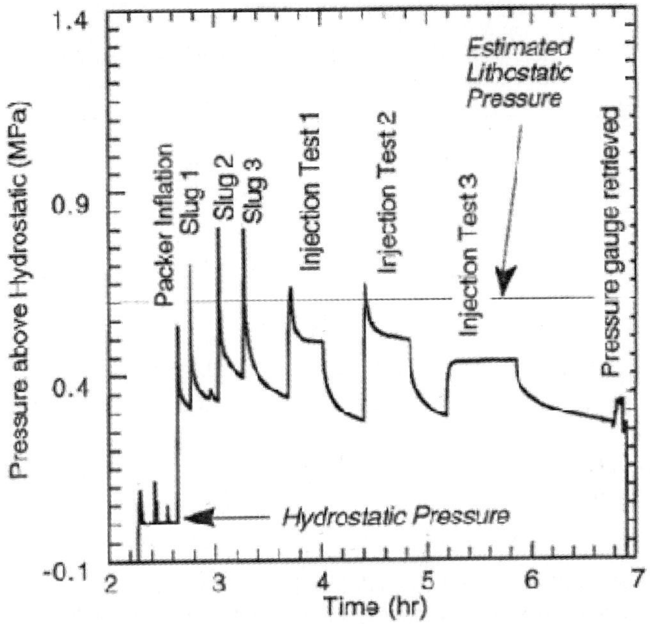

Figure 5. Slug test data.

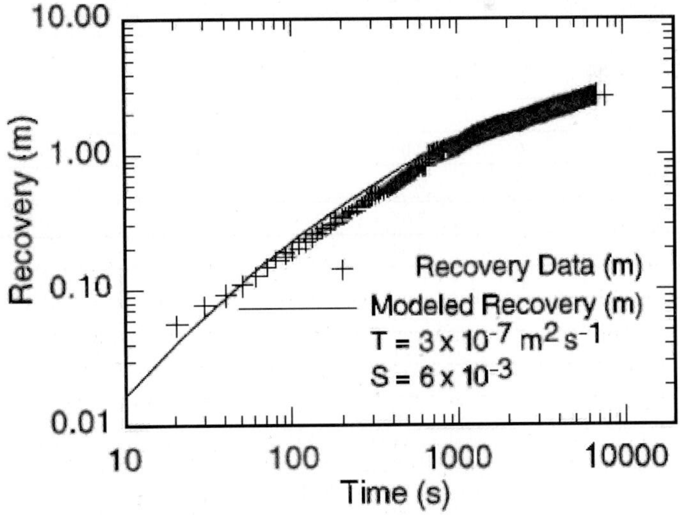

Figure 6. Match of pump test data to recovery curve for the Oct. 3, 1993.

Table 1. Estimated seepage velocity at measured points in the Kettle Creek embayment; transects 1, 2, and 3 are at the back, middle, and mouth of the embayment, respectively; stations are ordered from south to north (see Fig. 3). The depth of port insertion is 53 cm.

Transect	Station	h, cm	i	K, cm/s	v, m/s
1	1	0.5	0.01	$3.3 \cdot 10^{-6}$	$3.3 \cdot 10^{-8}$
	2	11.5	0.22	$1.8 \cdot 10^{-7}$	$4.0 \cdot 10^{-8}$
	3	3.3	0.06	$6.6 \cdot 10^{-8}$	$4.0 \cdot 10^{-9}$
	4	5.5	0.10	$1.3 \cdot 10^{-6}$	$1.3 \cdot 10^{-7}$
	5	13.3	0.25	$9.0 \cdot 10^{-8}$	$2.3 \cdot 10^{-8}$
	6	0.4	0.01	$2.4 \cdot 10^{-7}$	$2.4 \cdot 10^{-9}$
2	5	6.1	0.11	$7.2 \cdot 10^{-7}$	$8.0 \cdot 10^{-8}$
	4	8.7	0.16	$1.8 \cdot 10^{-7}$	$2.9 \cdot 10^{-8}$
	3	3.1	0.06	$1.8 \cdot 10^{-7}$	$1.1 \cdot 10^{-8}$
	2	5.6	0.11	$2.4 \cdot 10^{-7}$	$2.6 \cdot 10^{-8}$
	1	4.2	0.08	$4.0 \cdot 10^{-8}$	$3.2 \cdot 10^{-9}$
3	1	0.6	0.01	$4.2 \cdot 10^{-7}$	$4.2 \cdot 10^{-9}$
	2	1.3	0.02	$1.8 \cdot 10^{-7}$	$3.6 \cdot 10^{-9}$
	3	1.8	0.03	$1.0 \cdot 10^{-7}$	$3.0 \cdot 10^{-9}$

Table 2. Estimated seepage velocity at each measuring point in Silver Bay; transects 3, 2, and 1 are at the back, middle, and mouth of the embayment, respectively, and stations are ordered from north to south, as shown in Fig. 3.

Transect	Station	h, cm	i	K, cm/s	v, m/s
3	1	0.40	0.0085	$4.4 \cdot 10^{-7}$	$3.7 \cdot 10^{-11}$
	2*	0.89	0.005	$7.1 \cdot 10^{-9}$	$3.3 \cdot 10^{-13}$
	3	0.36	0.008	$1.3 \cdot 10^{-6}$	$1.0 \cdot 10^{-10}$
	4	4.49	0.096	$1.2 \cdot 10^{-7}$	$1.1 \cdot 10^{-10}$
	5	0.26	0.006	$3.2 \cdot 10^{-7}$	$1.8 \cdot 10^{-11}$
	6	0.04	0.001	$9.8 \cdot 10^{-8}$	$8.3 \cdot 10^{-13}$
	7	0.01	0.0002	$3.4 \cdot 10^{-7}$	$7.2 \cdot 10^{-13}$
2	1	2.03	0.043	$5.9 \cdot 10^{-7}$	$2.5 \cdot 10^{-10}$
	2	0.67	0.014	$3.0 \cdot 10^{-7}$	$4.3 \cdot 10^{-11}$
	3*	6.11	0.032	$2.2 \cdot 10^{-8}$	$7.1 \cdot 10^{-11}$
	4	2.85	0.061	$3.5 \cdot 10^{-7}$	$2.1 \cdot 10^{-10}$
	5	4.21	0.090	$5.1 \cdot 10^{-7}$	$4.6 \cdot 10^{-10}$
1	1	1.31	0.028	$6.6 \cdot 10^{-10}$	$1.8 \cdot 10^{-11}$
	2	0.70	0.015	$5.2 \cdot 10^{-10}$	$7.8 \cdot 10^{-13}$
	3	1.25	0.027	$1.4 \cdot 10^{-10}$	$3.8 \cdot 10^{-10}$

*A 190 cm long probe was used; other stations are based on a 47 cm port insertion depth.

Table 3. Estimated seepage velocity at each measuring point in the Toms River estuary; transects 3, 2, and 1 are at the back, middle, and mouth of the embayment, respectively, and stations are ordered from north to south as shown in Fig. 3. A dash indicates the data could not be interpreted.

Transect	Station (from south)	h, cm	i	K, cm s^{-1}	v, m s^{-1}
3	1	3.57	0.076	$2.0 \cdot 10^{-7}$	$1.5 \cdot 10^{-10}$
	2*	2.47	0.013	$6.6 \cdot 10^{-8}$	$8.9 \cdot 10^{-12}$
	3	2.03	0.043	$1.7 \cdot 10^{-6}$	$7.3 \cdot 10^{-10}$
	4	0.5	0.011	$4.6 \cdot 10^{-7}$	$4.9 \cdot 10^{-11}$
	5	2.5	0.053	$1.5 \cdot 10^{-6}$	$8.0 \cdot 10^{-10}$
2	1	2.84	0.060	$1.2 \cdot 10^{-7}$	$7.3 \cdot 10^{-11}$
	2	2.53	0.054	$1.3 \cdot 10^{-7}$	$7.0 \cdot 10^{-11}$
	3*	17.61	0.096	$5.2 \cdot 10^{-8}$	$5.0 \cdot 10^{-11}$
	4	2.49	0.053	$2.8 \cdot 10^{-7}$	$1.5 \cdot 10^{-10}$
1	1	—	—	—	—
	2*	0.82	0.004	$1.1 \cdot 10^{-7}$	$4.7 \cdot 10^{-12}$
	3	0.26	0.006	$2.0 \cdot 10^{-7}$	$1.1 \cdot 10^{-11}$
	4	0.53	0.011	$1.5 \cdot 10^{-7}$	$1.7 \cdot 10^{-11}$
	5	2.07	0.044	$3.5 \cdot 10^{-8}$	$1.5 \cdot 10^{-11}$
	6	0.02	0.0004	$2.4 \cdot 10^{-6}$	$1.0 \cdot 10^{-11}$
	7	2.13	0.045	$2.0 \cdot 10^{-6}$	$9.1 \cdot 10^{-10}$

*A 190 cm long probe was used; other stations are based on a 47 cm depth to port.

Acknowledgments

This work was supported by grant OCE-9202603 of the National Science Foundations, USA and grant R/S 95001 of the New Jersey Sea Grant Consortium, National Oceanographic and Atmospheric Administration, USA.

References

1. M. J. Kennish, and R. E. Lutz (eds.), *Ecology of Barnegat Bay, New Jersey, Lecture Notes on Coastal and Estuarine Studies 6*, Springer-Verlag, New York (1984).
2. D. A. Cherkauer, and J. M. McBride, A remotely operated seepage meter for use in large lakes and rivers, *Ground Water* **26(2)**, 165-171 (1988).
3. B. Carson, E. Suess, and J. Strasser, Fluid flow and mass flux determinations at vent sites on the Cascadia margin accretionary prism, *J. Geophysical Res.* **95(B6)**, 8891-8897 (1990).

4. M. Watt, M. Johnson, and P. J. Lacombe, Hydrology of the unconfined aquifer system, Toms River, Metedeconk River, and Kettle Creek basins, New Jersey, 1987-90, *U. S. Geological Survey Water Resources Investigations Report*, United States Geological Survey, Denver, Colorado (1994).
5. R. S. Nicholson, and M. K. Watt, Simulation of groundwater flow in the unconfined aquifer system of the Toms River, Metedeconk River, and Kettle Creek basins, New Jersey, *U. S. Geological Survey Water Resources Investigation Report 97-4066*, United States Geological Survey, Denver, Colorado (1997).
6. G. P. Lennon, D. Allen, C. L. McIlvaine, and B. Carson, Measurements of groundwater flow into Kettle Creek embayment, New Jersey, *Proceedings of the XXVII Congress of the Intern. Soc. of Hydr. Res.*, San Francisco California, 28-33 (1997).
7. E. J. Screaton, B. Carson, and G. P. Lennon, Hydrogeologic properties of a thrust fault within the Oregon Accretionary Prism, *J. Geophys. Res.* 20025-20035 (1995).
8. E. J. Screaton, B. Carson, and G. P. Lennon, In-situ permeability tests at ODP Site 892: Characteristics of a hydrogeologically active fault zone on the Oregon Accretionary Prism, *Proc. Ocean Drilling Program Scientific Results* **146(1)**, 291-297 (1995).
9. R. H. Bennett, J. T. Burns, H. Li, C. M. Percival, and J. Lipkin, Pore-water pressure events during the in situ heat transfer experiment simulation: Piezometer probe technology, *Sandia National Labs Report SAND86-7172*, Albuquerque, New Mexico (1987).
10. J. D. Bredehoeft, and S. S. Papadopulos, A method for determining the properties of tight formations, *Water Resources Res.* **6(1)**, 233-238 (1980).
11. S. S. Papadopulos, J. D. Bredehoeft, and H. H. Cooper, Jr, On the analysis of 'slug test' data, *Water Resources Res.* **9(4)**, 1087-1089 (1973).
12. M. D. Wetzel, In Situ Determination of Hydraulic Conductivity and Hydraulic Head in Shallow Coastal Sediments, Thesis submitted in Partial Fulfillment of a Master of Science in Civil Engineering, Lehigh University, 1994.

A SHORT REVIEW OF CONFORMAL GRID GENERATION IN AN IRREGULAR AREA

JOHN WANG*, TING-KUEI TSAY† and FU-RU LIN‡

Civil Engineering, National Taiwan University
No. 1, Sec. 4, Roosevelt Road
Taipei City, 10617 Taiwan
**E-mail: d89521008@ntu.edu.tw*
†E-mail: tktsay@ntu.edu.tw
‡E-mail: r93521318@ntu.edu.tw

In this paper, a short review of grid generation methods is proposed. Demonstrations of a conformal grid generation by the boundary integral element method and the complex mapping technique are presented in detail.

To elaborate the grid generation method, discussions of the associated problems about conformality quality are included. Indicators of MDO, ADO, MAR, and AAR are adopted to examine the deviation from orthogonality and the aspect ratio in a grid system, so as to estimate the quality of the resulting grids.

In Sec. 2, illustrations of the establishment of an conformal grid generation on a simply-connected region is presented in a step-by-step way. Forward transformation from an irregular region to a hyper-rectangular region, from the hyper-rectangular region to a rectangular domain, grid generation on the rectangular domain, the backward transformation from the rectangular to the hyper-rectangle and from the hyper-rectangle to the original irregular region are exhibited.

In Sec. 3, several examples of the conformal grid generation for specific applications or further studies are showed, therefore the validity and applicability of this grid generation system are confirmed.

Overall, a new conformal grid generation technique considering the effect of the mathematical singularity in BIEM is presented. Present results of excellent grid quality may serve as a powerful tool in the studies of the boundary layer in CFD. The grids maintain conformal property even when they are close to the boundary in a nano-scale.

1. Introduction

To generate a grid system for further use of computational purposes, two major methods are often considered: differential equation methods and algebraic methods. The former produces grid systems by solving one or more PDEs that relate the physical domain and computational domain. On the other hand, the latter generates grids by applying interpolating techniques between the boundaries of the physical domain.[1–3]

Conformal grids, one of the orthogonal grids, are most wanted for its superior applicability. When a computation in an irregular region is transformed onto a rectangular computational region seeking for faster calculations, the more complex governing equations due to the transformations hinder the purpose. However, conformal mapping can remove the drawback by remaining the governing equation forms in the computational domain the same as that in the physical domain.

In this section, overviews for the orthogonal or near orthogonal grid generation methods are made. Several benchmark problems discussed in these previous articles are also taken for further studies in Sec. 3.

When a two-dimensional physical domain represented by Cartesian coordinates X-Y is mapped to the computational domain ξ-η, the metric tensor of the transformation, g_{ij}, is defined as

$$A = \begin{bmatrix} g_{11} & g_{12} \\ g_{21} & g_{22} \end{bmatrix} = \begin{bmatrix} \left(\frac{\partial X}{\partial \xi}\right)^2 + \left(\frac{\partial Y}{\partial \xi}\right)^2 & \left(\frac{\partial X}{\partial \xi}\frac{\partial X}{\partial \eta} + \frac{\partial Y}{\partial \xi}\frac{\partial Y}{\partial \eta}\right) \\ \left(\frac{\partial X}{\partial \xi}\frac{\partial X}{\partial \eta} + \frac{\partial Y}{\partial \xi}\frac{\partial Y}{\partial \eta}\right) & \left(\frac{\partial X}{\partial \eta}\right)^2 + \left(\frac{\partial Y}{\partial \eta}\right)^2 \end{bmatrix} \tag{1}$$

and the Laplace equations in the physical domain, $\nabla^2\xi = 0$, $\nabla^2\eta = 0$, are transformed to the computational domain

$$g_{22}\frac{\partial^2 X}{\partial \xi^2} - 2g_{12}\frac{\partial^2 X}{\partial \xi \partial \eta} + g_{11}\frac{\partial^2 X}{\partial \eta^2} = 0 \tag{2}$$

$$g_{22}\frac{\partial^2 Y}{\partial \xi^2} - 2g_{12}\frac{\partial^2 Y}{\partial \xi \partial \eta} + g_{11}\frac{\partial^2 Y}{\partial \eta^2} = 0 \tag{3}$$

When $g_{12} = 0$, the mapping is an orthogonal mapping. That is, the orthogonality of the grids can be reserved. If another condition, $g_{11} = g_{22}$, is satisfied, the grid transformation belongs to a conformal mapping.

In the past studies, people used to say that the lack of robustness or "stiffness" of the conformal mapping makes it ill-suited for applications. Various kinds of orthogonal or nearly orthogonal grid generation methods are therefore developed to obtain the balance between the orthogonality and the smoothness of the grids.

1.1. *TTM system*

The pioneers of the orthogonal grid generation are thought to be Thompson, Thames and Mastin.[4] They developed the best known approach to boundary-fitted orthogonal curvilinear with a pair of covariant elliptic equations. The milestone of the article in the grid generation was called TTM system.[5]

$$\frac{\partial^2 \xi}{\partial X^2} + \frac{\partial^2 \xi}{\partial Y^2} = P(\xi, \eta) \qquad \text{and} \qquad \frac{\partial^2 \eta}{\partial X^2} + \frac{\partial^2 \eta}{\partial Y^2} = Q(\xi, \eta) \tag{4}$$

the partial differential equation relates the Cartesian coordinates X-Y with the boundary-fitted coordinate, ξ-η, where, P, Q are chosen (essentially by trial and error) to achieve the orthogonality and control the spacing of coordinate lines.[1,2,6]

Due to functions P and Q, Eq. (4) is no longer a conformal mapping from X-Y plane to ξ-η plane. The governing equations after the transformation from X-Y plane to ξ-η plane are then derived

$$g_{22}\frac{\partial^2 X}{\partial \xi^2} - 2g_{12}\frac{\partial^2 X}{\partial \xi \partial \eta} + g_{11}\frac{\partial^2 X}{\partial \eta^2} + J^2\left(P\frac{\partial X}{\partial \xi} + Q\frac{\partial X}{\partial \eta}\right) = 0 \tag{5}$$

$$g_{22}\frac{\partial^2 Y}{\partial \xi^2} - 2g_{12}\frac{\partial^2 Y}{\partial \xi \partial \eta} + g_{11}\frac{\partial^2 Y}{\partial \eta^2} + J^2\left(P\frac{\partial Y}{\partial \xi} + Q\frac{\partial Y}{\partial \eta}\right) = 0 \tag{6}$$

where, $J = \frac{\partial X}{\partial \xi}\frac{\partial Y}{\partial \eta} - \frac{\partial X}{\partial \eta}\frac{\partial Y}{\partial \xi}$. J is the Jacobian of the metric tensor g_{ij}.

The functions P and Q are called "control functions" for they are used to control the grid spacing and clustering. They can be homogenous or non-homogenous all over the domain. Different functions of P and Q are studied for different types of grids.[2,7–9] However, they are not orthogonal, not to mention conformal despite that they can generate smooth grids for domains of various kinds of geometry.

1.2. *RL system*

The conformal mapping is the most favored mapping method, which conserves dually the orthogonality and the equality of the scale factors (so that a small square can be mapped into a square on the other plane). It best simplifies the governing equation after the transformation from one plane to another plane. However, it has ill-condition that makes it uneasy to find the relationship.[6] In case of only orthogonality is to be conserved but not conformality, that is, g_{12}, the off-diagonal component of the covariant metric tensor, is equal to zero. Or, the Beltrami equations are satisfied

$$\frac{\partial X}{\partial \eta} = -f\frac{\partial Y}{\partial \xi}$$

$$\frac{\partial Y}{\partial \eta} = f\frac{\partial X}{\partial \xi} \tag{7}$$

Equation (4) can be simplified to be

$$\frac{\partial}{\partial \xi}\left(f\frac{\partial X}{\partial \xi}\right) + \frac{\partial}{\partial \eta}\left(\frac{1}{f}\frac{\partial X}{\partial \eta}\right) = 0$$

$$\frac{\partial}{\partial \xi}\left(f\frac{\partial Y}{\partial \xi}\right) + \frac{\partial}{\partial \eta}\left(\frac{1}{f}\frac{\partial Y}{\partial \eta}\right) = 0 \tag{8}$$

where

$$f = \frac{h_\eta}{h_\xi} \tag{9}$$

and

$$h_\eta = \sqrt{g_{22}} = \sqrt{\left(\frac{\partial X}{\partial \eta}\right)^2 + \left(\frac{\partial Y}{\partial \eta}\right)^2}$$

$$h_\xi = \sqrt{g_{11}} = \sqrt{\left(\frac{\partial X}{\partial \xi}\right)^2 + \left(\frac{\partial Y}{\partial \xi}\right)^2} \tag{10}$$

Equation (8) is called the "RL system" in memory of Ryskin and Leal, the authors of another article of milestone in grid generation.[6] It is noted that if the orthogonal factor g_{12} in Eqs. (5) and (6) is zero and the following control functions are set, Eqs. (5) and (6) can lead to the RL system.[6]

$$P = \frac{1}{h_\xi h_\eta} f_\xi, \quad Q = \frac{1}{h_\xi h_\eta}\left(\frac{1}{f}\right)_\eta \tag{11}$$

In Eq. (8), f is the so-called distortion function, which controls the space of grid-lines and the length scales of the transformation. Many studies focus on how to obtain the distortion function f. However, it is beyond the scope of this paper. The purpose of this section is to show that the establishment of a conformal grid system, with the aid of complex mapping, is in some cases not difficult at all. In addition, the two major types of singularities in the boundary integral element method: mathematical singularities and geometrical singularities, as mentioned in the previous studies,[10] playing important roles in the grid generations are discussed in this article.

1.3. *Other grid generation systems*

The other grid generation systems can be concluded to be either the type of TTM system or RL system mentioned above or the blending of the two systems. The difference of the available methods is in the way the distortion function f is obtained.

1.3.1. *2D orthogonal grid generation with boundary point distribution control by Luis (1996)*[11]

In this method, the governing equations are the same as RL system. The topic of this method is the calculation of the grid cell aspect ratio, the so-called distortion function f. To solve the f from the nonlinear Eq. (8), procedures are taken as follow

(1) Use algebraic interpolation to obtain the initial approximation of the grids. In this model, the linear interpolation is applied.
(2) Calculate the distortion function from the available grid using Eqs. (9) and (10).
(3) With fixed value f, grids are solved iteratively via Eq. (8).
(4) Define the convergence criteria. Repeat steps 2 and 3 if the criteria is not satisfied.

The above procedures are almost the same for the many studies on calculation of function f.[8,12] However, if the sliding boundary conditions (Neumann-Dirichlet) is applied, relocating of the boundary nodes to satisfy the orthogonality is necessary between steps 3 and 4. Sliding boundary conditions are the boundary conditions adopting Eq. (7) and the equation that describes the boundary (including Dirichlet boundary condition).

1.3.2. *Orthogonal grid generation with floating boundary points by Jeng (1999)*[12]

A fast trial and error method for the boundary grid distribution was suggested by Jeng in 1999.[12] A least square formulation of the floating-boundary-point method for two-dimensional orthogonal grid generation was proposed. The least square method minimizes the deviation from the Beltrami equation, so as to obtain the best orthogonality and smoothness in the grids.

In Jeng's paper, the procedures of the calculation for distortion function f is basically the same as that of Luis (1996).[11] However, when a sharp internal corner is met, the method developed in his method is applied to search for the best balance between the smoothness and the orthogonality at the corner.

1.3.3. *Nearly orthogonal grid system by Akcelik (2001)*[8]

In 2001, Akcelik developed a nearly orthogonal two-dimensional grid system by adding a force constant in the control function P and Q to prevent grid lines from collapsing onto each other.

$$\frac{\partial}{\partial\xi}\left(f\frac{\partial X}{\partial\xi}\right)+\frac{\partial}{\partial\eta}\left(\frac{1}{f}\frac{\partial X}{\partial\eta}\right)+P_X(h_\xi)+Q_X(h_\eta)=0$$

$$\frac{\partial}{\partial\xi}\left(f\frac{\partial Y}{\partial\xi}\right)+\frac{\partial}{\partial\eta}\left(\frac{1}{f}\frac{\partial Y}{\partial\eta}\right)++P_Y(h_\xi)+Q_Y(h_\eta)=0 \tag{12}$$

where

$$P(h_\xi)=c\left(h_\xi-\frac{\bar{h_\xi}^2}{h_\xi}\right)$$

$$Q(h_\eta)=c\left(h_\eta-\frac{\bar{h_\eta}^2}{h_\eta}\right) \tag{13}$$

$\bar{h_\xi}$ and $\bar{h_\eta}$, the mean scale factors, are the averaged h_ξ and averaged h_η over the entire domain. The subscript X and Y for P, Q represent the quantity in the X-direction and Y-direction.

It is noted that as the positive force constant c in Eq. (13) increases, the method resembles conformal mapping. When it is zero constant, the method reduces to the RL system.[8] The governing equations are solved by Finite Difference Method (FDM) and the Dirichlet or Neumann-Dirichlet boundary conditions are imposed. For the Neumann-Dirichlet boundary conditions, the positions of the nodes should be adjusted according to the orthogonality characteristics of the domain to maintain the grid quality on the boundaries.

1.3.4. *2D nearly orthogonal mesh generation by Zhang (2004)*[9]

Two methods for the generation of 2D nearly orthogonal grids were proposed by Zhang in 2004. They were developed to improve the distortion or overlapping at a complex geometry, particularly in those with sharp corners or strong curvatures. One method is developed by introducing control functions into the RL system. The method is formulated through the blending of the conformal mapping and the orthogonal mapping.

$$g_{22}\frac{\partial^2 X}{\partial\xi^2}-2g_{12}\frac{\partial^2 X}{\partial\xi\partial\eta}+g_{11}\frac{\partial^2 X}{\partial\eta^2}+J^2\left[(1-\gamma_p)P\frac{\partial X}{\partial\xi}+(1-\gamma_q)Q\frac{\partial X}{\partial\eta}\right]=0 \tag{14}$$

$$g_{22}\frac{\partial^2 Y}{\partial\xi^2}-2g_{12}\frac{\partial^2 Y}{\partial\xi\partial\eta}+g_{11}\frac{\partial^2 Y}{\partial\eta^2}+J^2\left[(1-\gamma_p)P\frac{\partial Y}{\partial\xi}+(1-\gamma_q)Q\frac{\partial Y}{\partial\eta}\right]=0 \tag{15}$$

where γ_p in $[0,1]$ and γ_q in $[0,1]$ are effect-control factors in the ξ and η directions, respectively. Compare the above equations with the RL system and the TTM system, definitions of γ_p and γ_q are settled

$$\gamma_p = \frac{|\bar{h_\xi} - h_\xi|}{\bar{h_\xi}}$$

$$\gamma_q = \frac{|\bar{h_\eta} - h_\eta|}{\bar{h_\eta}} \tag{16}$$

where $\bar{\gamma_p}$ and $\bar{\gamma_q}$ are the locally averaged scale factors in the ξ and η directions, respectively. It can be observed that if $\gamma_p = \gamma_q = 1$ $(h_\xi = h_\eta = 0)$, Eqs. (14) and (15) are reduced to the Laplace equations for the conformal mapping. On the other hand, if $\gamma_p = \gamma_q = 0$ $(h_\xi = \bar{h_\xi}, h_\eta = \bar{h_\eta})$, Eqs. (14) and (15) turn to the Poisson equation for the orthogonal equation. γ_p and γ_q can be viewed as the deviation from smoothness. As the values increase, the adjustments for smoothness also increase by means of Eqs. (14) and (15) to reach the balance.

The other method is that the contribution factors are introduced into the finite difference equation of a modified RL system (adding two sources in the RL equations). Through the contribution factors, the divergence of the modified RL equations, which correspond to the overlapping of grids, can be avoided. The factors are the exponent of distance of the grid point to the surrounding points. When the distortion of f at one neighbor is high, which implies that distance between them is small, the contribution factors of that neighbor is small as well. This will decrease the contribution of the high f from the neighbor node. The convergence of the FDM is then ascertained.

After establishment of conformal grid generation method in next section, four benchmark problems are selected to demonstrate the present conformal mapping. To illustrate the applicability and validity of present results, comparisons with the grids produced by Luis,[11] by Jeng,[12] by Akcelik,[8] and by Zhang[9] are made in the following section.

2. Illustrations for the establishment of a conformal grid generation on a simply-connected region

Numerical computations can be divided into three steps: preprocessing, calculation and post processing. In most cases, it is tedious and time-consuming to generate grids by manpower. Therefore, this study intends to develop a highly effective tool in numerical grid generation with the boundary-fitted orthogonal grid for a two-dimensional arbitrary domain. It can improve speed and accuracy of numerical calculations. This study also intends to improve a method of a boundary-fitted orthogonal coordinate system.[13]

In 1983, Liggett and Liu[14] solved the Laplace's equation with BIEM to obtain the potential and potential derivatives on the nodal points, and then applied the same boundary integral equation again with the calculated potential and potential derivatives to compute the unknown potential and potential derivatives in the domain. By doing this, the BIEM not only reduces the required solutions by one dimension to those on the boundaries, but also avoids the repetition of solving the Laplace equations if the generated co-ordinate system is not desirable. The linear shape functions of the boundary element has overcome the singular difficulty usually encountered in numerical boundary integration approach. However, they also reported that there exists a boundary layer in some numerical computations.

In the past, some studies showed that BIEM can be used to generate grids.[1] Combining the Cauchy-Riemann's condition, the Laplace equation can be used to perform a conformal mapping in a hyper-rectangular region, which is composed of four smooth boundaries and four right angles. The grids can be generated numerically by a set of forward and backward conformal mapping with BIEM.[2]

Furthermore, Tsay *et al.*[15,20] applied this technique in solving the mild-slope equation to describe the propagations of water-waves and tides. In 1997, Tsay and Hsu[13] produced a series of grid system in the irregular simply-connected region by adopting a sequential Z^n complex mapping to solve the problem encountered when the physical region is no longer hyper-rectangular and orthogonal conditions on the four corners are not satisfied.

Recent studies by Wang and Tsay,[10] based on the previous results by Liggett and Liu (1983),[14] successful resolved the fundamental numerical boundary layer problem in the traditional BIEM technique. In contrast to conventional numerical integration methods which suffer from the numerical boundary layer, the mathematical singularities can be shown to vanish when the integration limits are taken properly, Fig. 1. This re-examination of the limits of the angle terms derives the conclusion that the numerical boundary layer is in fact non-existent. Therefore, one can produce the orthogonal grids all over the domain, even the locations very close to the boundary (i.e. within 10^{-14} unit length), without losing its conformality.

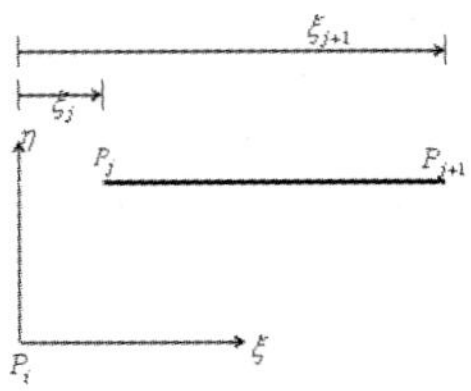

$$\left(\tan^{-1}\frac{\xi_{j+1}}{\eta_i} - \tan^{-1}\frac{\xi_j}{\eta_i}\right)$$

		Wang and Tsay(2005)	Liggett & Liu(1983)
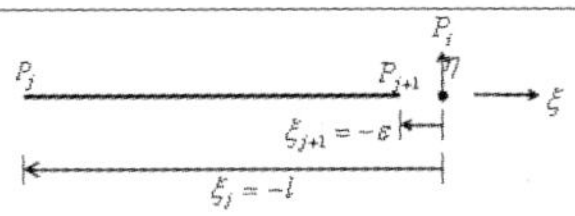	$\tan^{-1}\frac{\xi_{j+1}}{\eta_i}$ $\tan^{-1}\frac{\xi_j}{\eta_i}$	$-\frac{\pi}{2}$ $-\frac{\pi}{2}$	0 0
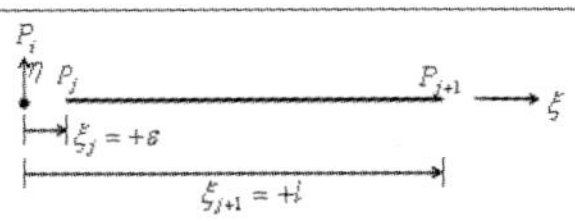	$\tan^{-1}\frac{\xi_{j+1}}{\eta_i}$ $\tan^{-1}\frac{\xi_j}{\eta_i}$	$\frac{\pi}{2}$ $\frac{\pi}{2}$	0 0
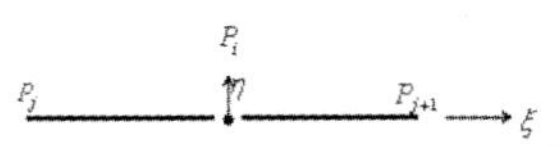	$\tan^{-1}\frac{\xi_j}{\eta_i}$ $\tan^{-1}\frac{\xi_{j+1}}{\eta_i}$	$\frac{\pi}{2}$ $-\frac{\pi}{2}$	0 0

Fig. 1. Comparisons of the results by Wang and Tsay (2005)[10] with those by Liggett and Liu (1983)[14] in the evaluations of the angle limits in BIEM.

To illustrate the grids generation in a simply-connected region, the following procedures are taken to describe this method step-by-step. Also, an example of grid generation at Dongshih of Taiwan is attached for illustration. Owing to the severe storm surge, Dongshih Harbor has been a benchmark example of flooding study in Taiwan. In Fig. 2, the upper and lower sides are the left-hand side, the artificial boundaries of the sea, while the right-hand side is the land. Dongshih Harbor is located at the south of a small peninsula.

2.1. *Forward transformation from an irregular region onto a hyper-rectangular region*

To play a role of a bridge between an irregular physical region and a hyper-rectangular computational region, a sequence of complex mapping is employed to serve for the purpose. A region between two straight lines with an angle of α in the Z-plane is to be mapped into an area with a desired

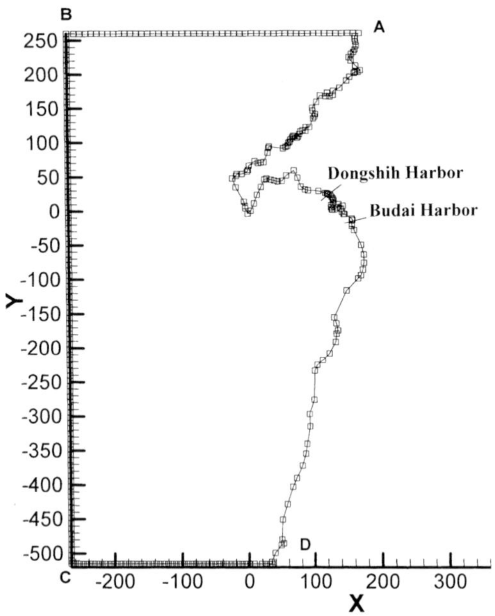

Fig. 2. Boundary-fitted orthogonal grids for a simply-connected domain: Dongshih Harbor case.

angle of β in the W-plane[13]

$$W - W_0 = \frac{1}{A}(Z - Z_0)^A, \quad A = \frac{\alpha}{\beta} \tag{17}$$

W_0 and Z_0 are the branch points on the complex plane of W and Z. The function of $\frac{1}{A}(Z - Z_0)^A$ is a multi-value function. A branch cut with the argument θ, $-\pi < \theta < \pi$, outside the domain is chosen so that the regions are continuous and the transformation function is a single-value function. By using the complex mapping as mentioned in Eq. (17), an irregular region on the physical domain (Fig. 2) is transformed to be a hyper-rectangular region with four right angles and four smooth curvilinear boundary lines (Fig. 3). One-by-one sequential transformation for angles on the domain of X-Y plane are implemented. Four angles on that domain are transformed to be $\frac{\pi}{2}$ and the other internal boundary angles are transformed to be π to shape a hyper-rectangular region on the U_1-V_1 plane as shown in Fig. 3. For comparison, the correspond footnotes of A, B, C, and D denoting the four right angles are shown on the two planes, respectively. It is noted that point A on the U_1-V_1 plane is really a right angle, though it is not easy

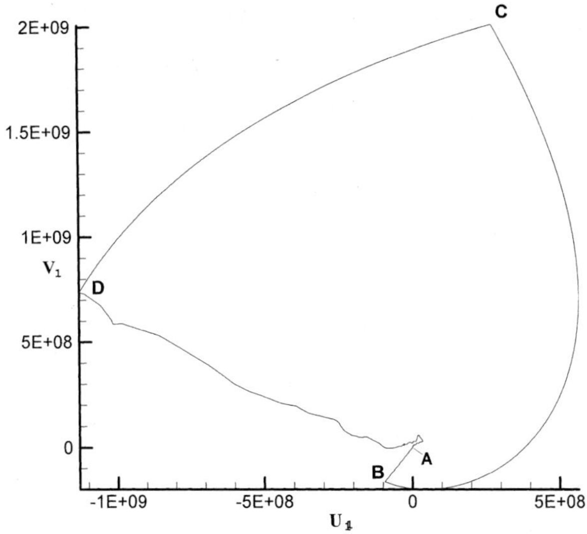

Fig. 3. A hyper-rectangular region after complex mapping: Dongshih Harbor case.

to be observed by naked eyes. At the meantime, in Fig. 3, the condense of the nodal distribution between curve A to D shows the geometrical singularity problem mentioned in the the paper by Wang and Tsay[10] should be overcome by adding points for both simulating the shape and smoothing the curve.

For a complex mapping from Z-plane to W-plane with a function $F(Z)$, the mapping will be conformal if the function $F(Z)$ is analytical,[17] and thus maintaining the angles during the mapping except the singular point Z_0. Therefore, one can perform the angle transformation one by one to produce four right angles on the transformed W-plane.

2.2. *Forward transformation from the hyper-rectangular region into a rectangular domain*

As shown in Fig. 4, a conformal transformation of a hyper-rectangular region, $ABCD$, in the X-Y plane into a rectangle, $A'B'C'D'$, in the ξ-η plane can be performed by solving

$$\frac{\partial^2 \xi}{\partial X^2} + \frac{\partial^2 \xi}{\partial Y^2} = 0 \qquad \text{and} \qquad \frac{\partial^2 \eta}{\partial X^2} + \frac{\partial^2 \eta}{\partial Y^2} = 0 \tag{18}$$

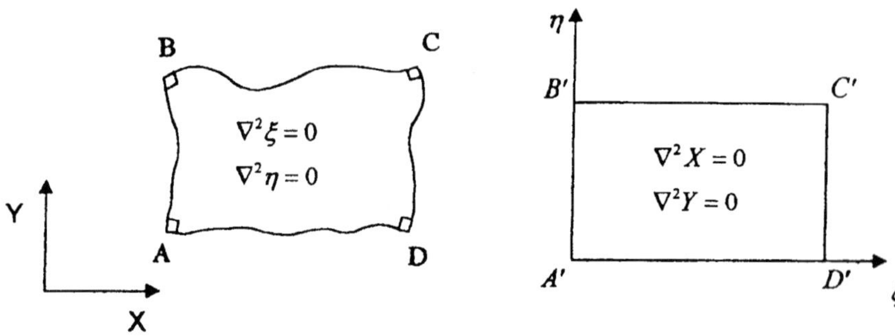

Fig. 4. Conformal mapping with BIEM from a hyper-rectangular domain to a rectangular domain.[13]

on the X-Y plane, with the boundary conditions derived from Cauchy-Riemann conditions.

$$\frac{\partial\xi}{\partial n}=0 \quad on \quad \eta=0 \quad \eta=\eta_0 \tag{19}$$

$$\frac{\partial\eta}{\partial n}=0 \quad on \quad \xi=0 \quad \xi=\xi_0 \tag{20}$$

where n is the outward normal direction of the boundaries.

Eqs. (19) and (20) correspond to those on the ξ-η plane imposed with 2 Dirichlet B.C. and 2 Neumann B.C. for solving $\nabla^2X=0$ and $\nabla^2Y=0$, respectively. To solve the Laplace equation as shown above, BIEM is applied

$$c(P)\Phi(P)=\int_\Gamma\left(\frac{\Phi(Q)}{r(P,Q)}\frac{\partial r(P,Q)}{\partial n}-\ln r(P,Q)\frac{\partial\Phi(Q)}{\partial n}\right)ds \tag{21}$$

P, Q denote the base point(source point) and the field point, respectively. Meanwhile, Φ can be either ξ or η. Also, $c(P)$ equals to π or $\frac{\pi}{2}$ in this case. In this way, the ξ and η for boundary nodal points of X-Y plane can then be obtained and so as to locate the corresponding nodal points on the ξ-η plane for the backward conformal mapping. The rectangular domain transformed from the hyper-rectangular domain of the example of the Dongshih harbor in the previous section is shown in Fig. 5. That is transformed by BIEM. In the figure, one can observe that the node distribution on the boundaries is not uniform. If the density of the nodal distribution is too samll, the adjustment of the nodal distribution on the original physical domain should be reconsidered to reduce the errors resulting from using linear elements in simulating irregular boundaries.

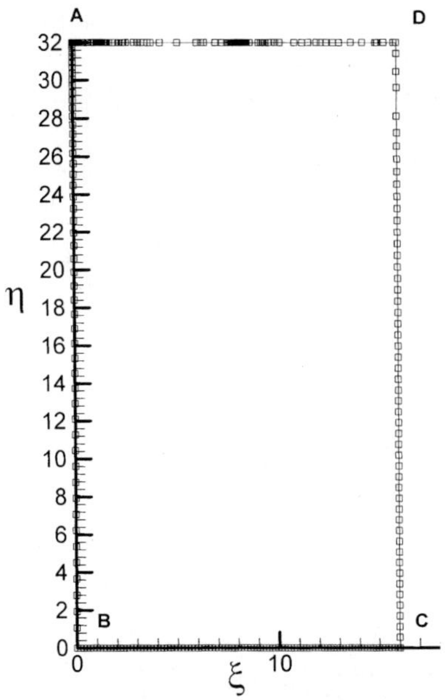

Fig. 5. Boundary nodal distribution on the rectangle in ξ-η plane after mapped by BIEM.

2.3. *Generation of grids on the rectangular domain*

Because the governing equations on Z- and W-plane are both Laplace equations, the conformal properties can then be expected. To generate an orthogonal grid system on Z-plane, we should generate orthogonal grids on the W-plane. It is very straightforward to generate grids on the rectangular region on W-plane. That is one of the reasons a rectangular region is chosen as the mapped region.

2.4. *Backward transformation from the rectangular region onto the physical domain*

Backward transformation from the rectangular region into the original physical domain can be achieved by solving the Laplace equation of X and Y with independent variables ξ and η. The governing equation is the same as Eq. (21) except the $\Phi = X$ and $\Phi = Y$. The Dirchlet boundary conditions on the nodal points along the boundaries of ξ-η plane are the corresponding values of X and Y on the X-Y plane. The derivatives on the boundary nodes can be obtained by using BIEM to solve $\nabla^2 X = 0$ and $\nabla^2 Y = 0$.

In this way, one can accomplish the calculation of X and Y for any point within the ξ-η domain. Then the grids on the original physical domain can be obtained by solving the grids on the ξ-η domain sequentially (Fig. 6).

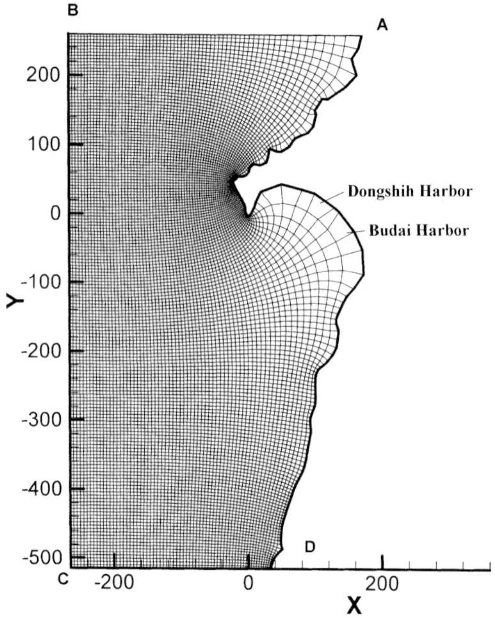

Fig. 6. Boundary-fitted curvilinear orthogonal grids on the waters aside Dongshih Harbor.

3. Studies of benchmark problems

In this section, several benchmark examples are selected to demonstrate the validity of present theory. To compare with the results by others, some indicators are adopted to examine the deviation from orthogonality and the aspect ratio. The maximum deviation from orthogonality (MDO), averaged deviation from orthogonality (ADO), maximum grid aspect ratio (MAR), and averaged grid aspect ratio (AAR) are used to study quality of the resulting grids.

$$MDO = max_{i,j}(|90^o - \vartheta_{i,j}|) \tag{22}$$

$$ADO = \frac{1}{m-2}\frac{1}{n-2}\sum_{i=2}^{m-1}\sum_{j=2}^{n-1}(|90^o - \vartheta_{i,j}|) \tag{23}$$

where the subscripts i and j denote the grid positions in the X- and Y-directions, respectively. The m and n are the maximum number of mesh lines in ξ and η directions, respectively; and ϑ is defined as

$$\vartheta_{i,j} = \cos^{-1}\left(\frac{g_{12}}{h_\xi h_\eta}\right) \tag{24}$$

$$MAR = max_{i,j}\left[max\left(f_{i,j}, \frac{1}{f_{i,j}}\right)\right] \tag{25}$$

$$AAR = \frac{1}{m-2}\frac{1}{n-2}\sum_{i=2}^{m-1}\sum_{j=2}^{n-1}\left[max\left(f_{i,j}, \frac{1}{f_{i,j}}\right)\right] \tag{26}$$

The $f_{i,j}$ is the value of the distortion function f at location (i, j). The MAR and AAR are used to evaluate the smoothness of a mesh.

When an orthogonal grid system is considered, that is, $g_{12} = 0$. It can be easily observed from Eqs. (24) and (22) that MDO and ADO are of the value 0. Moreover, in case of the conformal grid system is considered, another $g_{11} = g_{22}$, the MAR and AAR will go to the value of 1.

3.1. *Half donut*

As shown in Fig. 7, a region limited by two non-concentric half circles and the x-axis, $y = 0$, is presented. The small half circle has diameter of 1 and the larger one has a diameter of 3. In this case, Luis (1996)[11] pointed out that a particular choice of the boundary point distribution is required. The grids' collapse will be found except a different node distribution is specified along the half-circle of smaller diameter (Fig. 7). Also, Jeng[12] indicated that when uniform grid points were distributed throughout all boundaries, the floating-boundary-point method would generate a non-smooth grid system. Faced with this problem, Jeng (1999)[12] used floating-boundary-point method with 300 cycles of iterations by changing the distortion function f. However, in the present result, uniform boundary distribution on this example produces a smooth grid system (Fig. 8). It shows that the present method can be applied on a smooth boundary without difficulty. It is noted that the grids on the boundaries are generated and not necessarily coincide with the points used to describe the boundary geometry. To estimate the grid qualities of this case, in Table 1, one can find the four indicators are all optimal. Therefore the qualities of the grids are guaranteed.

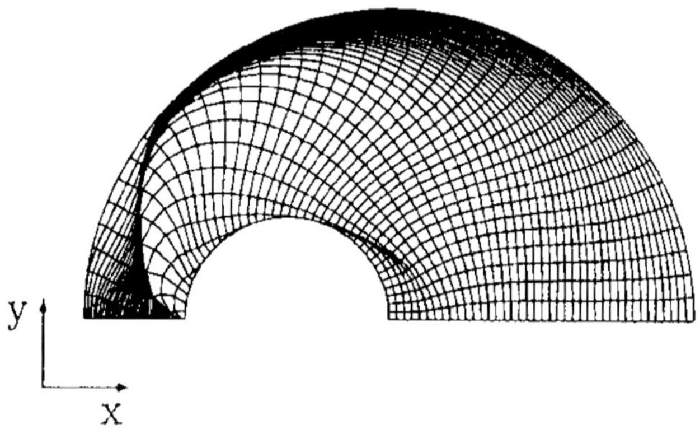

Fig. 7. Surface of discontinuity in a half donut (Luis, 1996).[11]

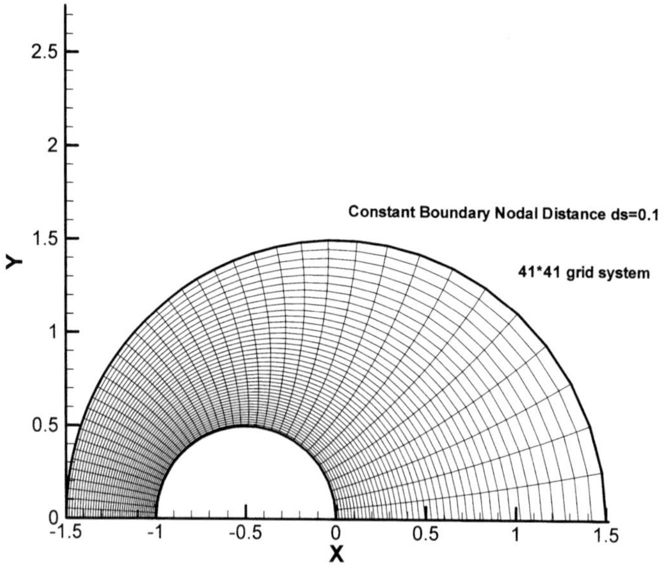

Fig. 8. 41 × 41 conformal grids for a half donut by present method.

Table 1. Evaluation of meshes of the half donut case.

	Figure	mesh size	MDO	ADO	MAR	AAR
Zhang[9]	Fig. 5 D(1)	41×41	3.70	0.62	5373.00	147.00
Zhang	Fig. 5 D(2)	41×41	5.10	1.32	17.2	5.24
Zhang	Fig. 5 D(3)	41×41	13.20	3.22	8.36	3.06
Zhang	Fig. 5 D(4)	41×41	13.90	4.24	8.40	3.04
Zhang	Fig. 5 D(5)	41×41	0.13	0.04	6.80	3.55
Akcelik[8]	Fig. 8 (a)	41×41	33.00	2.64	34.52	5.05
Akcelik	Fig. 8 (b)	41×41	1.02	0.35	7.66	3.44
Luis[11]	Fig. 2 (C)	41×41	71.20	1.44	N.A.	N.A.
Luis	Fig. 2 (D)	41×41	0.20	0.04	N.A.	N.A.
Present Results	Fig. 8	41×41	4.86×10^{-9}	4.86×10^{-9}	1.00	1.00

3.2. *A trapezoid-like region*

As shown in Fig. 9, a trapezoid-like region limited by two coordinate axes, $X = 0$, $Y = 0$, the lines $y = 1$ and $x = \frac{1}{2} + \frac{1}{6}\cos(\pi y)$ is presented.[8,11,18] Figure 9 presents 41×41 nodes in this domain. In 2001, Akcelik[8] calculated this benchmark problem and pointed out that the grids would overlap each other if the specified boundaries are used and grid points are equidistant. Applying the force constant to the grids will improve this particular situation, however, the orthogonality is worsen (Fig. 10). In present study, a nano-scale gap of 10^{-14} is built between the boundaries and the first

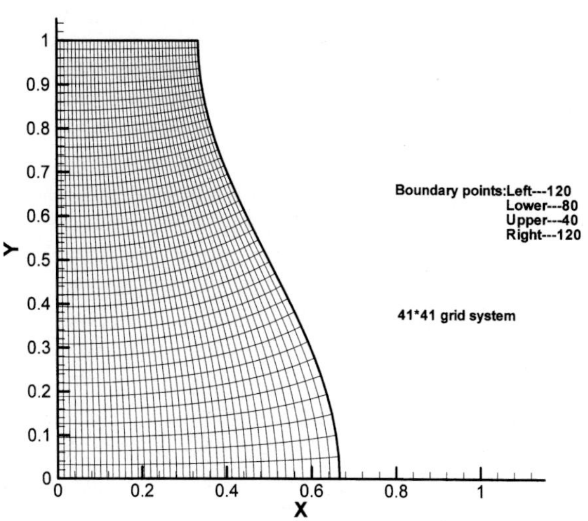

Fig. 9. A trapezoid-like region with 360 boundary points and 41×41 conformal meshes by present method.

inner grid. However, they are unlikely to be distinguished by naked eyes. In this paper, the numerical boundary-layer problems are not encountered and smooth, conformal grids are obtained (Fig. 9). It implies that present technique may be extended to nano-scale for some advanced studies.

In addition, the qualities of the grids can be observed from Table 2. In the table, four indicators are all the optimum, meaning the conformal grids are ensured.

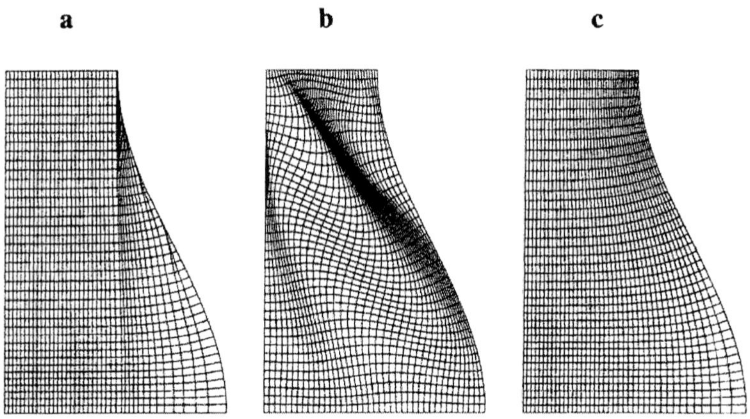

Fig. 10. Sliding boundaries are used for the straight sides in Grid a and c, while specified boundaries are used on the other sides. Force constants are 0 in Grid a, 0.01 in Grid b and c. Along the specified boundaries, grid points are equidistant, while at the left side $Y = \xi$.[8]

Table 2. Evaluation of meshes of the trapezoid-like region.

	Figure	mesh size	MDO	ADO	MAR	AAR
Zhang[9]	Fig. 4 C(1)	41×41	1.11	0.37	46.10	3.98
Zhang	Fig. 4 C(2)	41×41	1.41	0.65	12.4	2.88
Zhang	Fig. 4 C(3)	41×41	3.42	1.98	3.75	2.25
Zhang	Fig. 4 C(4)	41×41	3.28	2.02	3.76	2.25
Zhang	Fig. 4 C(5)	41×41	0.07	0.02	2.99	2.21
Akcelik[8]	Fig. 9 (b)	41×41	3.62	0.92	22.30	3.40
Akcelik	Fig. 9 (c)	41×41	0.50	0.04	6.13	2.26
Luis[11]	Fig. 9 (a)	41×41	0.20	0.04	N.A.	N.A.
Luis	Fig. 9 (b)	41×41	1.45	0.24	N.A.	N.A.
Luis	Fig. 9 (c)	41×41	5.67	0.34	N.A.	N.A.
Present Results	Fig. 9	41×41	4.86×10^{-9}	4.86×10^{-9}	1.00	1.00

3.3. *A four-semicircular region*

A unit square with a half circle on each side is proposed as a benchmark problem for discussion of grid generation. The grid nodes are equidistant along the boundaries. Luis (1996) stated that the maximum deviation from orthogonality occurs in the vicinity of the corners, where the angle between the grid lines is 270°.[11] Similarly, Zhang studied this problem in his paper in 2004.[9] As shown in Fig. 11, the mesh lines are contracted to the center-lines of domain B in X and Y directions. The grids generated by RL system as shown in Fig. 11(B1) are the worst smoothness. The other grid methods generate better grids by adding smoothness control function while sacrificing a little orthogonality. Nevertheless, the above mentioned methods are not able to resolve the predicament that the grids overlapping on the four corners.

In present method, adding elements around the four singular corners successfully minimize the errors efficiently and thus overcome this problem. After the improvement of the nodal distribution, no more overlapping at the four corners is found, Fig. 12.

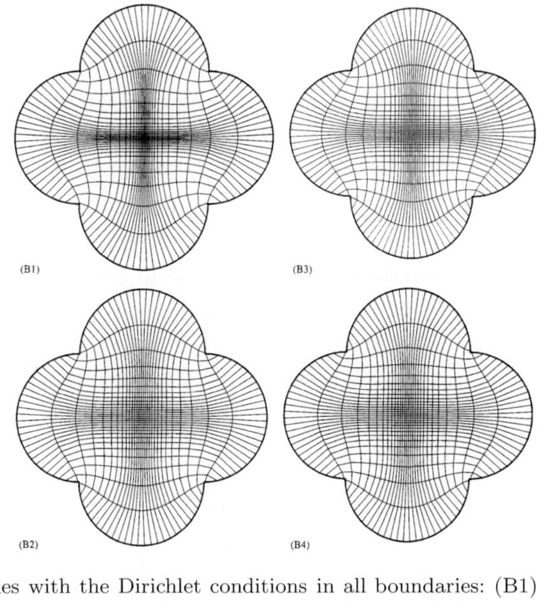

Fig. 11. Meshes with the Dirichlet conditions in all boundaries: (B1) the RL system; (B2) the RL system with smoothness control functions; (B3) the RL with contribution factors, $\alpha = 0.01$; and (B4) the hybrid RL system, $\alpha = 0.002$ and $r_c = 0.5$. (Zhang *et al.*, 2004)

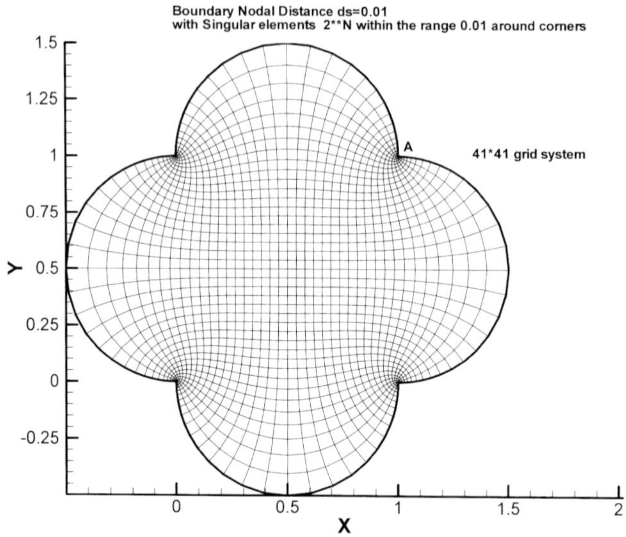

Fig. 12. Conformal grids on the physical X-Y domain with present method.

The four corners, which are 270°, generate singularities.[10] Nevertheless, the complex mapping method can be applied to overcome the geometrical singularity problems. After the mapping, however, it also causes nonuniform boundary nodal distribution on the transformed hyper-rectangle, thus it may derive great errors when boundary nodal points are not sufficient.

Four indicators in Table 3 are of the optimum. That implies the present grid system is really a conformal grid system. The MDO and the ADO are almost zeroes all over the domain, even on the four corners of Fig. 12.

Table 3. Evaluation of meshes of the four-semicircular region.

	Figure	mesh size	MDO	ADO	MAR	AAR
Zhang[9]	Fig. 3 B(1)	30×30	4.28	0.07	13.00	4.22
Zhang	Fig. 3 B(2)	30×30	9.57	0.33	7.50	2.70
Zhang	Fig. 3 B(3)	30×30	8.00	0.19	9.40	3.31
Zhang	Fig. 3 B(4)	30×30	7.89	0.23	8.30	2.96
Akcelik[8](2001)	Fig. 10 (b)	41×41	23.96	0.62	14.82	4.20
Luis[11](1996)	Fig. 9 (a)	41×41	12.50	0.18	N.A.	N.A.
Present Results	Fig. 12	41×41	4.86×10^{-9}	4.86×10^{-9}	1.00	1.00

4. Conclusions

A conformal grid generation method and the associated proper treatment of numerical boundary-layer problems within the BIEM[10] are presented in this paper.

The process of the conformal grid generation can be divided into five major steps: forward transformation from a physical domain to a hyper-rectangular region, forward transformation from the hyper-rectangular region to a rectangular domain, grid generation on the transformed rectangle, backward transformation from the mapped rectangle to the hyper-rectangle and backward transformation from the the hyper-rectangle to the original physical domain. Among the four stages, the first and the last stages are completed by a complex mapping to eliminate the geometric singularities,[10] whereas the intermediate stages are accomplished by introducing the analytic form of boundary integral element.[14] For the intermediate stages, grid lines can be generated directly on or close to the boundary to the low 10^{-14}.[10] A minor but important proper treatment of integration limits within the boundary integral element method has clarified the elimination of numerical boundary layer encountered before.[14] As far as the numerical singularity is concerned, the analytical integration approach (BIEM) may be a better choice over numerical integration method (BEM).

In the past, the generation of conformal grid system is deemed as very difficult. It has been reported that the grid lines will "collapse" or intersect near some boundary. In the present study, the overlapping of the grid lines near degenerated corners is analyzed to be the reason of geometrical singularity, further proofs and applications would be enforced in the next study.

In addition, using the method developed in the paper by Wang and Tsay,[10] the grid lines locating very near the boundary can be produced within a nano scale. This is thought to be important for the calculation of the boundary layer in the fluid dynamics.

Acknowledgments

With deep gratitudes, we present this research results based on the analytical approach of the element integrations in BIEM, a previous work developed by Professors J. A. Liggett and P. L.-F. Liu.

We hereby wish Professor Liu many Happy Birthdays to come.

References

1. Thompson, J. F. and Warsi, Z. U. A., Boundary-fitted coordinates systems for numerical solution of partial differential equations-a review, *Journal of Computational Physics*, **47**, 1 (1982).
2. Thompson, J. F., Warsi, Z. U. A. and Mastin, C. W., *Numerical Grid Generation, Foundations and Applications* (Elsevier Science Pub. Co., New York, North-Holland, 1985).
3. Robert, R. B., Hsieh, C. K. and Li, H., Grid generation in two dimensions using the complex variable boundary element method, *Appl. Math. Modelling*, **19**, 322 (1995).
4. Thompson, J. F., Thames, F. C. and Mastin, C. W., Boundary-fitted curvilinear coordinate system for sollutions of partial differential equations on field containing any number of arbitrary two dimensional bodies, in *NASA CR2729*, 1976.
5. Thompson, J. F., Thames, F. C. and Mastin, C. W., TOMCAT-A code for numerical generation of boundary-fitted curvilinear coordinate system on field containing any number of arbitrary two-dimensional bodies, *Journal of Computational Physics*, **24**, 274 (1977).
6. Ryskin, G. and Leal, L. G., Orthogonal Mapping, *Journal of Computational Physics*, **50**, 71 (1983).
7. Tamamidis, P. and Assanis, D. N., Generation of orthogonal grids with control of spacing, *Journal of Computational Physics*, **94**, 437 (1991).
8. Akcelik, V., Jaramaz, B. and Ghattas, O., Nearly orthogonal two-dimensional grid generation with aspect ratio control, *Journal of Computational Physics*, **171**, 805 (2001).
9. Zhang, Y., Jia, Y. and Wang, S. S. Y., 2D nearly orthogonal mesh generation, *International Journal for Numerical Methods in Fluids*, **46**, 685 (2004).
10. Wang, J. and Tsay, T. K., Analytical evaluation and application of the singularities in boundary element method, in *Engineering Analysis with Boundary Elements*, **29**, 241 (2005).
11. Luis, E., 2-D orthogonal grid generation with boundary point distribution, *Journal of Computational Physics*, **125**, 440 (1996).
12. Jeng, Y. N. and Chen, C. T., Two-dimensional orthogonal grid generation with floating boundary points, *Numerical Heat Transfer, Part B*, **36**, 207 (1999).
13. Tsay, T. K. and Hsu, F. S., Numerical grid generation of an irregular region, *International Journal for Numerical Methods in Engineering*, **40**, 343 (1997).
14. Liggett, J. A. and Liu, P. L-F., *The Boundary Integral Equation Method for Porous Media Flow* (Allen & Unwin Ltd, UK, 1983).
15. Tsay, T. K., Ebersole, B. A. and Liu, P. L., Numerical modelling of wave propagation using parabolic approximation with a boundary-fitted coordinate system, *International Journal for Numerical Methods in Engineering*, **27**, 37 (1989).
16. Tsay, T. K., Yeh, G. T., Wilson, G. V. and Toran, L. E., *Gridmaker: A Grid Generator for Two- and Three-dimensional Finite Element Subsurface Flow Models*(Oak Ridge National Laboratory, Tennessee, 1990).

17. Schinzinger, R. and Laura, P. A. A., *Conformal Mapping: Methods and Applications* (Elsevier, New York, 1991).
18. Duraiswami, R. and Prosperetti, A., Orthogonal mapping in two dimensions, *Journal of Computational Physics*, **98**, 254 (1992).
19. Lin, D. L., *Numerical Boundary-fitted Orthogonal Grid Generation* (M.S. thesis, National Taiwan University, Taiwan, 1998).
20. Tsay, T. K., Wang, J. and Huang, Y. T., Numerical generation and grid controls of conformal grids in multiply-connected regions, *International Journal for Numerical Methods in Engineering*, **67**, 1045 (2004).
21. Wang, J. and Tsay, T. K., A correct method in boundary element integration, in *Proceedings of the Fifth International Conference on Hydrodynamics* (Tainan, Taiwan, 2002).

SWASH MOTION DRIVEN BY BICHROMATIC WAVE GROUPS OVER SLOPING BOTTOMS

SHIH-CHUN HSIAO

Department of Hydraulic and Ocean Engineering
National Cheng Kung University, Tainan 701, Taiwan
E-mail: schsiao@mail.ncku.edu.tw

HWUNG-HWENG HWUNG

Department of Hydraulic and Ocean Engineering
Research Center of Ocean Environment and Technology
National Cheng Kung University, Tainan 701, Taiwan
E-mail: hhhwung@mail.ncku.edu.tw

YU-HSIEN LIN

Department of Hydraulic and Ocean Engineering
National Cheng Kung University, Tainan 701, Taiwan
E-mail: n8892112@ccmail.ncku.edu.tw

This paper examines the swash oscillation induced by bichromatic waves on sloping beaches. In particular, a series of physical experiments for three different sloping beaches of 1/10, 1/20 and 1/40 were conducted in a long wave flume (300 m × 5 m × 5.2 m) at Tainan Hydraulics Laboratory to investigate the swash motion. Ninety-three wave gauges along the wave flume and three run-up wires parallel to the sloping bottom were used. Detailed analyses of the swash motion dependence on beach slope, incident wave energy, and frequency difference are given. The cross-shore structures of long wave components are also examined.

1. Introduction

The swash zone is the portion of beach slope which is alternatively covered by the upsurge of water and exposed as the water retreats. Swash oscillations seriously influence cross-shore sediment transport as well as the deposition of sediment above the mean water level at the shoreline and play a significant role in beach stability. Consequently, it is essential to have a better understanding of the mechanism of swash motions. (e.g., Wright and Short, 1984; Hughes, 1992; Baldock et al., 1997).

The major source of infragravity wave energy (low frequency wave) in the inner surf zone and swash zone is still not fully understood. They may be from the incident free long wave offshore (Herbers et al., 1994), may be generated by propagating wave groups (Longuet-Higgins and Stewart, 1962) or even can be radiated by breakpoint forcing mechanism (Symonds et al., 1982; Schäffer, 1993). Furthermore, Watson et al. (1994) implied that the infragravity wave energy in the surf zone may also be forced by the interaction of individual bores, with the backwash resulting in an outgoing free long wave. Most field data show that the swash motion is consistent with the hypothesis of a free standing long wave due to small incident wave conditions and mild-slope topography (Guza and Thornton, 1982; Holland et al., 1995). However, if the wave condition becomes strongly nonlinear or the bottom slope is steep, the breakpoint forced long wave and bore interaction are not negligible.

The present paper addresses the influence of bichromatic wave grouping on the shoreline motion from the inner surf zone to swash zone. Particularly, the effects of various incident wave conditions and beach sloping gradients (1/10, 1/20 and 1/40) on the swash motions are discussed. From our experimental analysis, two different swash characteristics can be identified: One is the shoreline motion driven by low frequency waves and the other is that driven by the individual swash bores. In terms of the group frequency difference initiated by nonlinear wave groups, this paper aims to shed more lights on its effect on the magnitude at the group frequency, which controls the swash excursion on steeper slopes. Additionally, varying incident wave steepness is also a significant factor in determining the width of the swash zone, yet the magnitude of the shoreline motion is independent of incident wave height of monochromatic waves when the inner surf zone is fully saturated (Huntley et al., 1977). Furthermore, the measured amplitudes of low frequency motion at the group frequency induced by bichromatic waves for a range of primary mean frequencies and group frequencies are discussed to understand the influence of generated long waves on the swash zone for each slope.

This paper considers two-dimensional long wave forcing by wave grouping through accurately controlled laboratory experiments. Surf beat generation by bichromatic waves is analyzed via spectral analysis and the swash characteristics are further identified by surf parameter ξ. The laboratory study includes measurements of filtered long wave relative to digital signals recorded by fixed vertical gauges and that by run-up wire. Section 2 briefly reviews the previous literature; the experimental setup and data analysis are described in Section 3. More detailed results and discussions are further presented in Section 4. The conclusion and implication of this paper are drawn in Section 5.

2. Previous Literatures

Munk (1949) and Tucker (1950) were first to uncover infragravity waves with period ranges from one minute to several minutes based on the field measurements in California and British coasts, respectively. They showed that the infragravity wave, bound to the short wave groups, is generated by wave interaction. However, this incident bound long wave is inconsistent with the surf beat, which in general refers to free long waves. Longuet-Higgins and Stewart (1964) suggested that the surf beat was the component released through wave breaking and then subsequent reflection at the shoreline. Battjes (1974) conducted a series of experiments and implied that an increase in incident wave height contributes to wave breaking with complete dissipation in the inner surf zone and swash zone, which are therefore saturated. Symonds et al. (1982) subsequently proposed that a time-varying breaking point position induced by incident wave groups radiates long wave at the group frequency both onshore and offshore. While the shoreward propagating long wave reflects at the shoreline, an interference pattern is formed and the amplitude of the final seaward propagating wave would vary according to the group frequency and surf zone width. Kostense (1984) provided a series of laboratory experiments on 1/20 slope to support the Symonds et al. model qualitatively. Guza et al. (1984) suggested that the incident wave groups would be destroyed during the breaking process, and thereby infragravity wave energy is related to the degree of the saturation at the shoreline. List (1991) further indicated that incident nonlinear wave groups may exist after wave breaking, thus the wave groupiness does not completely reduce to zero in the inner surf zone.

Raubenheimer et al. (1995) simulated swash oscillations numerically with some field measurements on a mild beach (1/40) in comparison and found that infragravity wave energy dominates the run-up while the dissipation of wave groups within short wave frequencies is dependent on the sloping effect. Baldock et al. (1997) performed a series of regular waves, bichromatic waves and irregular waves to investigate the swash motion in inner surf zone and swash zone on 1/10 slope. A large portion of wave groups induced by bichromatic waves were found to remain at the still water shoreline, which is modulated at the group frequency. Another significant result is that the swash oscillation driven by bichromatic waves on steep slope (1/10) is principally dominated by the bound long waves initiated by nonlinear wave groups. The results of irregular waves also have the same trends. They concluded that the swash zone is saturated while the shoreline motion is governed largely by low frequency motions induced by wave grouping. Madsen et al. (1997) considered the cross-shore motion of wave groups (bichromatic waves) and irregular waves with emphasis on shoaling, breaking and run-up as well as the generation of surf beats using a time domain Boussinesq type model, which resolves the primary wave motion as well as the long waves. According to their studies, swash oscillations are largely governed by the type of wave breaking, and that the individual bores would dominate the shoreline motion if the breaker is of plunging type, whereas low frequency motion dominates for the spilling breaker.

Baldock et al. (2000) investigated bichromatic wave groups, which cover a broad range of wave amplitudes, short wave frequencies, group frequencies and modulation rates. They proposed that surf beat generation is dependent on the surf zone width and linearly dependent on the short wave amplitude. Baldock and Huntley (2002) implied that the surf beat is highly consistent with the surf zone width (time-varying breakpoint mechanism), rather than the release of incident bound long wave. Subsequently, the outgoing free long wave shows a positive correlation with incident short wave groups and is linearly dependent on short wave amplitude. In addition, the phase relationship between incident bound long waves and outgoing free long waves may be the predominant mechanism in the surf beat generation. Karunarathna et al. (2005) presented a systematic study of swash motions on 1/10, 1/30 and 1/60 slopes based on numerical experiments. The emphases of their study were to identify the shoreline motions under different driving mechanisms and to compare swash characteristics on varying slopes.

3. Experimental Set-Up and Data Analysis

3.1. *Wave flume*

The experiments were performed in a long wave flume (Fig. 1), which is 300 m long, 5.0 m wide and 5.2 m deep with a working water depth of 3.5 m, at Tainan Hydraulics Laboratory, National Cheng Kung University. The 1/10 slope starts at -35 m from the shoreline and the beginnings of 1/20 and 1/40 slopes are at -70 m and -140 m, respectively. The origin of the horizontal coordinate (x) is set at the intersection of the still water line with the beach face, positive onshore. There is a programmable, high resolution wave maker located at one end of the flume. The wave maker is a piston type paddle activated by a hydraulic cylinder. The optimal performance for wave periods of the wave maker is designed between 1 sec to 3 sec. In addition, the motion of hydraulic cylinder is

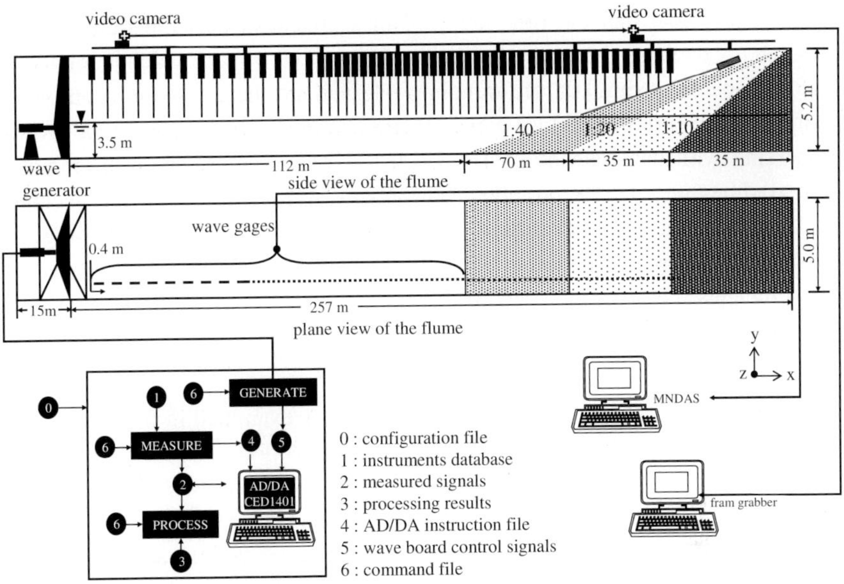

Fig. 1. The experimental set up of wave flume and instrumentation.

commanded by a programmable controller, which receives external input signal for motion.

3.2. *Instrumentation*

The temporal variations of surface elevation were measured using a total of ninety three wave gauges for 1/10, 1/20 slopes and eighty seven wave gauges for 1/40 slope distributed along the flume with three run-up wires located parallel to the sloping bottom. Each wave gauge has a dynamic range of 4.5 m at 12 bits digitization with noise typically less than 4 counts (~0.5 mm). Both the wave gauges and the run-up wires are calibrated through a series of experiments to ensure their linearity and stability. A typical calibration curve for a wave gauge indicates an R-square value of 0.9999 in the linearity of its response (e.g., Hwung and Chiang, 2005). The input signal at 25 Hz can be either generated in real time or read from a data file stored on the hard disk. The data acquisition is PC based Multi-Node-Acquisition-System (MNDAS), which is developed by Tainan Hydraulics Laboratory. It is designed specially to cope with mainly two problems. One is the synchronized problem for a large amount of parallel inputs and the other is the signal decay due to long distance transmission of data in the wave flume. Note that the time delay of measured data between first and last gauge in the data acquisition system approximates to 0.01 sec, which is relatively small compared to the 0.04 sec sampling interval and the 1.6 sec incident wave periods. Time series of surface elevation are obtained simultaneously for further processing with data rate of 25 Hz. Each test run is recorded for 10 minutes of real time in order to provide data samples sufficiently long for accurate calculation of the wave evolution. After finishing each test run, damping the possible long wave oscillation in the wave flume is necessary between successive measurements (Hwung et al., 2007).

3.3. *Wave generation*

The swash motion induced by bichromatic waves is examined in this study. The wave profiles were generated using a linear superposition of two closely spaced frequency components of equal amplitude, and have a mean wave period of 1.4, 1.6 and 2.0 sec, respectively (see Table 1). Note that the dimensionless water depths kh, where k is wave number and h is the water depth, range from 2.93 to 9.78 which are about or well within the deep water wave regime. Thus, the possible spurious free long wave due to mismatch of the boundary conditions at wave paddle surface and linear wave system is small (Rapp and Melville, 1990; Baldock et al., 1996). Specifically, the wave length L_0 is calculated based on mean period and surf similarity parameters for the 1/10 slope indicate a plunging breaker type and a spilling breaker type for the 1/20 and 1/40 slopes. In order to examine the effect of the frequency difference between two carrier waves in the swash zone, a total of fifty two typical cases were performed with full modulation rates ($a_2/a_1 = 1$) for each slope (1/10, 1/20 and 1/40). Emphases are placed on the effect of beach slope, incident wave heights and frequency difference on the hydrodynamics in swash zone.

3.4. *Analysis techniques*

Amplitude spectra were obtained directly by FFT using 4096 data points sampled by 25 Hz with Hanning window applied. The filtered long wave is separated digitally by the band-pass filter method from the amplitude spectra in the range of 0.1 f_c ~ f_c (f_c is the frequency at the minimum energy between gravity and infragravity wave) to provide the low frequency oscillation in the run-up and swash depth at the still water shoreline ($x = 0$) (Hwung et al., 2006).

Table 1. Experimental wave conditions of initial bichromatic wave trains.

case	d_0 (m)	T_1 (sec)	T_2 (sec)	H_0 (m)	f_m (Hz)	Δf (Hz)	ξ_0 (1/10)	ξ_0 (1/20)	ξ_0 (1/40)
B1	3.5	1.40	1.80	0.15	0.625	0.16	0.52	0.26	0.13
B2	3.5	1.40	1.80	0.13	0.625	0.16	0.55	0.28	0.14
B3	3.5	1.40	1.80	0.10	0.625	0.16	0.63	0.32	0.16
B4	3.5	1.40	1.80	0.08	0.625	0.16	0.71	0.36	0.18
B5	3.5	1.45	1.75	0.15	0.625	0.12	0.52	0.26	0.13
B6	3.5	1.45	1.75	0.13	0.625	0.12	0.55	0.28	0.14
B7	3.5	1.45	1.75	0.10	0.625	0.12	0.63	0.32	0.16
B8	3.5	1.45	1.75	0.08	0.625	0.12	0.71	0.36	0.18
B9	3.5	1.50	1.70	0.15	0.625	0.08	0.52	0.26	0.13
B10	3.5	1.50	1.70	0.13	0.625	0.08	0.55	0.28	0.14
B11	3.5	1.50	1.70	0.10	0.625	0.08	0.63	0.32	0.16
B12	3.5	1.50	1.70	0.08	0.625	0.08	0.71	0.36	0.18
B13	3.5	1.55	1.65	0.15	0.625	0.04	0.52	0.26	0.13
B14	3.5	1.55	1.65	0.13	0.625	0.04	0.55	0.28	0.14
B15	3.5	1.55	1.65	0.10	0.625	0.04	0.63	0.32	0.16
B16	3.5	1.55	1.65	0.08	0.625	0.04	0.71	0.36	0.18
B17	3.5	1.20	1.60	0.12	0.714	0.21	0.51	0.26	0.13
B18	3.5	1.20	1.60	0.10	0.714	0.21	0.55	0.28	0.14
B19	3.5	1.20	1.60	0.08	0.714	0.21	0.62	0.31	0.16
B20	3.5	1.20	1.60	0.05	0.714	0.21	0.78	0.39	0.20
B21	3.5	1.25	1.55	0.12	0.714	0.15	0.51	0.26	0.13
B22	3.5	1.25	1.55	0.10	0.714	0.15	0.55	0.28	0.14
B23	3.5	1.25	1.55	0.08	0.714	0.15	0.62	0.31	0.16
B24	3.5	1.25	1.55	0.05	0.714	0.15	0.78	0.39	0.20
B25	3.5	1.30	1.50	0.12	0.714	0.10	0.51	0.26	0.13

Table 1. (*Continued*)

B26	3.5	1.30	1.50	0.10	0.714	0.10	0.55	0.28	0.14
B27	3.5	1.30	1.50	0.08	0.714	0.10	0.62	0.31	0.16
B28	3.5	1.30	1.50	0.05	0.714	0.10	0.78	0.39	0.20
B29	3.5	1.33	1.48	0.12	0.714	0.08	0.51	0.26	0.13
B30	3.5	1.33	1.48	0.10	0.714	0.08	0.55	0.28	0.14
B31	3.5	1.33	1.48	0.08	0.714	0.08	0.62	0.31	0.16
B32	3.5	1.33	1.48	0.05	0.714	0.08	0.78	0.39	0.20
B33	3.5	1.35	1.45	0.12	0.714	0.05	0.51	0.26	0.13
B34	3.5	1.35	1.45	0.10	0.714	0.05	0.55	0.28	0.14
B35	3.5	1.35	1.45	0.08	0.714	0.05	0.62	0.31	0.16
B36	3.5	1.35	1.45	0.05	0.714	0.05	0.78	0.39	0.20
B37	3.5	1.80	2.20	0.22	0.500	0.10	0.53	0.27	0.13
B38	3.5	1.80	2.20	0.20	0.500	0.10	0.56	0.28	0.14
B39	3.5	1.80	2.20	0.15	0.500	0.10	0.64	0.32	0.16
B40	3.5	1.80	2.20	0.10	0.500	0.10	0.79	0.40	0.20
B41	3.5	1.85	2.15	0.22	0.500	0.08	0.53	0.27	0.13
B42	3.5	1.85	2.15	0.20	0.500	0.08	0.56	0.28	0.14
B43	3.5	1.85	2.15	0.15	0.500	0.08	0.64	0.32	0.16
B44	3.5	1.85	2.15	0.10	0.500	0.08	0.79	0.40	0.20
B45	3.5	1.90	2.10	0.22	0.500	0.05	0.53	0.27	0.13
B46	3.5	1.90	2.10	0.20	0.500	0.05	0.56	0.28	0.14
B47	3.5	1.90	2.10	0.15	0.500	0.05	0.64	0.32	0.16
B48	3.5	1.90	2.10	0.10	0.500	0.05	0.79	0.40	0.20
B49	3.5	1.95	2.05	0.22	0.500	0.03	0.53	0.27	0.13
B50	3.5	1.95	2.05	0.20	0.500	0.03	0.56	0.28	0.14
B51	3.5	1.95	2.05	0.15	0.500	0.03	0.64	0.32	0.16
B52	3.5	1.95	2.05	0.10	0.500	0.03	0.79	0.40	0.20

$0 \le \xi_0 \le 0.5$: **S**pilling breaker type; $0.5 < \xi_0 < 3.3$: **P**lunging breaker type.

Another relevant parameter used to describe the characteristics of wave groupiness is the time-domain groupiness factor GF (Huang, 2004), which is defined as

$$GF = \left\{ \frac{1}{T} \int_0^T \left[\frac{E_s(t) - \overline{E_s}}{\overline{E_s}} \right]^2 dt \right\}^{1/2} \tag{1}$$

where $E_s(t)$ is a moving average of instantaneous wave energy defined as

$$E_s(t) = \int_{-T_P/2}^{T_P/2} \left| \frac{\eta(t+\tau)}{T} \right|^2 Q(\tau) d\tau \tag{2}$$

and $\overline{E_s}$ is the mean value of $E_s(t)$. Moreover, $\eta(t)$ is a continuous real function of time series, T_p is the period of spectral peak and $Q(t)$ is a rectangular data window given as

$$Q(t) = \begin{cases} \dfrac{1}{T_p}, & |t| < \dfrac{T_p}{2} \\ 0, & |t| > \dfrac{T_p}{2} \end{cases} \tag{3}$$

In this study, the value for the frequency of the mother wavelet is chosen as 2π, which assures this wavelet as an admissible wavelet.

An existing technique for distinguishing incident bound long wave (IBLW), incident free long wave (IFLW), and reflected, offshore free long wave (OFLW), namely the three array method, was first presented by Kostense (1984), used by Madsen et al. (1997) and then further elaborated by Baldock et al. (2000). However, such a method does not take the wave nonlinearity and the influence of shoaling characteristics into account, and may produce substantial errors in the shallow water depth region ($d/L_0 > 0.1$, where L_0 represents the initial wave length). Therefore, in order to estimate the amplitudes of IBLW and OFLW in the shoaling zone of the flume ($d < 3.5$ m for 1/10, 1/20 and 1/40 slopes; $d < 2.5$ m for 1/60 slope), the time series of surface elevation for both IBLW and OFLW were separated by applying the phase shift and the amplification factor from the two-array method of Frigaard and Brorsen (1995) to modify the amplitudes of IBLW and OFLW at two different locations in the time domain. The digital signals of free surface elevation were then measured in the low-frequency band (frequency range: $0 < f_{lf} < f_c$; f_c is the minimum frequency between primary mean frequency and infragravity wave frequency, see Hwung et al., 2006), resulting in spectra of IBLW and OFLW.

4. Results and Discussion

4.1. *The beach slope effect on surf beat generation*

In order to avoid uncertainty in the experimental observation, the generation mechanisms of surf beat are limited to bichromatic wave groups in this study. Thus, the low frequency component should predominantly occur at a given frequency, leading to a regular pattern of long wave forcing. A series of previous works, including laboratory experiments (Kostense, 1984; Schäffer, 1993; Baldock et al., 1997, 2000) and numerical models (Symonds et al., 1982; Madsen et al., 1997; Karunarathna et al., 2005), have documented the aspect of long wave generation through bichromatic waves. However, due to the constraints of the experimental apparatus, the aforementioned laboratory analyses were mostly performed in wave flumes with a plane beach of steep slope (1/10). On the other hand, some numerical research driven to understand the mechanism of swash motion for both steep and mild beaches need to be further verified. Accordingly, in this study a series of experiments were performed to investigate the swash features on three typical types of slopes, with and without saturation in the inner surf zone. In Fig. 2, spectral analyses of swash motion and data of the first gauge station for three slopes (1/10, 1/20, 1/40) are presented. Both the long wave component (subharmonic) and short wave components (primary waves and superharmonic) decrease in their amplitudes with the decrease of the beach slope. It is noted that numerical results of Karunarathna et al. (2005) show that the low frequency component at the shoreline increases first and then decreases with decreasing slope for given incident wave conditions. This discrepancy may be partly due to different incident wave conditions used in our physical experiments and their numerical experiments. More specifically, our experiments were carried out under deep water wave conditions (dimensionless water depth $k_m h = 5.5$ where k_m is the wave number corresponding to the mean frequency of the bichromatic wave group) and their numerical experiments were conducted in the intermediate depth to shallow water depth regimes. Additionally, it is known that wave modulation in our long wave flume may result in energy transfer to lower frequency components (Hwung et al., 2006) and this may not happen in their numerical experiments.

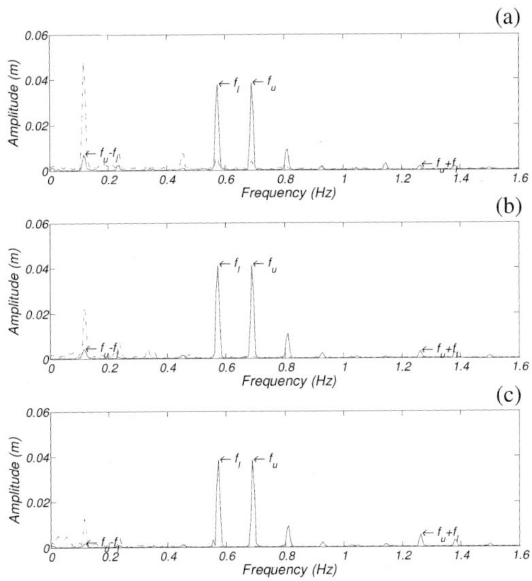

Fig. 2. The illustration of wave components in amplitude spectra at the initial stage, 15 m from wave board (solid line), and run-up (dotted line) for $H_0 = 0.1$ m, $f_m = 0.625$ Hz and $\Delta f = 0.12$ Hz. (a) 1/10 slope ($\xi_0 = 0.63$), (b) 1/20 slope ($\xi_0 = 0.32$) and (c) 1/40 slope ($\xi_0 = 0.16$) respectively. f_l: lower primary wave component; f_u: upper primary wave component; $f_u + f_l$: super-harmonic; $f_u - f_l$: sub-harmonic.

4.2. *The dependence of long wave motion on short wave periods*

The results of Symonds et al. (1982) implied that the long wave induced by time-varying breakpoint does not mainly depend on the individual primary wave frequencies, but on the frequency difference. To verify this hypothesis, the vertical swash excursions of three different primary mean frequencies $f_m = 0.71$, 0.625, and 0.5 Hz with the same group frequency and wave height are compared for the 1/10 slope as shown in Fig. 3. The selected ranges of these mean frequencies lead to plunging breaker type according to Battjes (1974). Clearly, the shoreline movement is seen to modulate at the group frequency irrespective of primary mean frequencies and a large portion of the wave groups still remain at the shoreline. Moreover, it is evident that the resultant shoreline motion covers individual bores riding on the low-frequency oscillation at the group frequency ($\Delta f = 0.08$ Hz). The corresponding spectral analyses of Figs. 3(a), 3(b) and 3(c) are presented in Figs. 4(a), 4(b) and 4(c) respectively. For the case of small surf parameter ($\xi_0 = 0.55$) as shown in Fig. 4(a), the magnitude of the short wave frequencies is relatively insignificant in driving the swash motion, therefore the short wave frequencies have less influence on surf beat mechanism. However, for the larger surf parameter, the magnitudes of the short wave frequencies

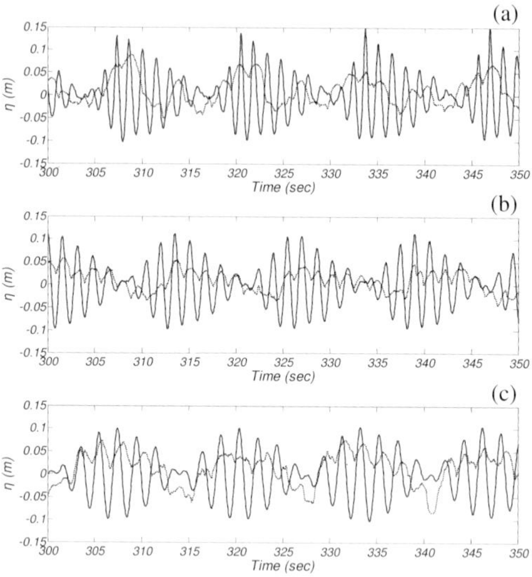

Fig. 3. Surface elevation measured by offshore gauge (solid line) and vertical swash excursion (dotted line) for H_0 = 0.1 m on 1/10 slope. (a) f_m = 0.71 Hz (T_1 = 1.325 sec and T_2 = 1.475 sec, ξ_0 = 0.55: plunging type); (b) f_m = 0.625 Hz (T_1 = 1.5 sec and T_2 = 1.7 sec, ξ_0 = 0.63: plunging type); (c) f_m = 0.5 Hz (T_1 = 1.85 sec and T_2 = 2.15 sec, ξ_0 = 0.79: plunging type).

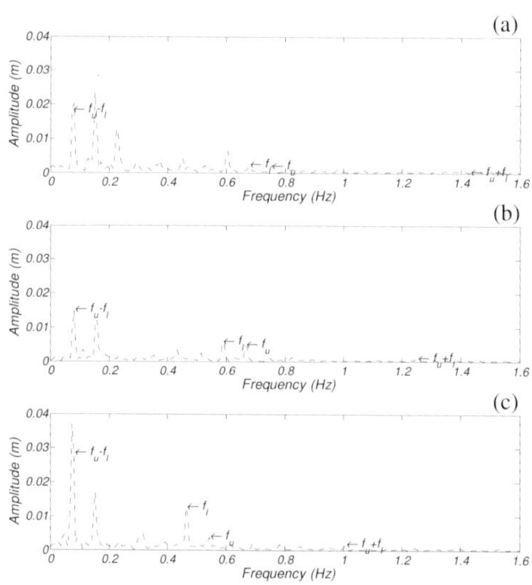

Fig. 4. The corresponding amplitude spectra measured by run-up wire (dotted line) for H_0 = 0.1 m on 1/10 slope. (a) f_m = 0.71 Hz (T_1 = 1.325 sec and T_2 = 1.475 sec, ξ_0 = 0.55: plunging type); (b) f_m = 0.63 Hz (T_1 = 1.5 sec and T_2 = 1.7 sec, ξ_0 = 0.63: plunging type); (c) f_m = 0.5 Hz (T_1 = 1.85 sec and T_2 = 2.15 sec, ξ_0 = 0.79: plunging type).

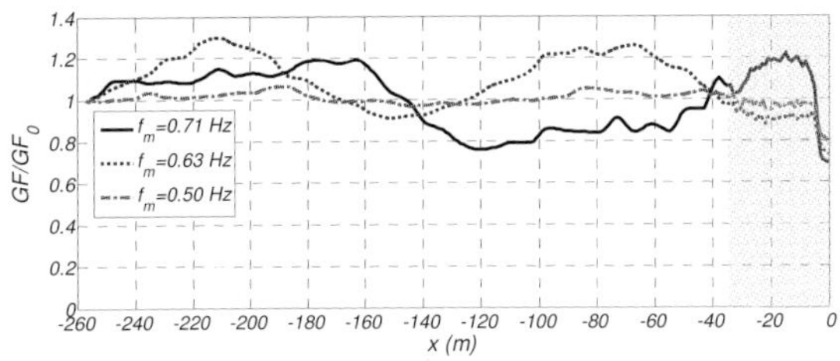

Fig. 5. Normalized groupiness factor against distance offshore from the still water shoreline for the given incident wave height $H_0 = 0.1$ m and group frequency $\Delta f = 0.08$ Hz on 1/10 slope ($x > -35$ m). (Solid line: f_m = 0.71 Hz; dotted line: f_m = 0.63 Hz; dashed-dotted line: f_m = 0.5 Hz.) The hatched area represents the plane beach.

gradually increase and play a significant role on the surf beat motion. Note that the subharmonic magnitude does increase monotonically with surf parameter when the waves are of spilling breakers. These results are consistent with numerical results by Karunarathna et al. (2005).

Furthermore, to refine our explanation, an indicator *GF* computed by equation (1) has been used to describe the cross-shore evolution of wave groupiness. This indicator is normalized by characteristic target GF0 with data being recorded offshore (x = -257 m). Fig. 5 shows the cross-shore evolution of grouping structure for three types of primary mean frequencies (f_m = 0.71, 0.625 and 0.55 Hz) on 1/10 slope. It can be seen that the grouping structure with higher primary mean frequency appears a sinusoidal pattern in the cross-shore region, accompanied with readily regrouping after group structure energy in decomposition. When the wave groupiness undergoes the initial breaking process across the surf zone, wave groupiness decreases readily with higher primary mean frequency. Finally, the grouping structure of lower primary mean frequency remaining at the shoreline is larger than that of higher ones. In fact, when the surf zone is unsaturated, low-frequency swash motions may be generated partly due to an incident breakpoint forced long wave (IBFLW) as suggested by Baldock et al. (1997), and partly due to direct forcing from swash bores. Consequently, it can be concluded that the incident wave groupiness rapidly decrease with increasing primary mean frequency from steep slope to gentle slope. Moreover, our measurement results are similar to the numerical calculations of Karunarathna et al. (2005).

4.3. *Swash profiles and spectral physics at different group frequencies*

As pointed out by Symonds (1982), group frequency may play a significant role in swash motion. Accordingly, four different group frequencies Δf for a given primary mean frequency (f_m = 0.5 Hz) are generated and compared to examine the characteristics of varying group frequencies on the swash motion with 1/10 slope as shown in Fig. 6(a). The results indicate that the magnitude of vertical swash excursion increases with increasing group frequency except for the case Δf = 0.03 Hz. The corresponding amplitude spectra are presented in Fig. 6(b), which appear the most evident low-frequency magnitude at the group frequency for Δf = 0.1 Hz. Additionally, substantial magnitudes appear at multiple times of Δf in the low frequency band, indicating a significant interaction occurs between long wave components in the swash zone.

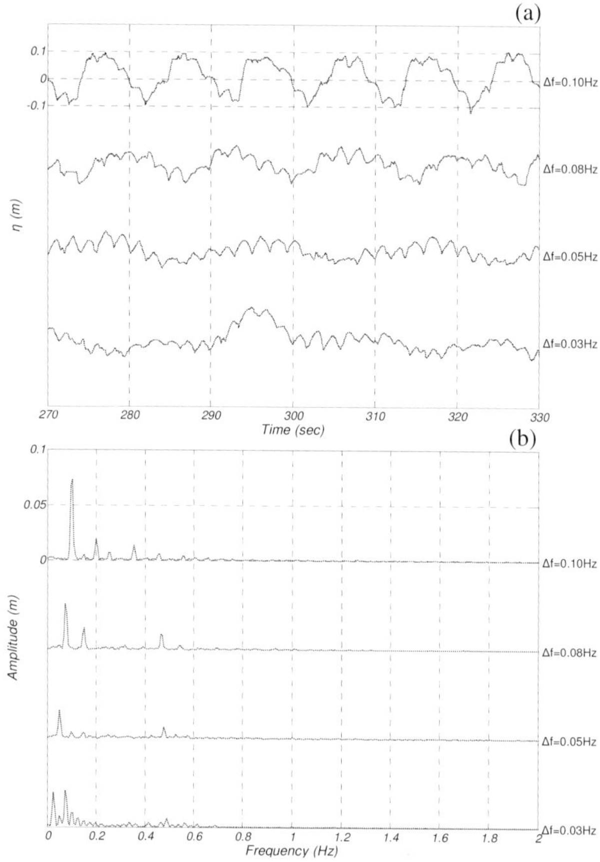

Fig. 6. (a) Vertical swash excursion; (b) Amplitude spectra at Δf = 0.03 Hz (T_1 = 1.95 sec and T_2 = 2.05 sec), Δf = 0.05 Hz (T_1 = 1.9 sec and T_2 = 2.1 sec), Δf = 0.08 Hz (T_1 = 1.85 sec and T_2 = 2.15 sec) and Δf = 0.1 Hz (T_1 = 1.8 sec and T_2 = 2.2 sec) for H_0 = 0.1 m and ξ_0 = 0.79 on 1/10 slope.

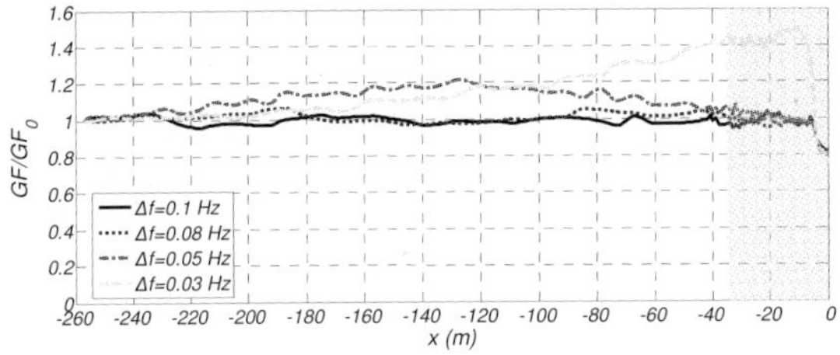

Fig. 7. Normalized groupiness factor against distance offshore from the still water shoreline for the given incident wave height H_0 = 0.1 m and primary mean frequency f_m = 0.5 Hz on 1/10 slope (x > -35 m). (Solid line: Δf = 0.1 Hz; dotted line: Δf = 0.08 Hz; dashed-dotted line: Δf = 0.05 Hz; dashed line: Δf = 0.03 Hz.) The hatched area represents the plane beach.

In the following, the normalized group factor is used to describe the relationship between group characteristics and the frequency difference. As shown in Fig. 7, four different types of group behavior are analyzed to characterize the evolution of wave groups due to frequency difference. It is interesting to know that the GF/GF_0 values of higher group frequency ones (Δf = 0.08, 0.1 Hz) keep nearly constant until the location where wave train commences to break, and then reduce to about 0.8. On the contrary, the group factors for lower group frequencies (Δf = 0.05, 0.03 Hz) show a gradual development at the initial stage and reach the maximum values at x = *-125* m and x = *-10* m, respectively. Moreover, a sharp gradient of wave groupiness is seen as the wave groups pass through the surf zone. The run-up data for each case present the same value on the aspect of grouping characteristic, which further shows almost no existence of short wave energies on the swash motion.

4.4. *Swash characteristics induced by varying energy levels*

In this section, the swash characteristics induced by varying energy levels are examined for three different bottom slopes. As shown in Fig. 8(a), vertical swash depths for four different incident wave heights on 1/10 slope are compared. The primary mean frequency of the wave groups is 0.5 Hz and group frequency is set at 0.08 Hz for these four cases. The swash profiles imply that the magnitude of swash oscillation increases with incident wave heights, suggesting that energy levels of initial wave structures actually influence the magnitude of the low frequency component at the group frequency with some individual bores riding on it. More detailed swash physics were analyzed via spectral analysis as shown in Fig. 8(b), which exhibits a positive proportionality between low frequency component at the group frequency and incident wave height. Although it is less apparent, the short wave energies have the same trend with those at the group frequency.

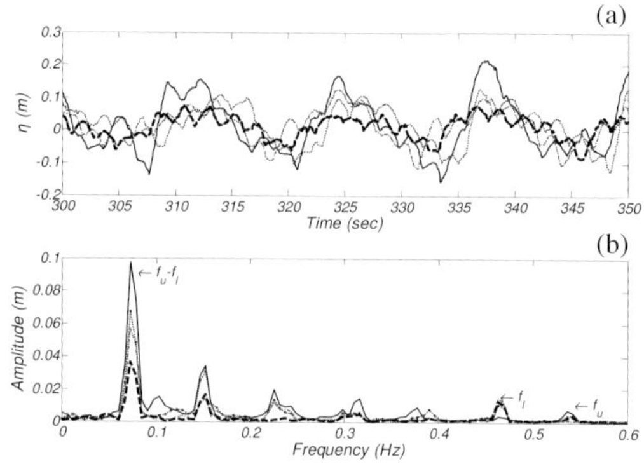

Fig. 8. (a) Vertical swash excursions measured for H_0 = 0.22 m (solid line, ξ_0 = 0.53), H_0 = 0.20 m (dotted line, ξ_0 = 0.56), H_0 = 0.15 m (dashed-dotted line, ξ_0 = 0.64) and H_0 = 0.1 m (dashed line, ξ_0 = 0.79) on 1/10 slope; (b) Amplitude spectra measured for H_0 = 0.22 m (solid line), H_0 = 0.20 m (dotted line), H_0 = 0.15 m (dashed-dotted line) and H_0 = 0.1 m (dashed line) on 1/10 slope.

The experimental results of the same wave conditions as Fig. 8 but on 1/20 slope indicate that energy carried by the low frequency component clearly dominates the width of the swash zone after wave breaking. The swash profiles in Fig. 9(a) appear to be almost the same response with some bores riding on the profile structures. Fig. 9(b) further demonstrates that milder slope reduces the difference between low frequency oscillations from varying incident wave heights while saturation just changes the height carried by primary short waves suggested by previous works (Madsen, 1997; Karunarathna, 2005).

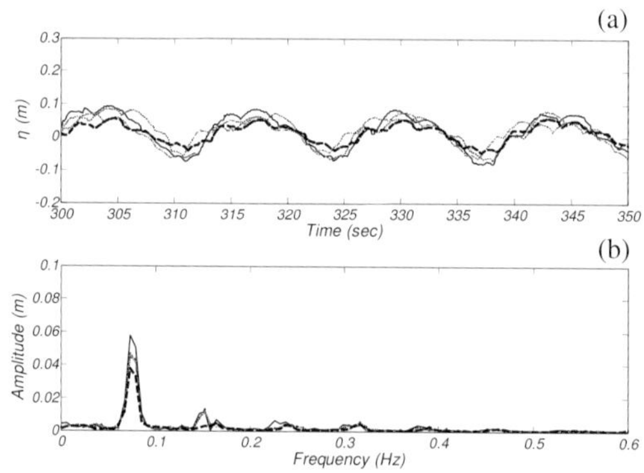

Fig. 9. (a) Vertical swash excursions measured H_0 = 0.22 m (solid line, ξ_0 = 0.27), H_0 = 0.20 m (dotted line, ξ_0 = 0.28), H_0 = 0.15 m (dotted-dashed line, ξ_0 = 0.32) and H_0 = 0.1 m (dashed line, ξ_0 = 0.40) on 1/20 slope; (b) Amplitude spectra measured for H_0 = 0.22 m (solid line), H_0 = 0.20 m (dotted line), H_0 = 0.15 m (dotted-dashed line) and H_0 = 0.1 m (dashed line) on 1/20 slope.

As the sloping gradient further decreases to 1/40, the low-frequency energy in the wave group plays a predominantly vital part in the shoreline excursion for all four different incident energy levels. The corresponding surf similarity parameter ranges from 0.13 to 0.20, which are of spilling breakers. Not surprising, the measured amplitude spectra shown in Fig. 10(b) reveals the magnitudes of short wave components are almost absent due to energy dissipation by wave breaking and bottom friction. In summary, while the individual swash bores riding on the low frequency swash will increase the width in the swash zone for steep slopes, the short wave components for milder slope almost vanish and make no contribution to the run-up. It is worth to note that the runup magnitudes do not monotonically increase with increasing initial amplitudes. As saturation is reached at the shoreline, the magnitude of shoreline motion is dominated by low-frequency components and thus may not be proportional to the short incident wave height.

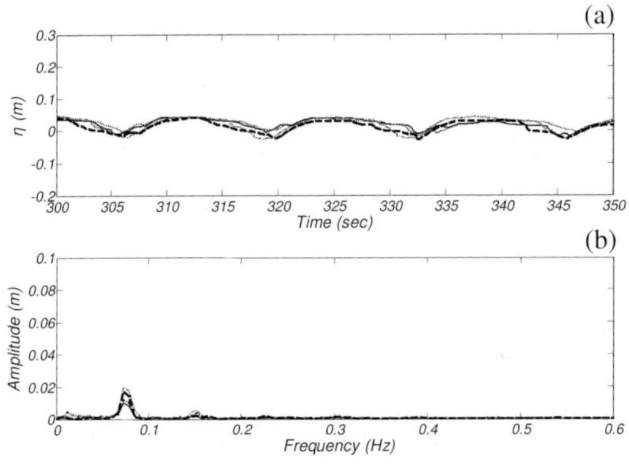

Fig. 10. (a) Vertical swash excursions measured for $H_0 = 0.22$ m (solid line, $\xi_0 = 0.13$), $H_0 = 0.20$ m (dotted line, $\xi_0 = 0.14$), $H_0 = 0.15$ m (dashed-dotted line, $\xi_0 = 0.16$) and $H_0 = 0.1$ m (dashed line, $\xi_0 = 0.20$) on 1/40 slope; (b) Amplitude spectra measured for $H_0 = 0.22$ m (solid line), $H_0 = 0.20$ m (dotted line), $H_0 = 0.15$ m (dashed-dotted line) and $H_0 = 0.1$ m (dashed line) on 1/40 slope.

4.5. *The effect of swash motion on long wave components*

Although the physical features of a time varying breakpoint mechanism have been well documented (e.g., Symonds et al., 1982; Schäffer, 1993; Baldock et al., 2000), most of their experiments were conducted in a single bottom slope. To verify and extend previous findings, the interference pattern of each long wave component based on our experimental data is examined in great details for three sloping bottoms. While the maximum of short wave groups meets the mean breakpoint at the instant in time, it is noted that the long wave forcing from radiation stress gradients emits both an incident breakpoint forced long wave (hereafter abbreviated as IBFLW) shoreward and an outgoing breakpoint forced long wave (OBFLW) seaward of the mean breakpoint. The reflected breakpoint forced long wave (RBFLW) is produced after the IBFLW propagates to the shoreline and then it reflects. If the OBFLW and RBFLW are in phase at the outer breakpoint, the constructive pattern takes place and the maximum amplitude of breakpoint forced long wave (BFLW) is attained. If this is the case, the mean breakpoint closely coincides with the node of a free standing long wave. On the other hand, the BFLW would reach the minimum response when the OBFLW and RBFLW are out of phase at the mean breakpoint, which approximates to the antinode for this long wave structure. In addition, if an incident bound long wave (IBLW) is assumed to be released during the breaking process and then to be reflected (RBLW), it would lead to further modification to the free standing long wave. More details of long wave phase relationships due to breakpoint mechanism can be seen in Fig. 1 by Baldock et al. (2000).

Figs. 11(a)–11(c) contrasts the cross-shore variation in the amplitudes of total long wave, IBLW and OFLW (the abbreviation of outgoing free long wave) at different group frequencies on 1/10 slope. In addition, the data are compared with the analytical solution of zeroth order Bessel function, which gives the most shoreward data measured by the run-up wire as an input. The steady setup of vertical swash excursion around the shoreline has been accounted for the real shoreline position in the x-axis. The data show the generation of partial standing long waves at discrete frequencies over the sloping bottom. Fig. 11(a) presents the case of the lower group frequency ($\Delta f = 0.039$ Hz) conducted in this experiment and the mean breakpoint is shoreward of the first nodal point for a free standing long wave. It is clearly seen that a node and an antinode is offshore of the breakpoint ($x \approx$ -5.4 m) and at the shoreline, respectively. In comparison, the mean breakpoint reaches the shoreline, which is an antinode. It should be noted that there is no existence of the nodal structure for a free standing long wave in the nearshore region. The measured data by the run-up wire appears as a small shoreline amplitude. It is also obvious that both the IBLW and the OFLW are dominant around the breakpoint. It could be expected that the strong phase cancellation between these two waves, which causes the suppression of the shoreline amplitude.

For the case in Fig. 11(b), the short wave breaking meets the node ($x_b \approx$ -3.8 m) of the nodal structure for a free standing long wave. Although the previous literatures (e.g. Mase, 1995; Baldock et al., 2004) exhibit that short wave swash-swash interactions may also contribute to the shoreline motion, but

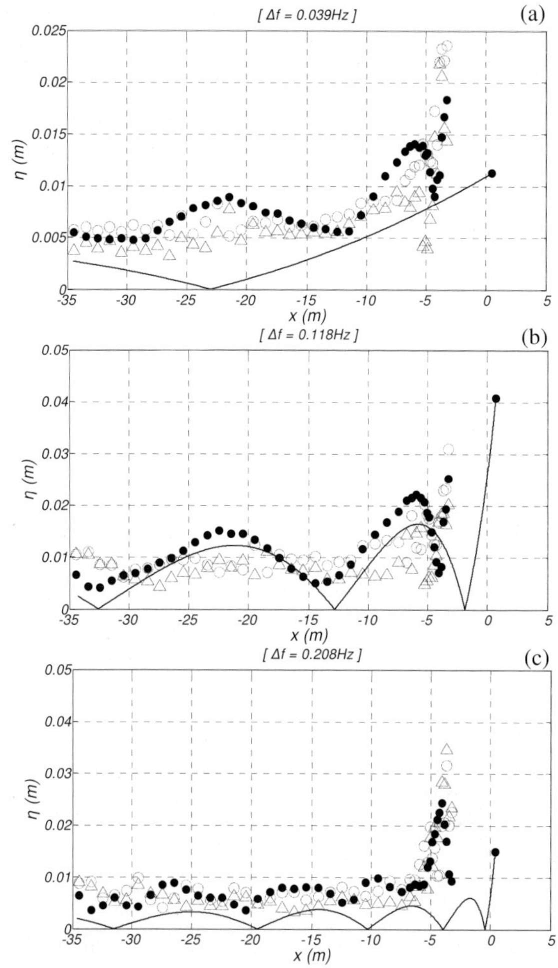

Fig. 11. The amplitudes of IBLW (triangle), OFLW (circle) and total low-frequency component (solid circle) for (a) $\Delta f = 0.118$ Hz, $x_b = -5.4$ m; (b) $\Delta f = 0.118$ Hz, $x_b = -3.8$ m; (c) $\Delta f = 0.208$ Hz, $x_b = -6.8$ m respectively, with $H_0 = 0.1$ m and $f_m = 0.625$ Hz on 1/10 slope. J_0: free standing wave solution (solid line).

there is little evidence of these interactions directly associated with the main source of OFLW. The maximum shoreline amplification observed occurs when the nodal point is closely in accordance with the mean breakpoint, which is in agreement with Baldock et al. (2000) and Lin et al. (2007). In contrast, the

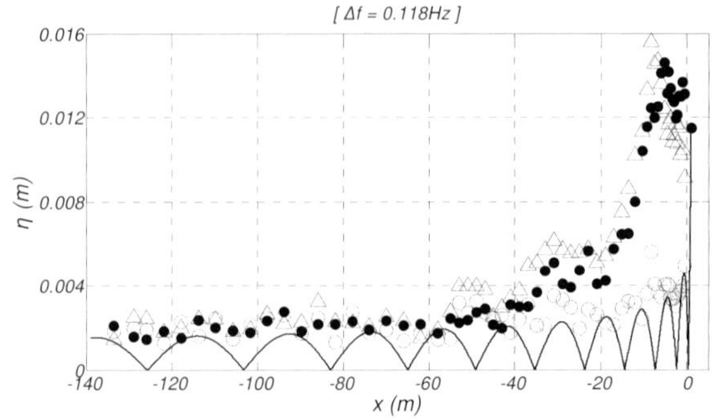

Fig. 12. The amplitudes of IBLW (triangle), OFLW (circle) and total low-frequency component (solid circle) for $\Delta f = 0.118$ Hz with $H_0 = 0.1$ m and $f_m = 0.625$ Hz on 1/40 slope. J_0: free standing wave solution (solid line).

minimum amplification of shoreline motion occurs at the higher group frequency ($\Delta f = 0.208$ Hz) in Fig. 11(c) while the mean breakpoint reaches the antinode. It could be conjectured that the surf zone width observed from the visual camera seems to influence the shoreline motion, which is consistent with the hypothesis of breakpoint forcing mechanism suggested by Symonds et al. (1982). The results indicate that the maximum shoreline amplification would correspond to a peak frequency of resonant response between the IBLW and the OFLW. Furthermore, the longer breaking excursion is found to contaminate the nodal structure for a free standing long wave seaward of the surf zone, which is due to the destructive interference of phase relationship between IBLW and OFLW.

Fig. 12 illustrates the identical cases of Fig. 11(b) but on 1/40 experimental slope. It is of interest to note that the breakpoint forcing mechanism is not applicable to the present cases due to the obscure nodal structure by the combination of the IBLW and OFLW in different phase relationship (see details in Lin et al., 2007), which show the predominance of the IBLW in the cross-shore structure of long wave components. The obscure nodal structure is therefore occurred in the nearshore region because there is no obvious resonant interaction between IBLW and OFLW. In addition, the measurement close to the shoreline indicates that a pronounced dissipation is occurred in the inner surf zone and swash zone, and that the swash-swash interaction would evidently produce additional outgoing long waves regardless of strong dissipation on the mild slopes.

5. Conclusion

A series of elaborate experiments to study the influence of sloping gradient, incident wave height and frequency difference on the swash motion for bichromatic wave groups were conducted and analyzed. The results mainly confirm and extend previous work, as summarized below:

(1) The surf similarity parameter can be used to characterize the type of shoreline motion for both monochromatic and bichromatic waves, and that plunging breakers result in obvious swash bores at the shoreline whereas the spilling breakers generate low frequency dominating shoreline motion. More specifically, for given incident wave the swash oscillation is dominated by both swash bores and low frequency motion for steep slope, whereas it is predominant by long wave motion only for mild slope.

(2) For the cases of bichromatic waves, the magnitude of long wave component diminishes as the mean primary frequency increases for given slope and frequency difference. In addition, according to swash profiles with subsequent spectral analysis, the swash height is proportional to group frequency (Δf), except for the case $\Delta f = 0.03$ Hz. The offshore wave height influences the swash depth, especially on steep slope (1/10), which suggests an unsaturated situation.

(3) Based on our analysis, the long wave generation in the swash zone is highly correlated to the width of breaking region and its magnitude is found to be linearly proportional to incident short wave. Moreover, if the cross-shore structure of the long wave component is of standing wave type, the location of mean breaking point corresponds well to nodal point of standing wave.

(4) On a steep slope, the maximum amplification of the swash motion corresponds well with the resonant construction between the IBLW and the OFLW. On the other hand, the minimum amplification of the shoreline motion occurs due to the phase cancellation of these two waves at higher or lower group frequency than the peak one. This verifies that the larger breaking excursion would lead to a destructive interference pattern on the dynamic set-up (Symonds et al., 1982; Baldock et al., 2000; Lin et al., 2007).

(5) On a mild slope, the breakpoint forcing mechanism is found to be ineffective for the generation of OFLW. Besides, the large dissipation of the IBLW in the nearshore region (especially in the swash zone) is considered to provide less energy to the OFLW.

Acknowledgments

The authors gratefully acknowledge the Ministry of Education, Taiwan for the support of the project: Program for Promoting University Academic Excellence under Grant No. A-91-E-FA09-7-3. Special thanks also go to the research staff of Tainan Hydraulics Laboratory. In addition, we sincerely thank Dr. Baldock's kindness for use of his code to analyze the long wave components.

References

1. Baldock, T. E., Holmes, P., Horn, D. P., 1997. Low frequency swash motion induced by wave grouping. Coastal Eng. 32, 197-222.
2. Baldock, T. E., Holmes, P., 1999. Simulation and prediction of swash oscillation on a steep beach. Coastal Eng. 36, 219-242.
3. Baldock, T. E., Huntley, D. A., Bird, P. A. D., O'Hare, T., Bullock, G. N., 2000. Breakpoint generated surf beat induced by bichromatic wave groups. Coastal Eng. 39, 213-242.
4. Baldock, T. E., Huntley, D. A., 2002. Long wave forcing by the breaking of random gravity waves on a beach. Proc. R. Soc. Lond. A 458, 2177-2201.
5. Battjes, J. A., 1974. Surf similarity. Proc. 14th Int. Conf. Coastal Eng., ASCE, 466-480.
6. Battjes, J. A., 1988. Surf zone dynamics. Ann. Rev. Fluid Mech. 20, 257-293.
7. Elgar, S., Herbers, T. H. C., Okihiro, M., Oltman-Shay, J., Guza, R. T., 1992. Observations of infragravity waves. J. Geophys. Res. 97, 15573-15577.
8. Guza, R. T., Thornton, E. B., 1982. Swash oscillation on a natural beach. J. Geophys. Res. 87, 483-491.
9. Guza, R. T., Thornton, E. B., Holman, R. A., 1984. Swash on steep and shallow beaches. Proc. 19th Int. Conf. Coastal Eng., ASCE, 708-723.
10. Herbers, T. H. C., Elgar, S., Guza, R. T., 1995. Generation and propagation of infragravity waves. J. Geophys. Res. 100, 24863-24872.
11. Holland, K. T., Raubenheimer, B., Guza, R. T., Holman, R. A., 1995. Run-up kinematics on a natural beach. J. Geophys. Res. 100(3), 4985-4993.
12. Huang, M. C., 2004. Wave parameters and functions in wavelet analysis. Ocean Eng. 31, 111-125.
13. Hughes, M. G., 1992. Application of a nonlinear shallow water theory to swash following bore collapse on a sandy beach. J. Coastal Res. 8, 562-578.
14. Huntley, D. A., Guza, R. T., Bowen, A. J., 1977. A universal form for shoreline run-up spectra? J. Geophys. Res. 82, 2577-2581.

15. Hwung, H. H., Chiang, W. S., 2005. The measurements on wave modulation and breaking. Meas. Sci. Technol. 16, 1921-28.
16. Hwung, H. H., Lin, Y. H., Hsiao, S. C., 2006. The investigation on the generation of infragravity wave. (in press). doi:10.1016/j.oceaneng.2006.08.012.
17. Hwung, H. H., Chiang, W.S., Hsiao, S.-C., 2007. Observations on the evolution of wave modulation, 463, 85-112, Proceedings of Royal Society.
18. Karunarathna, H., Chadwick, A., Lawrence, J., 2005. Numerical experiments of swash oscillations on steep and gentle beaches. Coastal Eng. 52, 497-511.
19. Kostense, J. K., 1984. Measurement of surf beat and set-down beneath wave groups. Proc. 19th Int. Conf. Coastal Eng., ASCE, 724-740.
20. Lin, Y. H., Hsiao, S. C., Hwung, H. H., 2007. The generation and dissipation of low-frequency waves. J. Geophys. Res. (under review)
21. List, J. H., 1991. Wave groupiness variation in the nearshore. Coastal Eng. 15, 475-496.
22. List, J. H., 1992. A model for the generation of two dimensional surf beat. J. Geophys. Res. 97, 5623-5635.
23. Longuet-Higgins, M. S., Stewart, R. W., 1962. Radiation stress and mass transport in gravity wave, with application to "surf beats". J. Fluid Mech. 13, 481-504.
24. Longuet-Higgins, M. S., Stewart, R. W., 1964. Radiation stress in water wave: A physical discussion with applications. Deep-Sea Res. 11, 529-562.
25. Madsen, P. A., Sorensen, O. R., Schäffer, H. A., 1997. Surf zone dynamics simulated by a Boussinesq type model: Part 2. Surf beat and swash oscillations for wave groups and irregular waves. Coastal Eng. 32, 289-319.
26. Mansard, E. P. D., Funke, E. R., 1980. The measurement of incident and reflected spectra using a least squares method. Proc. 17th Int. Conf. Coastal Eng., ASCE, New York, 154-172.
27. Munk, W. H., 1949. Surf beats. Trans. Am. Geophys. Union 30, 849-854.
28. Raubenheimer, B., Guza, R. T., Elgar, S., Kobayashi, N., 1995. Swash on a gently sloping beach. J. Geophys. Res. 100(5), 8751-8760.
29. Ruessink, B. G., 1998. Bound and free infragravity waves in the nearshore zone under breaking and nonbreaking conditions. J. Geophys. Res. 103, 12795-12805.
30. Schäffer, H. A., 1993. Infragravity waves induced by short wave groups. J. Fluid Mech. 247, 551-588.

31. Symonds, G., Huntley, D. A., Bowen, A. J., 1982. Two-dimensional surf beat: Long wave generation by a time-varying breakpoint. J. Geophys. Res. 87, 492-498.
32. Tucker, M. J., 1950. Surf beats: Sea wave of 1 to 5 min. period. Proc. R. Soc. London, Ser. A 202, 565-573.
33. Watson, G., Peregrine, D. H., 1992. Low frequency waves in the surf zone. Proc. 23rd Int. Conf. Coastal Eng., ASCE, 818-831.
34. Watson, G., Barnes, T. C. D., Peregrine, D. H., 1994. The generation of low frequency waves by a single wave group incident on a beach. Proc. 24th Int. Conf. Coastal Eng., ASCE, 776-790.
35. Wright, L. D., Short, A. D., 1984. Morpho-dynamic variability of surf zones and beaches: A synthesis. Mar. Geol. 56, 93-118.

SAND TRANSPORT UNDER NEARSHORE WAVE AND CURRENT AND ITS IMPLICATION TO SANDBAR MIGRATION

TIAN-JIAN HSU and XIAO YU

Center for Applied Coastal Research, Civil and Environmental Engineering
University of Delaware, Newark, DE 19716, USA

A process-based prediction of nearshore morphological evolution requires reliable parameterization of the wave-current induced sediment transport rate. Hence, a two-phase model for sediment transport (Hsu et al. 2004; Amoudry et al. 2007) is utilized in this paper to study sheet flow sand transport driven by wave and current. The model is first demonstrated to predict the measured sediment transport rate under nonlinear wave forcing in an oscillating water tunnel for two different wave periods (O'Donoghue and Wright 2004). Motivated by an earlier study of Dohmen-Janssen et al. (2001), the model is further used to study the effect of mobile bed and moving sediment on the time-averaged mean current in the wave-current boundary layer. Model results suggest that the mobile bed effect further increases the apparent roughness of the mean current in the wave-current boundary layer by a factor of two. Finally, the model is driven by a field measured near bed wave-current cross-shore velocity time series of 250 sec at Duck, NC (Gallagher et al. 1998). The model predicts a net onshore sediment transport flux consistent with field observation. More importantly, the model reveals the highly heterogeneous vertical structure of sediment flux and complex interplay among wave skewness/asymmetry, undertow current and sediment transport. For typical wave conditions at Duck, an offshore undertow current of magnitude smaller than about 0.2 m/s is not sufficient to overcome the effect of wave-induced onshore transport.

1. Introduction

Sediment transport has many important applications in land preservation, coastal zone management and hazard mitigation. Within the context of nearshore sediment transport, an outstanding issue that caught many researchers' attention has been the prediction of surf zone sandbar migration. In earlier years, much attention in this area focuses on understanding the surf zone wave-current hydrodynamics (e.g., Thornton and Guza 1983; Stive and Wind 1986; Ting and Kirby 1994; Lin and Liu 1998; Scott et al. 2005; see Svendsen 2006 for a comprehensive review) and bottom boundary layer processes (e.g., Grant and Madsen 1979; Soulsby et al. 1993; Cox and Kobayashi 1999; Liu 2006).

In the past decades, there have been increasing amounts of cross-shore sediment transport studies that emphasize the importance of sediment responses to nearshore waves and currents due to for example, effects of wave group and higher-order wave statistics (Thornton et al. 1996; Gallagher et al. 1998; Ruessink et al. 1998; Elgar et al. 2001; Hsu et al. 2006), nonlinear boundary layer processes (e.g., Trowbridge and Young 1989; Henderson et al. 2004), and the effects of breaking wave turbulence on sediment transport (e.g., Butt et al. 2004; Scott et al. 2007).

The most critical component in predicting cross-shore beach profile evolution is an accurate estimate of local sediment transport rate under wave-current forcing. Due to the presence of irregular seabed and moving sediment, the near bed boundary layer processes (velocity, turbulent statistics and bottom stress) under wave and current can be highly complex. The responses of sediment particles, namely the distribution of sediment concentration and its temporal variation in the bottom boundary layer are not well-understood. Several existing sediment transport formulae for predicting beach profile evolution are based on straightforward extension from fluvial sediment transport that assumes local equilibrium (e.g., Bailard 1981). However, the interactions among wave, current, and sediment may be highly nonlinear and such simple parameterization of sediment transport rate in the nearshore environment becomes questionable.

While there has not yet been a conclusive and universal agreement regarding the relative importance of surf zone processes on sandbar migration (specifically the onshore migration) due to the high complexity of the problem, these studies have recently encouraged a significant amount of research efforts toward understanding the small-scale sediment transport processes in the surf zone. These small-scale efforts include the measurement of concentrated region of sediment transport within the wave boundary layer (e.g., Ribberink and Al-Salem 1995; Dohmen-Janssen et al. 2001; Dohmen-Janssen and Hanes 2002, 2005; O'Donoghue and Wright 2004) and the development of various detailed models to capture the intermittent turbulence (e.g., Zedler and Street 2006; Chang and Scotti 2006), two-phase fluid-sediment interaction (Dong and Zhang 1997; Hsu et al. 2004; Amoudry et al. 2007), granular flow dynamics (e.g., Jenkins and Hanes 1998; Drake and Calantoni 2001; Hsu et al. 2004) and heterogeneity of discrete sediment particles (Calantoni et al. 2004).

One great advantage of using a two-phase flow approach with granular dynamics is that the concentrated regime of sediment transport can be modeled directly through a careful consideration of the intergranular stresses and fluid-sediment interaction. This provides a considerable improvement upon existing suspended load models which must rely on empirical sediment reference concentration or pick-up function. In this paper, we study detailed wave-current boundary layer sediment dynamics under laboratory and field measured wave forcing using a two-phase model. The model adopted here is revised from that of Hsu et al. (2004) originally developed for massive particles. These new revisions appropriate for typical beach sand and model validation with U-tube data are recently reported in Amoudry et al. (2007).

In order to push forward the two-phase model as a standard tool for predicting sediment transport rate under typical field measured random wave train and current, the two-phase model reported in Amoudry et al. (2007) is further extended with a fully-implicit numerical scheme for calculating the coupled fluid phase and sediment phases equations. Because the wave-current statistics in the typical field condition are often obtained with a realization of few minutes to an hour, in order to study realistic net sediment transport rate driven by wave-current forcing in the field, a realization of near-bed sediment transport of at least several minutes is also necessary. The newly enhanced numerical scheme is more numerically stable and can produce results driven by field measured random wave and current for a duration of 10-20 minutes within a reasonable amount of CPU time.

This paper is organized as follow. A brief model formulation, including the new implicit numerical scheme is discussed in Section 2. The model is further utilized as a tool to study several crucial sediment transport processes. In Sections 3, we focus on the effect of wave period and the effect of wave-current interactions on sediment-laden boundary layer and the resulting sediment transport. The model results are first validated with experimental data and further used to interpret the "apparent roughness" of the current due to the effects of waves, mobile bed and transported sediment. In Section 4, the model is driven by field measured wave-current forcing at Duck, NC (Elgar et al. 2001) during typical onshore sandbar migration condition. The two-phase model calculates detailed sediment flux distribution near the bottom seabed under realistic forcing and reveals critical mechanisms possibly responsible for the observed onshore and offshore transport. This paper is concluded in Section 5.

2. Model Formulation

We consider noncohesive sand transport in the sheet flow condition where the seabed can be considered locally flat without the presence of ripples (e.g., Nielsen 1992). In typical sandy beach with grain size around 0.2-0.3 millimeter, sheet flow occurs when the nondimensionalized bottom shear stress (i.e., the Shields parameter) is greater than about 1.0. Therefore, sheet flow is often observed near the sandbar crest or around the swash zone where peak wave orbital velocity amplitude is greater than about 0.8 m/s.

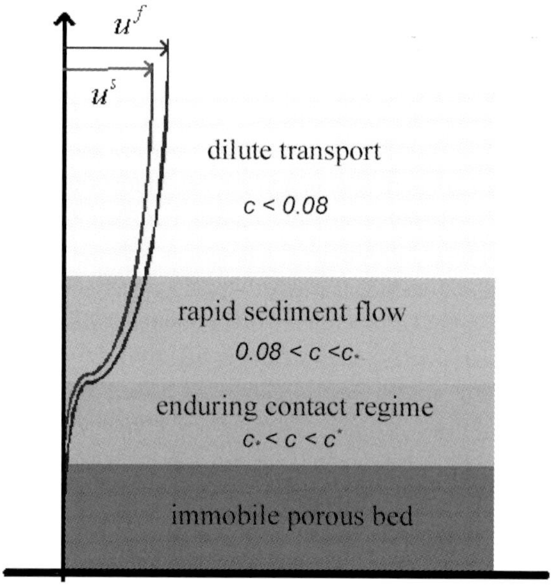

Figure 1: A schematic description of sheet flow based on dominant transport mechanisms. c^*=0.635 is the random-close-packing concentration while c_*=0.57 is the random-loose-packing concentration.

Sediment transport is conventionally considered as bedload and suspended load. Bedload is part of the load that is continuously in contact with the bed (or adjacent particles) during transport (Fredsoe and Deigaard 1992). According to this definition, the suspended load consists of particles that transport without being directly in contact with the bed or the adjacent particles. For less energetic conditions, the concentration in the boundary layer decreases dramatically from stationary bed concentration (about 60%) to a few percent at 5-10 grain diameter above the bed. The suspended load in this case is mainly driven by carrier flow turbulence, which can be called dilute suspension. However, for typical sheet flow conditions, the amount of suspended load is not significant. There exists a layer with thickness about 10-20 grain diameters in which the sediment concentration is between 10% and 50%. When transport is this concentrated, sediments are driven by intergranular processes through short-lived collisions or interaction with interstitial fluid (Bagnold 1954). Additionally, the flow turbulence may also be affected by the presence of particles (Gore and Crowe 1989). This layer is called rapid transport regime (see Figure 1).

In order to continuously model sediment transport processes from the bedload regime dominanted by contact forces to the less concentrated rapid transport regime and to the dilute regime dominated by turbulent suspension (see Figure 1), a two-phase flow theoretical framework with appropriate closures on intergranular interactions and carrier flow turbulence modulation is warranted.

Following Hsu et al. (2004), we consider an ensemble-averaged flow field where the sheet flow boundary layer is assumed to be fully developed in the streamwise (x) direction with fluid and sediment velocity denoted by u^f and u^s respectively. We focus on the vertical (z) distribution of fluid and sediment quantities across the boundary layer with fluid and sediment vertical velocity denoted by w^f and w^s, respectively. The carrier fluid is of density ρ^f, viscosity ν and the sediment is of diameter d, density ρ^s with concentration denoted by c. The streamwise fluid pressure gradient is assumed to be uniform across the bottom boundary layer and is prescribed according to given free-stream wave-current forcing. The vertical gradient of fluid pressure p^f is solved as part of the solution of the model. The mass and momentum equations of the fluid phase are written as

$$\frac{\partial(1-c)}{\partial t}+\frac{\partial(1-c)w^f}{\partial z}=0 \tag{1}$$

$$\frac{\partial\rho^f(1-c)u^f}{\partial t}+\frac{\partial\rho^f(1-c)w^f u^f}{\partial z}=-(1-c)\frac{\partial p^f}{\partial x}+\frac{\partial\tau^f_{xz}}{\partial z}-\frac{\rho^s}{T_P}c\left(u^f-u^s\right) \tag{2}$$

$$\begin{aligned}\frac{\partial\rho^f(1-c)w^f}{\partial t}+\frac{\partial\rho^f(1-c)w^f w^f}{\partial z}&=-(1-c)\frac{\partial p^f}{\partial z}+\frac{\partial\tau^f_{zz}}{\partial z}-\rho^f(1-c)g\\&\quad-\frac{\rho^s}{T_P}c\left(w^f-w^s\right)+\frac{\rho^s}{T_p}\frac{\nu_t}{\sigma_c}\frac{\partial c}{\partial z}\end{aligned} \tag{3}$$

The mass and momentum equation of the sediment phase are

$$\frac{\partial c}{\partial t}+\frac{\partial cw^s}{\partial z}=0 \tag{4}$$

$$\frac{\partial \rho^s c u^s}{\partial t} + \frac{\partial \rho^s c w^s u^s}{\partial z} = -c\frac{\partial p^f}{\partial x} + \frac{\partial \tau^s_{xz}}{\partial z} + \frac{\rho^s}{T_P} c\left(u^f - u^s\right) \tag{5}$$

$$\frac{\partial \rho^s c w^s}{\partial t} + \frac{\partial \rho^s c w^s w^s}{\partial z} = -c\frac{\partial p^f}{\partial z} + \frac{\partial \tau^s_{zz}}{\partial z} - \rho^s c g + \frac{\rho^s}{T_P} c\left(w^f - w^s\right) - \frac{\rho^s}{T_p}\frac{\nu_t}{\sigma_c}\frac{\partial c}{\partial z} \tag{6}$$

One of the unique features of the two-phase formulation is the inter-phase momentum transfer terms (the last term in (2) and (5) and the last two terms in (3) and (6)) modeled here as drag force with T_p the particle response time. The particle response time is a measure of the particle inertia in the fluid. Coarse and heavy particles have large T_p and the differences between the fluid and sediment velocities are expected to be large. On the other hand, fine and light particles are expected to more closely follow the fluid velocity. Due to the ensemble-averaging, detailed interaction between fluid turbulent eddies and sediment particles through drag is not resolved and is modeled here by turbulent diffusion through eddy viscosity ν_t and a Schmidt number σ_c (the last term in equations (3) and (6)).

The fluid stresses τ^f_{xz} and τ^f_{zz} represent various small-scale processes that are not resolved in the present model, including the viscous stress and Reynolds stress. In this model the eddy viscosity with k_f-ε_f closure is adopted to calculate the Reynolds stress. In multiphase flow, it is well-known that the presence of the dispersed phase (i.e., sediment phase) modifies the carrier flow turbulence (e.g., Gore and Crowe 1989). Using a fluid turbulent kinetic energy k_f equation derived from the two-phase theory, Hsu et al. (2004) obtain extra terms compared to the single-phase k-equation that are responsible for the damping and generation of turbulence due to the presence of sediment. The k_f and ε_f equations are calculated with numerical coefficients that are empirically determined for dilute condition. Amoudry et al. (2007) find that for typical beach sand of diameter ≈0.2 mm, the numerical coefficients adopted by Hsu et al. (2004) for dilute flow are inappropriate. Guided by Direct Numerical Simulation (DNS) results for fluid-particle interaction of Squire and Eaton (1994), Amoudry et al. (2007) adopt a modified numerical coefficient in the ε_f equation that explicit depends on concentration and the ratio of particle response time T_p to fluid turbulence timescale T_l and is able to predict reasonably well the sheet flow sand transport measured in oscillatory water tunnel (Dohmen-Janssen et al. 2001;

O'Donoghue and Wright 2004). In this paper, we adopt the turbulence closure detailed in Amoudry et al. (2007).

The sediment stresses τ_{xz}^{s} and τ_{zz}^{s} in equations (5) and (6) parameterize various small-scale intergranular interaction and sediment Reynolds stress due to the ensemble-averaging. Hsu et al. (2004) derived a balance equation of sediment fluctuation energy k_s from the two-phase theory. In this k_s equation, the production, diffusion and dissipation are calculated based on kinetic theory of collisional granular flow (Jenkins and Hanes 1998). Due to the drag force in the sediment phase momentum equation, the fluid turbulence kinetic energy k_f appears in the k_s equation as a source term representing an enhancement of sediment fluctuation due to fluid turbulence. This formulation is adopted here for the closure of sediment stresses.

In the highly concentrated regime just above the immobile bed (see Figure 1), where sediment concentration is in between random-close-packing (c^*=0.635) and random-loose-packing (c_*=0.57), sediment stresses are generated due to enduring contact and frictional force at the surface of the particles. The overall behavior of this sediment transport regime is like an elastic solid or a transition between solid and fluid. A particle on average is in contact with at least another particle and the kinetic theory must be modified (which assume binary collision). In this study, we adopt the formulation suggested by Hsu et al. (2004) in which an enhanced collision viscosity and a Herzian contact relation are utilized when sediment concentration is greater than random-loose-packing. A Coulomb Failure criterion is used to relate the shear and normal stresses at the immobile bed in order to determine whether a new layer of sediment in the immobile bed is mobilized and can be suspended (Hsu et al. 2004).

Notice that Hsu et al. (2004) further consider sediment Reynolds stress as an additional sediment stress due to large-scale turbulence averaging and adopt a one-equation closure. Because this closure is ad hoc and does not change the model results much, it is neglected both in Amoudry et al. (2007) and also in this paper.

3. Monochromatic Waves in Oscillating Water Tunnel

The two-phase model has been validated with oscillating water tunnel data measured by Dohmen-Janssen et al. (2001) and O'Donoghue and Wright (2004). The model is able to predict reasonably well the time-dependent sand concentration and flow velocity under wave-current forcing. Details of the model-data comparisons are reported in Amoudry et al. (2007). In this paper, we focus on further analyzing the model results under simple wave-current forcing to study the effects of wave period on sediment transport and the effects of waves on the current boundary layer in sediment-laden condition. These two interesting effects are critical to further study of sediment transport driven by realistic field forcing (Section 4).

3.1. *Effect of Wave Period on Sediment Transport*

O'Donoghue and Wright (2004) report a series of oscillating water tunnel experiments driven by Stokes 2nd-order wave velocity (u_{0max}=1.5 m/s and u_{0rms}=0.9 m/s) of wave period T=7.5 sec and 5.0 sec. Sediment concentration in the concentrated regime is measured by Conductivity Concentration Meter (CCM). The net sediment transport averaged over one wave period $\overline{q}$ for each case is also reported. Here, we focus on the cases with grain size $d=0.28$ mm. Detailed model-data comparison on sediment concentration for the cases of wave period 7.5 sec with different grain sizes has been presented in Amoudry et al. (2007). To demonstrate the model capability in predicting the concentrated regime of sheet flow transport, model-data comparison for time-dependent concentration within the first 10 grain diameters above the immobile bed are shown for two different wave periods ($T=7.5$ sec in the left panel and $T=5.0$ sec in the right panel) in Figure 2 with z_b representing the vertical coordinate relative to the initially undisturbed bed level. The model is able to predict time-dependent concentration reasonably well. The phase of the time-dependent concentration is captured well for the case of larger wave period. However, for the case of smaller wave period there is some discrepancy regarding the phase of suspended sediment concentration at z_b=1.05 mm. Notice that in the pick-up layer (the region with z_b<0), sediment is eroded from the bed and concentration decreases as the magnitude of forcing increases. However, in the suspension layer (z_b>0), concentration increases as the magnitude of forcing increases. Typical dilute suspended load models (e.g., Li and Davies 1997) only calculate the suspension layer (z_b>0) and parameterize empirically the pick-up layer (z_b<0) using a reference concentration or a pick-up function (Engelund and Fredsoe 1976; van Rijn 1984).

The measured and predicted net sediment transport rate averaged over one wave $\overline{q}$ is shown in Table 1. Overall the model predicts reasonably well the measured transport rate. Notice that the measured $\overline{q}$ is about 45% larger than that of the case with smaller wave period. This trend is captured well by the two-phase model.

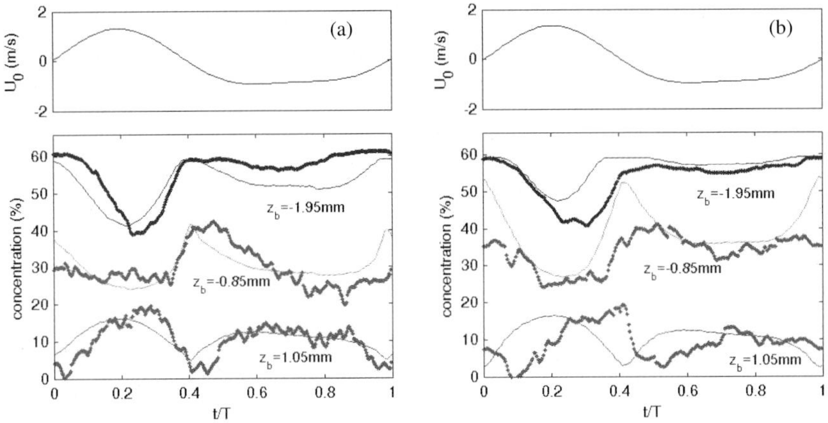

Figure 2: Time-dependent concentration within the first several millimeters above the immobile bed. The left panel is the results with period T=7.5 sec and the right panel is the results with T=5.0 sec. Measured concentration is represented by thick curve and model results are the thin curves. z_b represents the vertical coordinate relative to the initially undisturbed bed level.

Table 1. Measured and modeled net sediment transport rate averaged over one wave period.

	T = 5.0 sec	T = 7.5 sec
Model $\bar{q}$ (m^2/s) × 10^{-5}	4.8	2.0
Measured $\bar{q}$ (m^2/s) × 10^{-5}	5.3	3.6

3.2. *Effect of Wave-Current on Sediment Transport*

In the coastal environment, sediment transport is driven by both waves and currents. The nonlinear wave-current interactions significantly modify the boundary layer flow structure compared to that of current or wave alone (e.g., Grant and Madsen 1979; Fredsøe 1984; Soulsby et al. 1993). It is well-known that the effect of waves on the mean current velocity profile can be parameterized by an enhanced roughness, often called "apparent roughness" (Grant and Madsen 1979). Physically, the wave boundary layer significantly enhances the vertical mixing of flow momentum and hence modifies the shape of the mean current velocity profile. The "apparent roughness" concept is now widely used in phase-averaged circulation models. However, the effect of mobile bed and transported sediment on the mean current velocity profile under wave-current-sediment interactions has received less attention.

Recently, Dohmen-Janssen et al. (2001) demonstrated that due to the mobile bed effect, a much larger roughness is required for sheet flow transport. This conclusion is made by fitting directly the measured mean current velocity profiles to a logarithmic law. Using an earlier version of the present two-phase model, Hsu and Raubenheimer (2005) make similar suggestion that for coarse sand the time-dependent bottom stress in a sediment boundary layer can be predicted by a turbulent wave boundary layer (WBL) model based on Reynolds-averaged Navier-Stokes (RANS) equation (with only closures on turbulent Reynolds stress), provided that the roughness height K_s is elevated to a value ($K_s \approx 10\sim 20d$) much larger than typically used for clear fluid flow ($K_s \approx 2d$). The enhanced roughness is believed to compensate for the additional energy dissipation due to particle intergranular interaction not directly modeled in a single-phase approach. In this paper, we study the roughness of the current velocity profile in sediment-laden wave-current boundary layer using the two-phase model, a wave boundary layer model (Hsu et al. 2006), and measured data reported by Dohmen-Janssen et al. (2001).

Sheet flow sediment transport with sand grain size d=0.32 mm is measured by Dohmen-Janssen et al. (2001) in an oscillating water tunnel. The flow forcing is of a sinusoidal wave velocity of amplitude u_{0max}=1.47 m/s, period T=7.2 sec with a mean current superimposed. The magnitude of the mean current velocity is U=0.25 m/s at 10 cm above the bed based on direct Acoustic Doppler Velocimetry (ADV) measurement. In order to appropriately incorporate the wave and current forcing in the bottom boundary layer, measured velocity time series is first separated into a pure oscillatory component and a mean component. The oscillatory component is prescribed as time-dependent streamwise pressured gradient in the two-phase model. Then, a constant streamwise pressured gradient is added as mean current forcing so that model results for time averaged fluid velocity matches measured value at a given location above the bed (in this case it is 0.25 m/s at 0.1 m above the bed). An iteration procedure is required as the constant streamwise pressure gradient required to achieve the desired current velocity at given location above the bed is not known as a priori (Davies et al. 1988).

Detailed model-data comparisons for sediment concentration are reported in Amoudry et al. (2007). Model results for time-averaged mean current velocity profile agrees very well with the measured data (compare the solid curve with symbols in Figure 3). Specifically, the two-phase model captures the detailed shape within the first 2-3 cm above the bed. On the other hand, a clear fluid RANS wave-current boundary layer model with standard k-ε closure (dash-dotted curve) cannot predict the shape of the measured velocity profile near the bed.

Both the two-phase model and the clear fluid RANS model predict a logarithmic mean current velocity profile (it become a straight line in the semi-logarithmic plot shown in Figure 3) above the wave boundary layer. The wave boundary layer is about 5 cm in thickness. As demonstrated by Dohmen-Janssen et al. (2001), let us consider the mean current boundary layer velocity above the wave boundary layer following the rough wall logarithmic law:

$$u(z)=\frac{U_*}{\kappa}\ln\left(\frac{z}{z_0}\right) \tag{7}$$

with κ=0.4 the Karman constant, U_* the friction velocity of the mean current, and $z_0=K_s/30$. One can obtain z_0 (and hence K_s) for a given mean current velocity profile by extrapolation (i.e., extending the logarithmic profile and find the interception with the log-z axis). Using this procedure, the clear fluid RANS model predict a roughness height K_s=172 mm (i.e., z_0=5.75 mm or $K_s=540d$) for combined wave-current boundary layer without the consideration of sediment effects. Notice that in the clear fluid RANS model, a physical roughness $K_{s0}=2.5d$ is used as the bottom boundary condition for the flow velocity (Jensen et al. 1989) to represent the actual roughness elements at the bed. Therefore, the roughness due to wave-current interaction in the bottom boundary layer is enhanced by about 216 times. Interestingly, using the velocity profile calculated by the two-phase model, which further includes the effects of mobile bed and transported sediment, the roughness height is K_s=354 mm (i.e., z_0=11.8 mm or $K_s=1106d$). The presence of sediment and mobile bed in wave-current boundary layer further enhances the resulting current roughness by about two times. In summary, when both wave-current interaction and mobile-bed effects are considered, the roughness experienced by the mean current in this case is about 440 times larger than the pure current over a fixed bed.

Another way to exam the effect of waves and sediment transport on the mean currents is to calculate the friction velocity U_*. The friction velocity U_* required to generate a mean current velocity 0.25 cm/s at 10 cm above the bed with physical roughness $K_s=2.5d$ can be easily calculated by equation (7) to be 1.22 cm/s. We can further obtain the friction velocity for the cases with waves and sediment transport by directly examining the slope of the velocity profiles in Figure 3 and multiplying by the Karman constant. The friction velocity under wave-current interaction without sediment transport is 3.55 cm/s, which is about

2.9 times larger than that for the pure current condition. When considering sediment transport in wave-current boundary layer, the friction velocity further increases to 4.63 cm/s. In summary, the effect of wave-current interaction is very important in coastal sediment transport and hydrodynamic energy dissipation. Moreover, the effect of sediment transport may further increase the bottom roughness or bottom stress ($\sim U_*^2$) by a factor of 2.

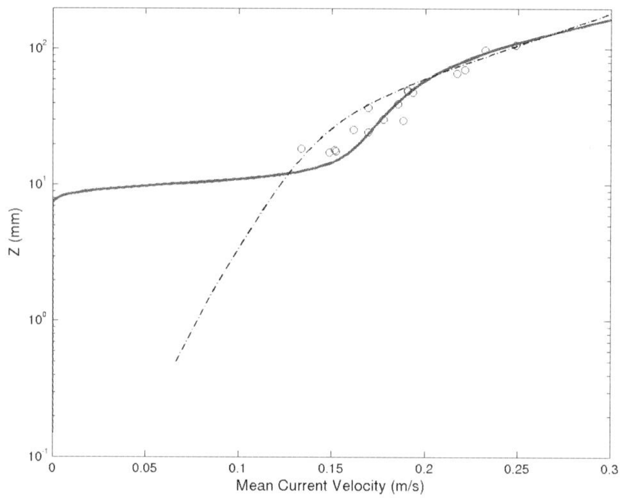

Figure 3: Comparisons of mean current velocity profiles from the measurement (circle), two-phase model results (solid curve), and clear fluid RANS model results (dashed-dot curve). The forcing condition is of a sinusoidal wave velocity of amplitude u_{0max}=1.47 m/s, period T=7.2 sec with a mean current (U=0.25 m/s at 10 cm above the bed) superimposed. The grain size is of d=0.32 mm. In the clear fluid RANS model, a physical roughness of K_s=2.5d is used. The two-phase model does not require specification of physical roughness.

Table 2. Apparent roughness and friction velocity under different conditions. Sand grain diameter is d=0.32 mm. K_s/d=2.5 is suggested by Janssen et al. (1989) based on fixed rough bed results.

	K_s/d	U_* (cm/s)
Pure current with fixed bed roughness	2.5	1.25
Wave + current with fixed bed roughness (clear fluid RANS model)	540	3.55
Wave + current with sediment transport (two-phase model)	1106	4.63

4. Sediment Transport During Onshore Sandbar Migration

Sandbar migration is an interesting morphodynamic feature in the surf zone of several meters water depth. It serves to encourage wave breaking and energy dissipation offshore and therefore reduces wave impact at the shoreline. During storm conditions, a sandbar is often driven offshore or completely washed out due to breaking wave induced undertow current. During calm weather the sandbar slowly reappears and moves onshore. It is generally agreed that the offshore-directed undertow current is the major cause for offshore bar migration while the onshore sandbar migration is mainly driven by waves. However, according to prior observations, the mechanisms that the waves drive the bar onshore are not well-understood (e.g., Thornton et al. 1996; Gallagher et al. 1998; Elgar et al. 2001; Capobianco et al. 2004). These critical observations have motivated many studies on the mechanics of wave-induced sediment transport.

Using more detailed modeling approaches that focus on the near bed bottom boundary layer processes, several possible mechanisms which encourage onshore sediment transport rate and sandbar migration have been proposed. For example, the unsteady response of bottom stress (i.e., the effects of phase-lag between bottom stress and free-stream wave forcing) and the resulting sediment transport due to the wave forcing have been attributed as one of the dominate mechanisms for onshore transport (Drake and Calantoni 2001; Hoefel and Elgar 2003; Nielsen and Callaghan 2003; Hsu and Hanes 2004). On the other hand, Trowbridge and Young (1989) and Henderson et al. (2004) emphasize the nonlinear bottom boundary layer effects (e.g., streaming) on onshore sediment transport. Recently, Hsu et al. (2006) demonstrated that by carefully choosing the mean current friction factor under waves (Grant and Madsen 1979; Soulsby et al. 1993), simple sediment transport parameterization that assume both bottom stress and resulting sediment transport rate are in-phase with the free-stream wave forcing can predict the observed onshore sandbar migration.

These small-scale studies provide very insightful information on bottom sediment transport. However, the theoretical and modeling formulations utilized in these studies are either not entirely consistent with the realistic field condition during sandbar migration or are limited by their assumptions on sediment transport mode. For example, the models utilized by Drake and Calantoni (2003) and Hsu and Hanes (2004) are of sophisticated multiphase models, but the closure assumptions used in their models are more appropriate for coarser sand grains. The analysis presented by Trowbridge and Young (1989), Nielsen and Callaghan (2003) and Hsu et al. (2006) utilizes power-law type formulae (e.g.,

Ribberink 1998) for sediment transport rate, which already assumes time-dependent sediment transport is in-phase with bottom stress. The model developed by Henderson et al. (2004) is of dilute suspended load model, which by definition does not incorporate sediment transport close to the bed and may suffer from the inaccuracy in sediment bottom boundary conditions. Finally, most of these models, except that of Henderson et al. (2004), do not explicitly incorporate wave-current interaction in their analyses or simulations. The fact that these models are not general enough for realistic sediment transport may be the cause of the seemingly inconsistent conclusions for the dominant mechanism of onshore sediment transport. Therefore, there is a need to utilize a more complete sediment transport model to study sediment transport rate during typical onshore sandbar migration.

The two-phase model presented in the previous sections has been validated with typical beach sand (~0.2 mm) under both nonlinear wave and wave-current forcing conditions (Figures 2 and 3. See also Amoudry et al. 2007). More importantly, the model calculates both the concentrated and dilute regimes of sediment transport so priori assumption on bedload and suspended load is not required. Therefore, the two-phase model is utilized here to study sediment transport during typical onshore sandbar migration events.

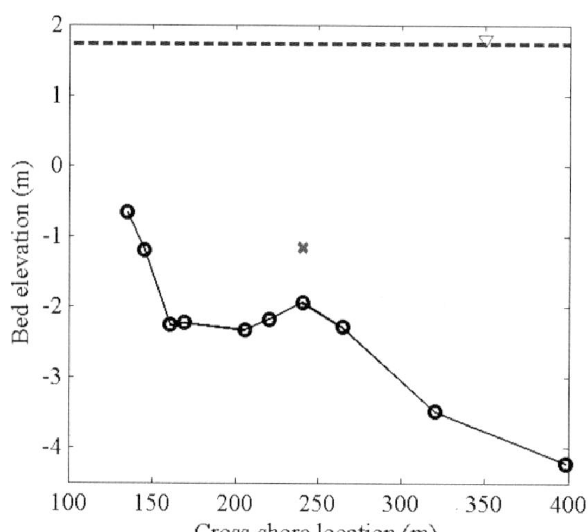

Figure 4: Measured cross-shore bathymetry (circles) at Duck, NC on September 23rd 1pm, 1994 (Gallagher et al. 1998). The mean free-surface level (dashed line) is 1.78m above the mean sea level. The cross symbol represent the location of the measured cross-shore velocity used to drive the two-phase model in this paper.

Measured time-series of near bed cross-shore flow velocity during the Duck 94 field experiment (Gallagher et al. 1998; Elgar et al. 2001) is used to drive the two-phase model. The sampling frequency of flow velocity measured by ADV is 2 Hz; hence all the short wave information is resolved. For the case discussed here, a 250 second time series of cross-shore flow velocity measured at 0.78 m above the seabed at cross-shore location x=240.6 m (i.e., around a sandbar crest, see Figure 4) of water depth 3.62 meter is used. This 250 sec time series (see Figure 5(a)) is adopted during a period in which the bar is observed to move onshore and the net sediment transport fluxes is also expected to be onshore directed.

The measured 250 sec cross-shore velocity time series is of R.M.S. wave velocity 0.564 m/s with a weak offshore directed undertow current of 0.103 m/s. The wave skewness is 1.03 which is positive and high. As can be seen from Figure 5(a), the onshore near bed peak wave velocity often exceeds 0.8 m/s and sheet flow condition is expected. With the measured wave and current velocity utilized as forcing to drive the two-phase model, the model calculates time-dependent wave-current boundary layer flow and sediment transport processes within the first 0.78 meter near the bed. The mean current velocity, concentration and sediment flux profiles are then time-averaged from the wave-phase resolving model results. The structure of the mean current velocity (Figure 5(b)) is heterogeneous. Within the first 5 cm near the bed, the mean current velocity is onshore directed. This onshore directed velocity is due to nonlinear fluid-sediment interactions resulted from skewed oscillatory forcing (According to numerical experiments, this onshore directed near bed flow is even stronger when offshore mean current is removed). Further away from the bed, the mean current becomes offshore directed due to applied mean undertow current forcing. Because most of the sediment transport occurs within the first two centimeter near the bed (see Figure 5(c) for time-averaged concentration profile within the first 10 cm above the bed), the time-averaged sediment flux is also onshore directed (see Figure 5(d)). The calculated net sediment transport rate is $\overline{q}=2.74\times10^{-5}$ m^2/s (onshore), which is consistent with the field observation.

However, further numerical experiments suggest that when increasing the offshore mean current forcing so that the calculated mean current profile is -0.25 m/s at 0.8 cm above the bed, the onshore directed mean current velocity near the bed cannot exist (Figure 6(b)). Sediment is suspended slightly higher due to stronger current, and more importantly the time-averaged sediment flux has both onshore (slightly weaker) and offshore (slightly stronger) directed components depending on the vertical location above the bed. The net sediment transport rate

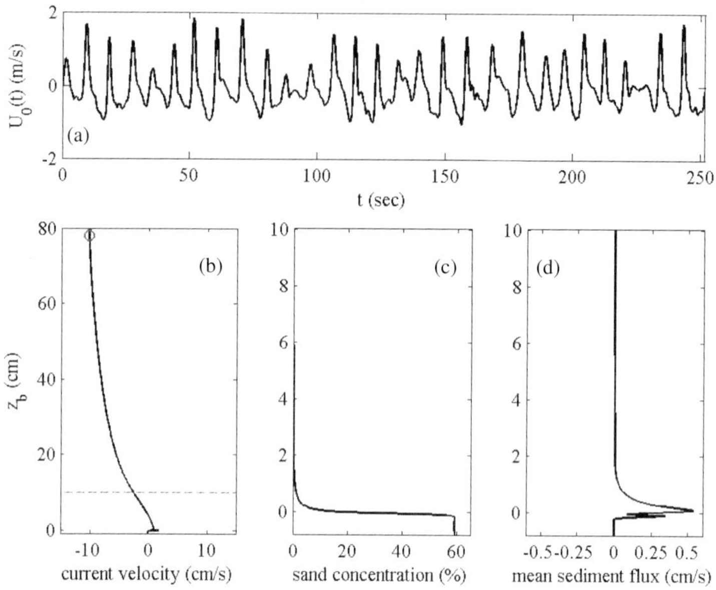

Figure 5: Field measured cross-shore velocity time series of 250 sec at 0.78 m above the seabed (a) is used to drive the two-phase model. Model results for (b) time-averaged mean current velocity profile with the circle represents the measured mean current velocity at 78 cm above the bed; (c) time-averaged sand concentration profile; (d) time-averaged sediment flux profile.

is $\overline{q}$=-2.76×10^{-6} m^2/s (offshore). It can be expected that when the applied offshore current is even stronger, larger offshore transport rate is obtained.

Our numerical results shown here are consistent with prior conjecture based on field observation that the wave skewness/asymmetry causes onshore transport and the mean undertow current causes offshore transport. However, model results further reveal that detailed onshore/offshore sediment flux distribution and interplay among wave, current and sediment in the water column are highly complex. It is not straightforward to quantify net transport rate. For the given wave forcing studied here, an undertow current of about 10 cm/s is not sufficient to cause net offshore transport because wave-induced onshore flux is dominant. However, when the undertow current is greater than about 0.25 m/s, the undertow-current-induced offshore flux dominates wave-induced onshore flux. As suggested by Gallagher et al. (1998), during the entire 2.5 month field observation, an onshore sandbar migration event is observed when the offshore undertow current is smaller than about 0.2 m/s. On the other hand, two offshore sandbar migration events are observed when the offshore undertow currents are

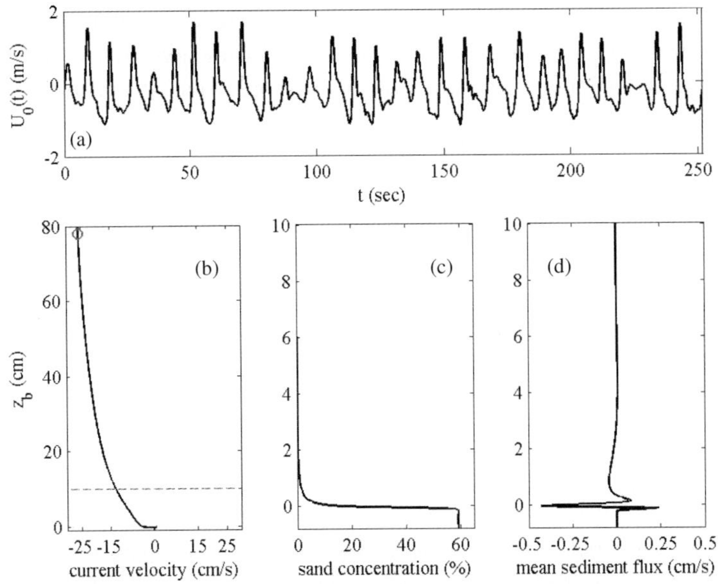

Figure 6: The same oscillatory wave condition as shown in Figure 5 but with a stronger offshore current of 0.25 m/s at 0.78 m above the bed. Model results for (b) time-averaged mean current velocity profile; (c) time-averaged sand concentration profile; (d) time-averaged sediment flux profile.

greater than about 0.3 m/s (some undertow currents of more than 0.5 m/s are observed). Our model results reported here are consistent with the field observation.

5. Conclusion

A two-phase flow model for boundary layer fluid flow and sediment transport is developed with appropriate closure of flow turbulence that accounts for the effect of sediment on turbulence modulation and closure of sediment stresses based on kinetic theory and the regime of enduring contact (Hsu et al. 2004; Amoudry et al. 2007). This model is utilized here to study several interesting questions of sheet flow sand transport in coastal environment. Model results suggest that the mean current apparent roughness (e.g., Grant and Madsen 1979) under wave-current interactions is further enhanced by a factor of two due to the mobile bed effects. When driving the model with field measured near bed cross-shore velocity (Gallagher et al. 1998), the model predicts strong interplay among nonlinear wave, undertow current and the resulting sediment transport.

Wave-induced sand transport occurs within the first several centimeters near the bed, which is onshore directed due to nonlinear interaction between the waves and sediment responses. The addition of offshore directed undertow current causes expected offshore transport. However, during typical less energetic breaking wave condition (typical onshore transport events), the mean undertow current is not strong enough to overturn the direction of sediment transport caused by wave-induced onshore transport. Numerical experiments suggest that the undertow current must exceed about 0.3 m/s to result in net offshore sediment transport. Detailed field data-model study for more events is warranted in order to provide a reliable quantitative parameterization for sediment transport in the surf zone.

Acknowledgments

Prof. Philip L.-F. Liu has always been a great mentor and source of inspiration. Tian-Jian Hsu likes to express his most sincere gratitude to Prof. Liu. Laboratory and field data provided by Marjolein Dohmen-Janssen, Tom O'Donoghue and Steve Elgar are greatly appreciated. Funding of this research is supported by U.S. Office of Naval Research (N00014-04-10217, N00014-05-10082) and National Science Foundation (OCE-0644497, CTS-0426811).

References

1. L. Amoudry, T.-J. Hsu, and P. L.-F. Liu, submitted, (2007).
2. R. A. Bagnold, *Proc. R. Soc. Lond. A*, 225, 49-63, (1954).
3. J. A. Bailard, *J. Geophys. Res.*, 86, 10938-10954, (1981).
4. T. Butt, P. Russell, J. Puleo, J. Miles, and G. Masselink, *Cont. Shelf Res.*, 24, 754-771, (2004).
5. J. Calantoni, K. T. Holland, and T. G. Drake, *Phil. Trans. R. Soc. Lond. A*, 262, 1987-2001, (2004).
6. M. Capobianco, H. Hanson, M. Larson, H. Steetzel, M. J. F. Stive, Y. Chatelus, S. Aarninkhof, and T. Karambas, *Coast. Eng.*, 47(2), 113-135, (2004).
7. Y. S. Chang, and A. Scotti, *J. Geophys. Res.*, 111(C7), C07001, (2006).
8. A. G. Davies, R. L. Soulsby, and H. L. King, *J. Geophys. Res.*, 93(C1), 491-508, (1988).
9. D. T. Cox, and N. Kobayashi, *J. Geophys. Res.*, 105, C6, 14223-14236, (1999).
10. C. M. Dohmen-Janssen, W. N. Hassan, and J. S. Ribberink, *J. Geophys. Res.*, 106(C11), 27103-27115, (2001).

11. C. M. Dohmen-Janssen, D. F. Kroekenstoel, W. N. Hassan, and J. S. Ribberink, *Coast. Eng.*, 46, 61-87, (2002).
12. C. M. Dohmen-Janssen, and D. M. Hanes, *J. Geophys. Res.*, 107, 3149, (2002).
13. C. M. Dohmen-Janssen, and D. M. Hanes, *Cont. Shelf Res.*, 25, 333-347, (2005).
14. P. Dong, and K. Zhang, *Coast. Eng.*, 36, 87-109, (1999).
15. T. G. Drake, and J. Calantoni, *J. Geophys. Res.*, 106(C9), 19,859-19,868, (2001).
16. F. Engelund, and J. Fredsoe, *Nordic Hydrology*, 7, 293-306, (1976).
17. S. Elgar, E. L. Gallagher, and R. T. Guza, *J. Geophys. Res.*, 106, 11623-11627, (2001).
18. J. Fredsøe, *J. Hydr. Eng.*, ASCE, 110, 1103-1120, (1984).
19. J. Fredsøe, and R. Deigaard, *Mechanics of Coastal Sediment Transport*, World Scientific, (1992).
20. E. L. Gallagher, S. Elgar, and R. T. Guza, *J. Geophys. Res.*, 103(C2), 3203-3215, (1998).
21. R. A. Gore, and C. T. Crowe, *Int. J. Multi-phase Flow*, 15, 279-285, (1989).
22. W. D. Grant, and O. S. Madsen, *J. Geophys. Res.*, 84, 1797-1808, (1979).
23. S. M. Henderson, J. S. Allen, and P. A. Newberger, *J. Geophys. Res.*, 109(C6), C06024, (2004).
24. F. Hoefel, and S. Elgar, *Science*, 299, 1885-1887, (2003).
25. T. J. Hsu, and D. M. Hanes, *J. Geophys. Res.*, 109(C5), C05025, (2004).
26. T. J. Hsu, J. T. Jenkins, and P. L.-F. Liu, *Proc. Roy. Soc. Lond. (A)*, 460, (2004).
27. T. J. Hsu, and B. Raubenheimer, *Cont. Shelf Res.*, 26, 589-598, (2006).
28. T. J. Hsu, S. Elgar, and R. T. Guza, *Coast. Eng.*, 53, 817-824, (2006).
29. J. T. Jenkins, and D. M. Hanes, *J. Fluid Mechanics*, 370, 29-52, (1998).
30. B. L. Jensen, B. M. Sumer, and J. Fredsøe, *J. Fluid Mech.*, 206, 265-297, (1989).
31. Z. Li, and A. G. Davies, *J. Waterw. Port Coastal Ocean Eng.*, 122, 157-164, (1997).
32. P. L.-F. Liu, *Proc. Royal. Soc. Lond. (A)*, 462(2075), 3481-3491, (2006).
33. P. Nielson, *Coastal Bottom Boundary Layers and Sediment Transport*, World Scientific, (1992).
34. T. O'Donoghue, and S. Wright, *Coast. Eng.*, 50, 117-138, (2004).
35. J. S. Ribberink, and A. A. Al-Salem, *Coast. Eng.*, 25, 205-225, (1995).
36. J. S. Ribberink, *Coast. Eng.*, 34, 59-82, (1998).

37. B. G. Ruessink, K. T. Houwman, and P. Hoekstra, *Marine Geology*, 152, 295-324, (1998).
38. C. P. Scott, D. T. Cox, T. B. Maddux, and J. W. Long, *Measurement Science and Technology*, 16(10), 1903-1912, (2005).
39. C. P. Scott, D. T. Cox, S. Shin, and T. B. Maddux, *Proc. 30th Int. Conf. Coast. Eng.*, in press, (2007).
40. R. L. Soulsby, L. Hamm, G. Klopman, D. Myrhaug, R. R. Simons, and G. P. Thomas, *Coast. Eng.*, 21, 41-69, (1993).
41. K. D. Squires, and J. K. Eaton, *J. Fluid Eng.*, 116, 778-784, (1994).
42. M. J. F. Stive, and H. G. Wind, *Coastal Eng.*, 10, 325-340, (1986).
43. I. A. Svendsen, *Introduction to Nearshore Hydrodynamics*, World Scientific, (2006).
44. E. B. Thornton, and R. T. Guza, *J. Geophys. Res.*, 88(C10), 5925-5938, (1983).
45. E. Thornton, R. Humiston, and W. Birkemeier, *J. Geophys. Res.*, 101, 12097-12110, (1996).
46. F. C. K. Ting, and J. T. Kirby, *Coast. Eng.*, 24, 51-80, (1994).
47. J. Trowbridge, and D. Young, *J. Geophys. Res.*, 94, 10971-10991, (1989).
48. L. C. van Rijn, *J. Hydraul. Eng.*, 110, 1494-1502, (1984).
49. E. A. Zedler, and R. L. Street, *J. Hydraulic Eng.*, 132(2), 180-193, (2006).

WAVE ATLAS FOR THE ARABIAN GULF

K. RAKHA, S. NEELAMANI, K. AL-BANAA* and K. AL-SALEM

Environment and Urban Development Division, Coastal and Air Pollution Department
Kuwait Institute for Scientific Research, P.O. Box: 24885, 13109 Safat, Kuwait
**E-mail: kbanaa@kisr.edu.kw*

Long term wave data is important for the design of coastal structures and for the calculation of sediment transport rates. Such data is usually obtained by hindcasting waves from wind data, when long term measured data is not available. The use of simplistic hindcasting formulas will not provide realistic data. In this study a third-generation wind driven wave model WAM is used to generate 12 years of wave data for the Arabian Gulf. The WAM model is a two dimensional model that uses non-stationary and non-homogeneous wind fields in predicting wind waves. The data showed that the maximum significant wave height is 5.5 m for data covering a period of 12 years. Extreme wave analysis of the data showed that the 100 Year wave height can reach 7.0 m in the deeper southern part of the Arabian Gulf. The corresponding mean spectral wave period reaches 8.0 sec.

1. Introduction

The ROPME (Regional Organization for the Protection of Marine Environment) Sea (Arabian Gulf) is a marginal sea in a typical arid zone. It lies between the latitude of 24°-30° N. The Arabian Gulf covers an area of 226,000 km^2. It is 990 km long and its width ranges from 56 to 338 km. It has a total volume of 7000 to 8400 km^3 of seawater (El-Gindy and Hegazi [1]). The entire basin lies upon the continental shelf. The average water depth of the Arabian Gulf is about 35.0 m. However, water depths of more than 107 m exist in some places in the south-eastern part of the Arabian Gulf (see Fig. 1). The Gulf's water depth increases in the south east direction. It is connected to the Gulf of Oman and the Arabian Sea through the Strait of Hormuz, which is 56 km wide with an average water depth of 107 m and allows water exchange between the Arabian Gulf and Arabian Sea.

The Arabian Gulf is surrounded by some of the main oil and gas exporting countries like Oman, UAE, Qatar, Bahrain, Kuwait, Iraq and Iran. Many marinas, harbors, ports, power plants, and other projects are being constructed or planned in the near-shore area, bordering the Arabian Gulf. Such projects require

* Corresponding author.

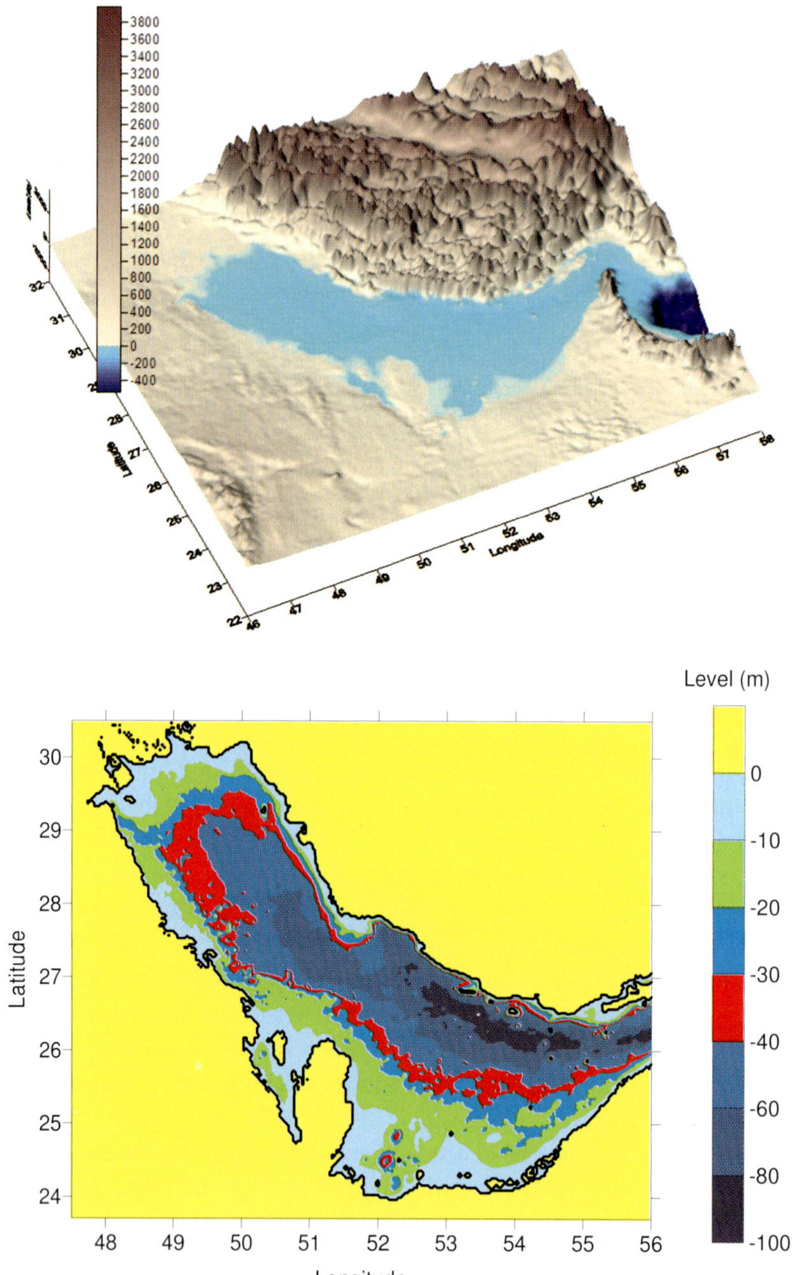

Figure 1. Topographic and bathymetric maps of the Arabian Gulf region.

both long term wave data as well as extreme wave conditions for operation and maintenance of marine projects and for their life time safe design. Yearly average wave conditions are also important for the prediction of wave induced currents, which are essential for the prediction of sediment transport rates in the surf zone. The calculation of littoral drift is needed for the prediction of shoreline changes due to any proposed structures in the nearshore area.

For most locations around the world, it is difficult to find long term wave data. Hindcasted wave data are usually used for such purposes. Long term wave data are usually calculated based on simplistic empirical wave hindcast formulas. These formulas assume a stationary and homogeneous wind speed over the Gulf. Such formulas can not provide realistic data.

El-Gindy and Hegazi [1], provided a hydrographic Atlas for the Arabian Gulf. They focused however on tidal and density driven currents. The Atlas also provided useful data on the salinity and temperature of the Arabian Gulf. No data on the wave conditions was provided.

An oceanographic data Atlas was also provided by Al-Yamani *et al.* [2], which is a valuable source of data for Kuwaiti territorial waters. This Atlas, however, does not include sufficient data on the hydrodynamics of the Kuwaiti territorial waters and does not cover the Arabian Gulf.

DHI Water and Environment together with Oceanweather Inc. produced a comprehensive metocean study of the Arabian/Persian Gulf (PERGOS). PERGOS utilizes hindcast models of high resolution. The wind and wave hindcasts were all made with basin-wide grids of spacing 0.0625° in latitude and longitude (7 km), nested within coarser grid systems. Two dimensional water levels (including surge and tide) and currents were also simulated using DHI's MIKE 21 flexible mesh model using a mesh resolution less than 7 km. The data covers a period of 20 years from 1983 (www.dhigroup.org). No data is available however, on the model validation and results.

2. Model Description

The WAM model (WAMDI Group [3]) is a third-generation wind driven wave model (WAMDI Group [3]). This is a two dimensional model, which uses non-stationary and non-homogeneous wind fields for predicting the wind waves.

The model is based on the discrete spectral action balance equation that is a generalization of the energy balance equation in the presence of currents. The physical processes included in the model are: wind-wave interaction (generation), wave-wave interaction (quadruplet), dissipation (white capping) and shallow water bottom dissipation (bottom friction). For the calculation of waves in the surf zone and in shallow water a nearshore wave model with a fine grid should be used to transform the offshore wave roses generated by this study.

The model was modified to operate on a personal computer and was setup for the Arabian Gulf (see Al-Salem *et al.* [4]). The model covered the Gulf to the 56.8° longitude with a grid spacing of 0.1° (Fig. 1). The model was applied using 24 wave directions (15° resolution) and 25 frequency bands and a time step of 300 sec. A 12 year database is generated from January 1, 1993 until December 31, 2004. This database is similar to the study of Benoit and Lafon [5].

2.1. *Input Data*

For the WAM model, the bathymetry was obtained from the admiralty charts for the study area. This data was digitized on the 0.1° grid as shown in Fig. 2.

The WAM model requires the spatial and the temporal values for the wind over the water body. Wind data obtained from the European Center for Medium Range Weather Forecast (ECMWF) was used. This wind data has a spatial resolution of 0.5° and a temporal resolution of 3.0 hours. As mentioned by Al-Salem *et al.* [4], the ECMWF wind data tends to under predict some of the storms. This conclusion was confirmed by comparing the ECMWF data with measurements at the Kuwait International Airport. In the Arabian Gulf, in general the dominant wind direction is northwesterly (Elshorbagy *et al.* [6]).

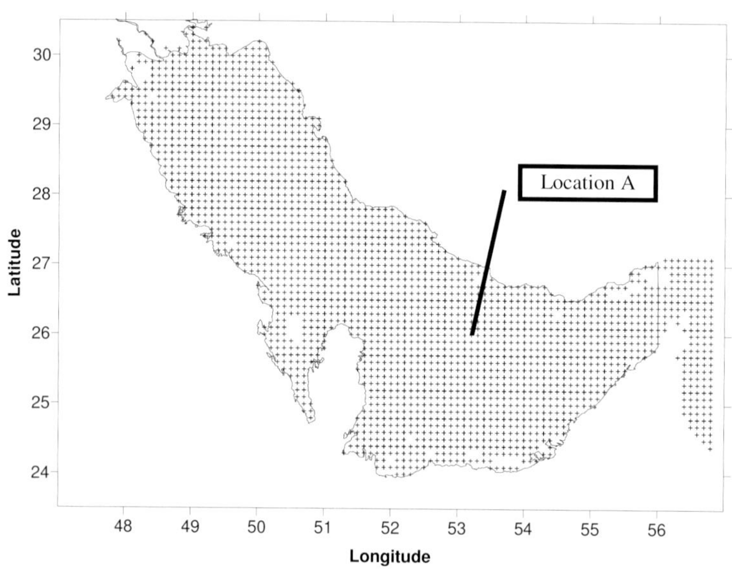

Figure 2. Topographic and bathymetric maps of the Arabian Gulf region.

3. Model Validation

The WAM model was validated using measured wave data at several locations in the Kuwaiti territorial waters (see Al-Salem *et al.* [4] and Rakha *et al.* [7]). The measurements covered a period of about four years where the wave height and period were recoded using a wave buoy and two towers. Figure 3 provides the location of these stations.

The model results showed some under prediction in the wave heights for storm events. This was due to the under prediction in the wind speed for such events. Using regression analysis, a suitable correction was proposed for waves exceeding 1.0 m. Figure 4 provides a sample of the corrected model results for the significant wave height H_s and the mean wave period T_m at the wave buoy. It can be seen that the model predicts H_s and T_m very well.

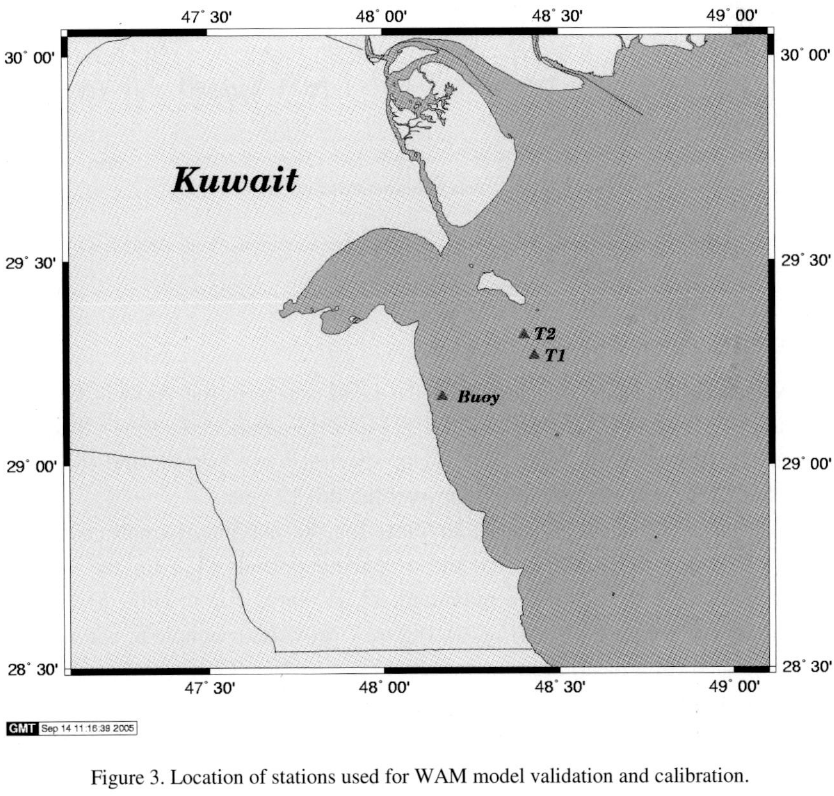

Figure 3. Location of stations used for WAM model validation and calibration.

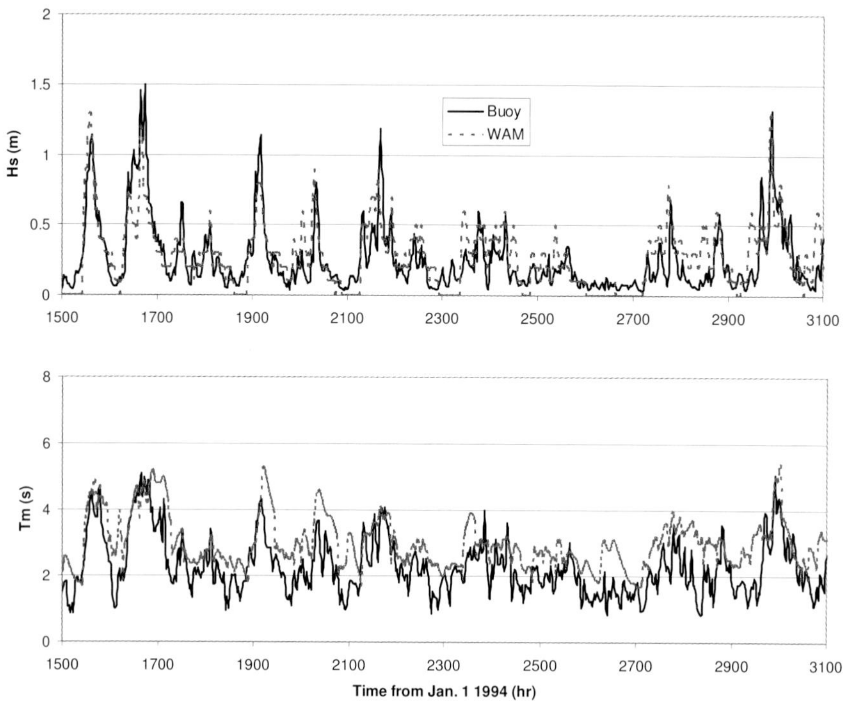

Figure 4. Comparison between measured and predicted wave conditions (Year 1994, at wave buoy).

4. Hydrodynamic Atlas

The WAM Model was used to model the wind waves in the Arabian Gulf for duration of 12 years (from January 1, 1993 until December 31, 2004). The data on the significant wave height, the mean spectral wave period, and the mean wave direction were stored every hour over the full 12 years.

Figures 5 and 6 provide contour plots for the maximum significant wave heights (H_s) and maximum of the mean spectral periods (T_m) for the 12 year period. It can be seen that the maximum H_s is above 5.0 m (Fig. 5) and the Maximum T_m reaches 8.0 sec (Fig. 6). Figure 7 provides a sample of a wave rose plot generated for a location in the middle of the Arabian Gulf with Longitude of 53.2° and Latitude 25.8° (Location A of Figs. 2 and 5). Figure 8 provides the wind and wave conditions over the Arabian Gulf for the extreme storm predicted at Location A (see Fig. 5). This extreme condition occurred on January 14, 1993 at 15:00 Hour.

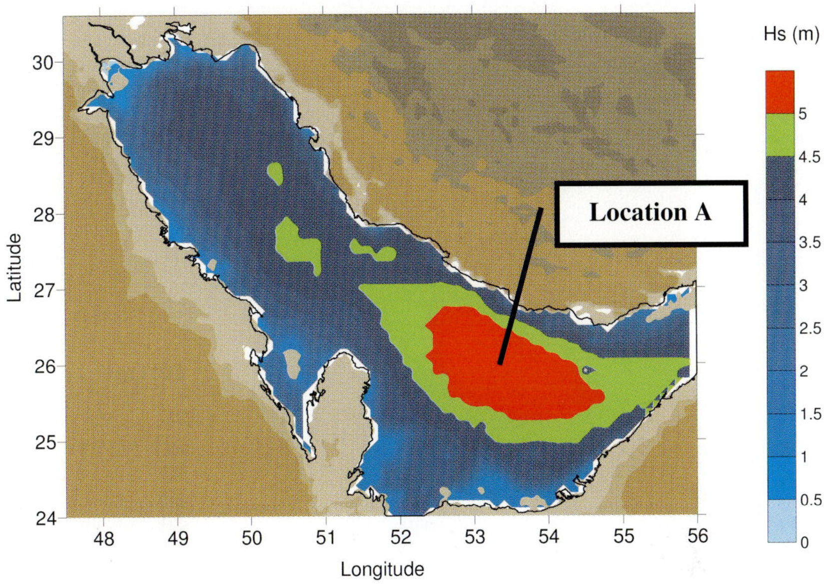

Figure 5. Maximum significant wave heights over 12 years for the Arabian Gulf.

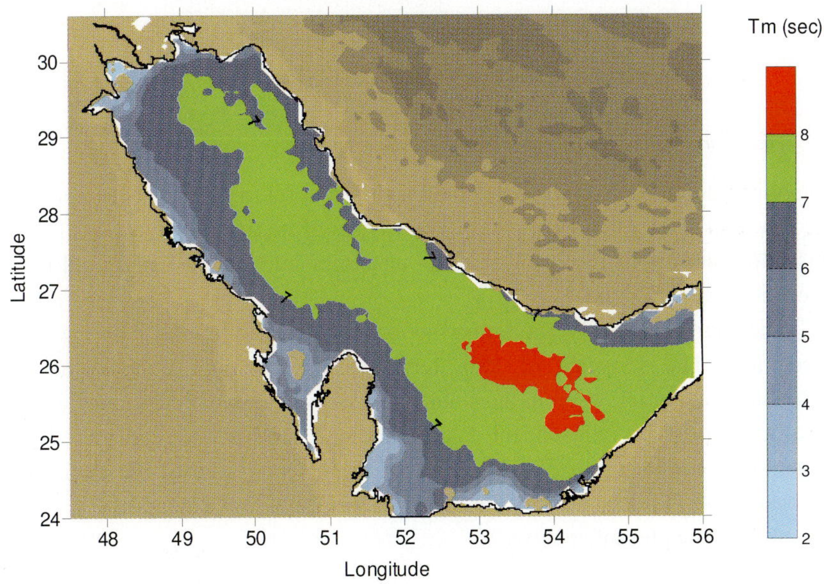

Figure 6. Maximum mean spectral wave period over 12 years for the Arabian Gulf.

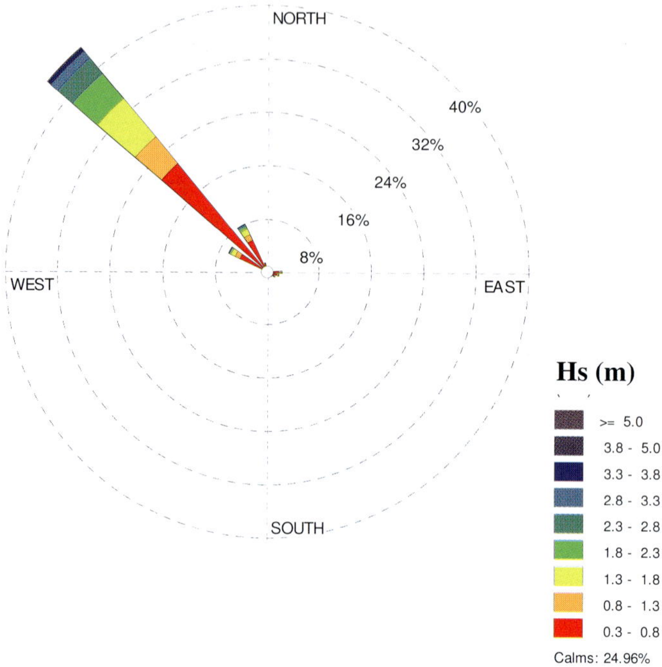

Figure 7. Wave rose at Location A in the Arabian Gulf.

4.1. *Extreme Waves*

Establishing the extreme wave heights for different return periods is essential for the safe and economic design of different types of marine structures. A lack of this information will result either in an unsafe or an over designed (and hence uneconomical) structure.

The Gumbel and Weibull distributions were used for the extreme value prediction. The statistics of long term wave prediction requires that the individual data points used in the statistical analysis be statistically independent. In order to produce independent data points, only storms were considered. The commonly used method to separate wave heights into storms, the Peak Over Threshold (POT) was used (Coles [8]). A threshold wave height of 1.0 m is selected for the present analysis.

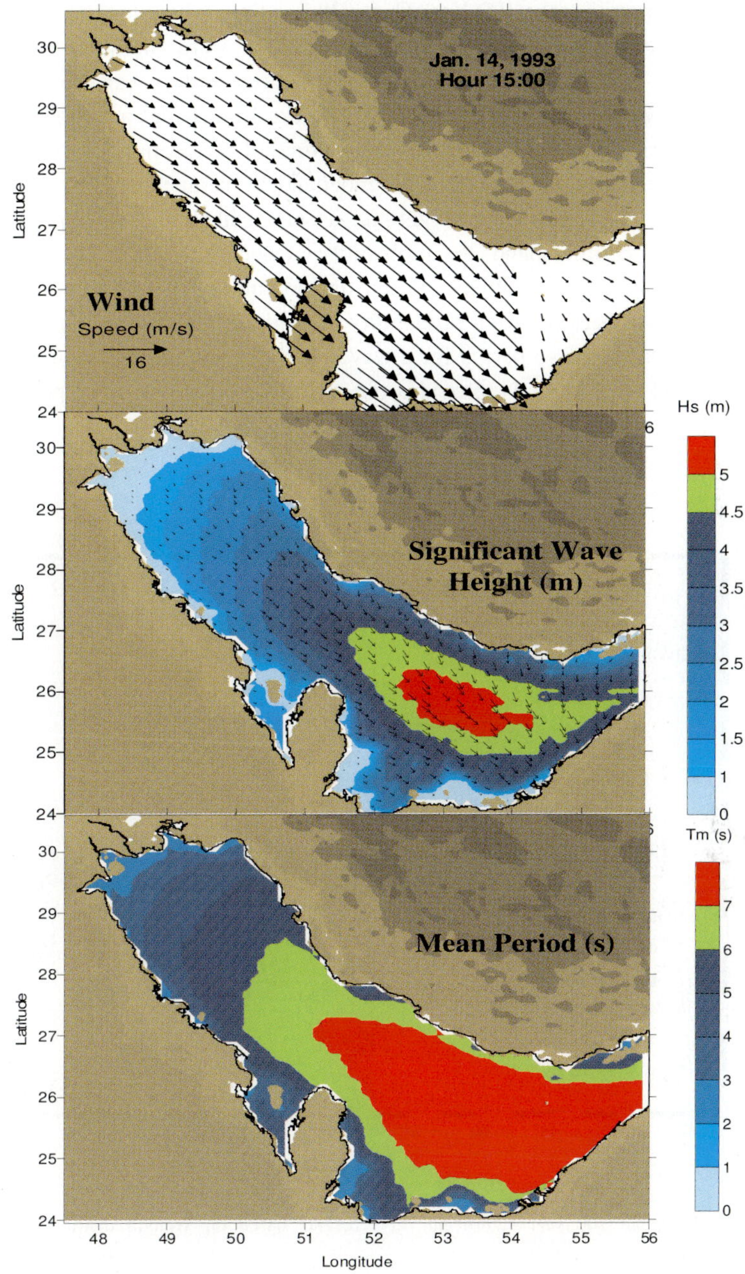

Figure 8. Wind and wave conditions over the Arabian Gulf January 14, 1993 Hour 15:00.

Figure 9 provides the 100 Year H_s for 38 locations where long term analysis was conducted. Furthermore, additional locations were analyzed along the Kuwaiti and the United Arab Emirates territorial waters can be found in Neelamani *et al.* [9] and [10] respectively). Figure 10 provides a contour plot for the 100 year H_s using all the data available (a total of 82 locations).

Wave height and wave periods are independent parameters. However it can be seen in general that as wave height increases, it is likely that wave period also increase. On the other hand, the probability of occurrence of high waves and long periods are more pronounced than the probability of occurrence of high waves and short periods. Joint probability of wave height and wave period was used to predict the wave period for a wave height of any desired return periods (Kamphuis [11]). Figure 11 provides the mean wave periods corresponding to the 100 year wave heights provided in Fig. 10.

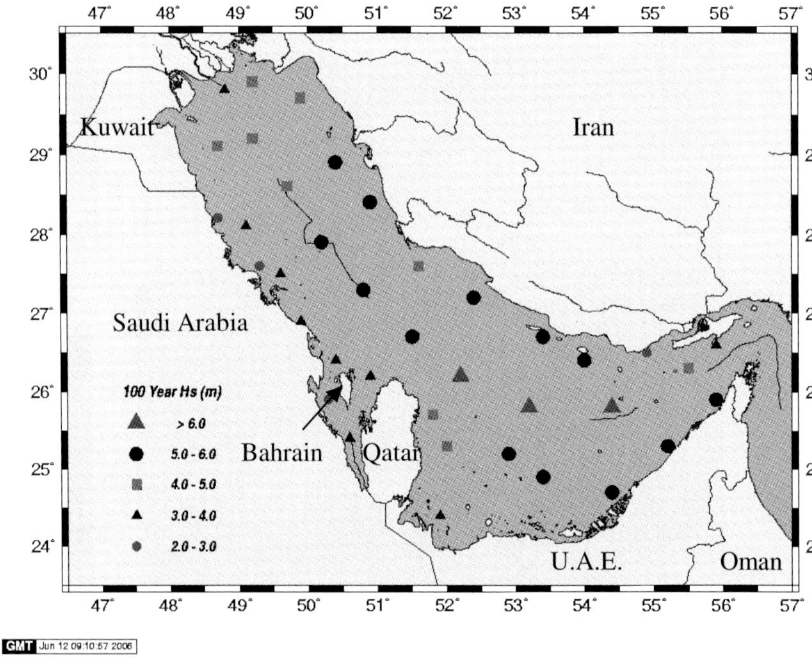

Figure 9. Predicted 100 year significant wave heights based on Weibull distribution (at 38 locations).

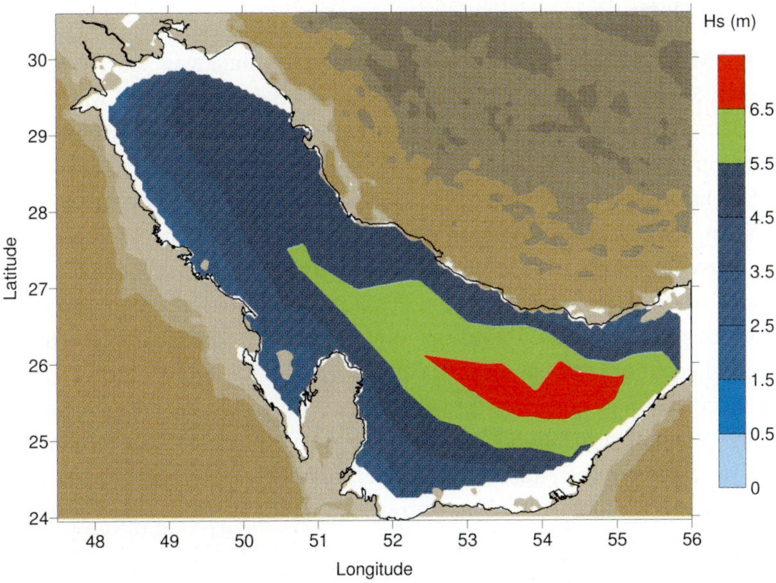

Figure 10. Predicted 100 Year significant wave heights in the Arabian Gulf.

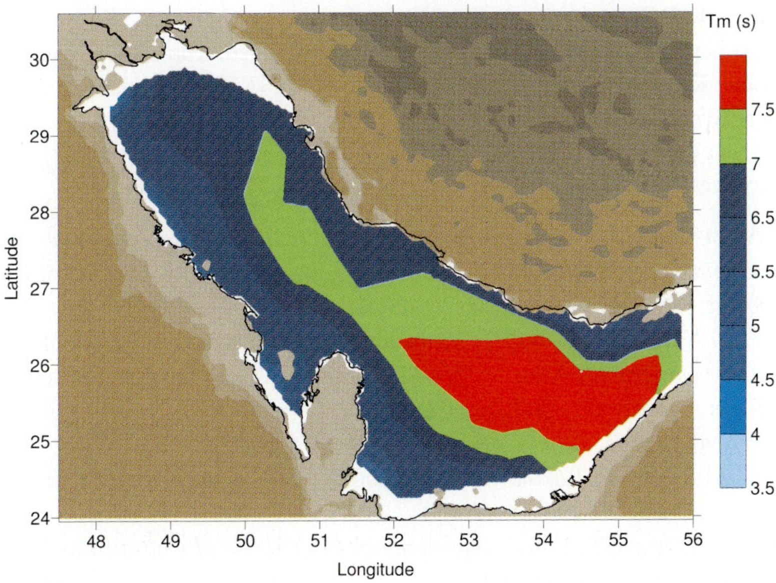

Figure 11. Predicted mean wave period corresponding to the 100 Year wave condition.

5. Conclusions

The WAM model was used to predict the wave climate in the Arabian Gulf (ROPME Sea) for 12 years using ECMWF wind data. The model was validated using data measured at the upper part of the Gulf and found to predict the wave climate very well.

The maximum significant wave height hindcasted for a period of 12 years was 5.3 m with a mean wave period of 8.0 sec. Gumbel and Weibull extreme value distributions are used to predict the significant wave heights at 82 different locations in the Arabian Gulf waters. Peak over threshold of 1.0 m was used for synthesizing the raw data. Based on Joint probability analysis, polynomial type predictive equations for the mean wave period from the significant wave heights of intended return periods were obtained.

The Weibull distribution was found to be very suitable for extreme wave height prediction in the Arabian Gulf waters. The 100 Year wave height reached up to 7.0 meters in the Gulf with a corresponding mean wave period reaching up to 8.0 sec.

Acknowledgments

This study was funded in part by Kuwait Foundation for the Advancement of Science (KFAS) and in part by Warba Insurance Company, Kuwait. Their support is greatly appreciated and highly acknowledged. The authors would like to thank Kuwait Institute for Scientific Research (KISR) for their invaluable assistance and support during this study.

References

1. A. El-Gindy and M. Hegazi, in *Atlas on Hydrographic conditions in the Arabian Gulf and the upper layer of the Gulf of Oman*, (University of Qatar, Qatar, 1996), p. 170.
2. F. Al-Yamani, J. Bishop, E. Ramadhan, M. Al-Husaini and A. Al-Ghadban, in *Oceanographic Atlas of Kuwait's Waters*, (Kuwait Institute for Scientific Research, Kuwait, 2004), p. 203.
3. WAMDI group, *J. Phys. Oceanogr.*, **18**, 1775 (1998).
4. K. Al-Salem, K. Rakha, W. Sulisz and W. Al-Nassar, *Arabian Coast 2005 Conf.*, Dubai, 27-29 Nov. (2005).
5. M. Benoit and F. Lafon, *International Conference of Coastal Engineering (ICCE)*, pp. 714-726 (2004).

6. W. Elshorbagy, M. Azam and K. Taguchi, *Journal of Waterway, Port, Coastal and Ocean Engineering*, **132**, 47 (2006).
7. K. A. Rakha, K. Al-Salem and S. Neelamani, *Kuwait Journal of Science and Engineering*, **34**(1A), 143 (2007).
8. S. Coles, in *An introduction to statistical modeling of extreme values*, (Springer-Verlag, 2001), p. 208.
9. S. Neelamani, K. Al-Salem and K. Rakha, *Emirates Journal for Engineering Research*, **11**, 37 (2006).
10. S. Neelamani, K. Al-Salem and K. Rakha, accepted by *Ocean Engineering Journal*, (2007).
11. J. Kamphuis, in *Introduction to Coastal Engineering and Management*, (World Scientific, 2000), p. 472.

NUMERICAL STUDY ON THE THREE-DIMENSIONAL DAM-BREAK BORE INTERACTING WITH A SQUARE CYLINDER

TSO-REN WU

Institute of Hydrological Sciences, National Central University
300 Jhongda Rd, Jhongli, Taoyuan 32048, Taiwan

PHILIP L.-F. LIU

School of Civil and Environmental Engineering, Cornell University
Ithaca New York 14853, USA

The interaction between a dam-break bore and a vertical square cylinder is studied by using a Reynolds Averaged Navier-Stokes (RANS) model with a $k-\varepsilon$ turbulence closure, a large-eddy simulation (LES) model, and direct numerical simulation (DNS). In all three numerical models the free-surface is tracked by the volume-of-fluid (VOF) method. Numerical results are compared with laboratory measurements in terms of the total force acting on the cylinder and the time history of velocity measured at a fixed point upstream of the cylinder. Overall, all three numerical models predict the force and velocity reasonably well, while the RANS model and LES appear to provide more accurate prediction during the period when strong turbulence is present. Based on the RANS model results, the turbulent intensity (TI), defined as the velocity fluctuation divided by the mean velocity, is in the range of 0.2 ~ 0.4 in the wake zone behind the cylinder and 0.4 ~ 0.8 in the plunging waves reflected from the end wall. The results from the RANS model and LES are compared with each other. The LES results reveal finer structure compared to the RANS results in terms of free-surface profiles and vorticities. The distribution of the vorticity field is similar to TI indicating that they are strongly correlated. The thin layer of water in front of the dam initially prevents the direct interaction between the bore and the bed. However, a series of vortexes generated by the edge of the square cylinder transports the water from the near bed region to the free surface.

1. Introduction

Videos taken during the 2004 Indian Ocean tsunami have shown that tsunami waves behaved like turbulent bores in many coastal areas. As these bores rushed towards land, they destroyed buildings and structures and in many places, carried a huge amount of debris. Field surveys also showed that scour holes appeared at the foot of many buildings, which are the result of interactions between buildings and runup and downrush flows (Liu et al., 2005). To reduce the tsunami damage in the future, it is important to develop a new design guideline for coastal structures subject to tsunami force loading.

However, our knowledge on tsunami-structure interaction is still very limited. In the past, the objectives of most of research work on wave-structure interaction have been towards evaluation of the functionality of a designed structure. Therefore, only the global characteristics, such as the transmission and reflection coefficients, are the focus of research. Because of the complexity of coastal structure, earlier research work has relied heavily on laboratory experiments and empirical or semi-empirical correlations (Irschik et al., 2003). Laboratory experiments often suffer from constraints on the range of testing physical parameters and scale effects. In theoretical approach, the potential flow assumption has been commonly adopted (Ting and Kim, 1994; Zhuang and Lee, 1996; Tang and Chang, 1998; Huang and Dong, 1999, Abul-Azm, 1994; Ertekin and Becker, 1996; Losada et al., 1996). Strictly speaking, the potential flow assumption is easily violated in the complex flow field associated with wave-structure interaction, especially when wave breaking occurs.

In recent years, because of the availability of computational resources, researchers have begun to develop 2D and 3D numerical models based on Navier-Stokes equations (NSE) to study wave-structure interaction. These models include the Lagrangian approach, e.g., the Smoothed Particle Hydrodynamics (SPH) method (Gomez-Gesteira and Dalrymple, 2004; Dalrymple and Rogers, 2005), the Eulerian approach, e.g., the Direct Numerical Simulation (DNS) (Chen et al., 1999), the Reynolds Averaged Navier-Stokes (RANS) with the volume-of-fluid (VOF) method (Lin and Liu, 1998a,b), as well as the Large-Eddy Simulation (LES) with the VOF method (Liu et al., 2005). Although the applications of the 3D models are still limited, evidence has demonstrated that these models are promising and deserve continuing development and refinement.

In this paper, our goal is to investigate interaction between a bore generated by a dam-break, and a slender square cylinder. To achieve this objective, we adopt both the RANS model and the LES model with a well-developed VOF method to calculate the velocity field, including the turbulent intensity, wave forces, and of course free surface elevations. Numerical results will be checked with the experimental data provided by Yeh, H. and Petroff, C. (http://engr.smu.edu/waves/solid.html). The inter-model comparisons will be discussed in terms of the importance of the turbulence modeling.

We will introduce the numerical models including RANS, LES and VOF in Section 2. The numerical setup and simulation information will be included in Section 3. The simulation results and validations will be presented in Section 4. In Section 5, we will discuss the spatial distribution of *TI*, the difference between RANS and LES in terms of the free-surface profiles, velocity fields, and vorticity structure, and the effects of the wet floor. The conclusion will be made on Section 6.

2. Numerical Models

In this section, we shall briefly review the RANS and LES model.

2.1. *Reynolds Averaged Navier-Stokes Equations*

The RANS model consists of continuity and momentum equations for the ensemble averaged flow quantities (Pope, 2001). A turbulent closure model must be selected in order to include the effects of turbulence on the mean flow. In this paper, we adopt the conventional eddy viscosity model with the $k-\varepsilon$ equations as the closure (Rodi, 1980; Lemos, 1992; Pope, 2001). We shall present only the essential equations to describe the model for completeness. More detailed description of the model can be found in Wu (2004) and Liu et al. (2005).

The governing equations for the RANS model are:

$$\frac{\partial \langle u_i \rangle}{\partial x_i} = 0 \tag{2.1}$$

$$\frac{\partial \langle u_i \rangle}{\partial t} + \frac{\partial \langle u_i \rangle \langle u_j \rangle}{\partial x_j} = -\frac{1}{\rho}\frac{\partial \langle P \rangle}{\partial x_i} + \frac{1}{\rho}\frac{\partial}{\partial x_j}\left[\mu_{eff}\left(\frac{\partial \langle u_i \rangle}{\partial x_j} + \frac{\partial \langle u_j \rangle}{\partial x_i}\right)\right] + g_i \tag{2.2}$$

where the bracket $\langle\ \rangle$ denotes the Reynolds averaging, i and $j = 1,2,3$, $\mu_{eff} = \mu + \mu_t$ is the effective viscosity with μ being the molecular viscosity and μ_t the turbulent viscosity, ρ the fluid density, u_i the *i-th* component of fluid particle velocity, *P* the pressure, and *g* the gravitational acceleration.

The turbulent viscosity is characterized by two parameters: the turbulent kinetic energy (TKE), *k*, or the velocity scale $\vartheta = \sqrt{2k}$, and a length scale ℓ. The dimensional analysis shows that: $\mu_t = \rho C_\mu \vartheta \ell$, where C_μ is a dimensionless constant (Rodi, 1980; Pope, 1975). The length scale can be further dimensionally related to the energy dissipation rate, ε, as $\ell = k^{3/2}/\varepsilon$. Thus, the turbulent viscosity can be expressed as (Rodi, 1980; Pope, 1975)

$$\mu_t = \rho C_\mu \sqrt{k}\ell = \rho C_\mu \frac{k^2}{\varepsilon} \tag{2.3}$$

To close the problem, one still needs to find information for k, ε and C_μ.

2.1.1. $k-\varepsilon$ model

The standard and commonly used balance equations for k and ε are (Launder and Spalding, 1974; Rodi, 1980):

$$\frac{\partial k}{\partial t}+\frac{\partial\langle u_j\rangle k}{\partial x_j}=\frac{1}{\rho}\frac{\partial}{\partial x_j}\left[\left(\mu+\frac{\mu_t}{\sigma_k}\right)\frac{\partial k}{\partial x_j}\right]+P_k-\varepsilon \tag{2.4}$$

$$\frac{\partial \varepsilon}{\partial t}+\frac{\partial\langle u_j\rangle \varepsilon}{\partial x_j}=\frac{1}{\rho}\frac{\partial}{\partial x_j}\left[\left(\mu+\frac{\mu_t}{\sigma_\varepsilon}\right)\frac{\partial \varepsilon}{\partial x_j}\right]+\frac{\varepsilon}{k}\left(C_{\varepsilon 1}P_k-C_{\varepsilon 2}\varepsilon\right) \tag{2.5}$$

in which σ_ε, $C_{\varepsilon 1}$, and $C_{\varepsilon 2}$ are empirical coefficients and the production term, P_k, is expressed as

$$P_k=-\left\langle u_i^{'}u_j^{'}\right\rangle\frac{\partial\langle u_i\rangle}{\partial x_j}\approx\frac{\mu_t}{\rho}\left(\frac{\partial\langle u_i\rangle}{\partial x_j}+\frac{\partial\langle u_j\rangle}{\partial x_i}\right)\frac{\partial\langle u_i\rangle}{\partial x_j} \tag{2.6}$$

where $u_i^{'}=u_i-\langle u_i\rangle$ denotes the velocity fluctuation (or the residual velocity).

The $k-\varepsilon$ model has been widely used and contains five universal empirical constants C_μ, σ_k, σ_ε, $C_{\varepsilon 1}$, and $C_{\varepsilon 2}$. The standard $k-\varepsilon$ model employs values for the constants that are calibrated by a wide range of turbulent flows (Rodi, 1980):

$$C_\mu=0.09;\sigma_k=1.00;\sigma_\varepsilon=1.30;C_{\varepsilon 1}=1.44;_{\varepsilon 2}=1.92$$

2.1.2. *Boundary conditions for* $k-\varepsilon$ *model*

In general two types of boundary conditions are required for solving $k-\varepsilon$ equations. On the free-surface the boundary condition employed in the present model is the zero flux condition: $\partial k/\partial n=\partial\varepsilon/\partial n=0$. On the other hand, along a solid boundary the wall boundary conditions ("law of the wall") are applied (Pope, 2001):

$$\varepsilon=P_k=-\left\langle u^{'}v^{'}\right\rangle\frac{d\langle u\rangle}{dy}=\frac{u_\tau^3}{\kappa y},k=\frac{u_\tau^2}{\sqrt{C_d}} \tag{2.7}$$

where y is the distance from the wall, κ (= 0.41) is the von Karman constant, and u_τ is the friction velocity given by: $u_\tau=\sqrt{|\tau_w|/\rho}$. The eddy viscosity is simply a linear function of the distance from the boundary, i.e., $\nu_t=\kappa u_\tau y$.

2.2. *Large Eddy Simulation*

For the purpose of comparison, the spatially filtered LES model is also used (Pope, 2001). The unresolved small scale motions are modeled by the simple Smagorinsky sub-grid scale (SGS) model (Deardroff, 1970). Thus, the filtered continuity and momentum equations are (Pope, 2001; Wu, 2004; Liu et al., 2005):

$$\frac{\partial \bar{u}_i}{\partial x_i} = 0 \tag{2.8}$$

$$\frac{\partial \bar{u}_i}{\partial t} + \frac{\partial \bar{u}_i \bar{u}_j}{\partial x_j} = -\frac{1}{\rho}\frac{\partial \bar{p}}{\partial x_i} + g_i + \frac{1}{\rho}\frac{\partial}{\partial x_j}\left[\bar{\mu}_{eff}\left(\frac{\partial \bar{u}_i}{\partial x_j} + \frac{\partial \bar{u}_j}{\partial x_i}\right)\right] \tag{2.9}$$

in which the over-bar denotes the spatial filtering, $\bar{u}_i$ and $\bar{p}$ are the filtered velocity and pressure, and $\bar{\mu}_{eff}$ is the effective viscosity in LES, which is defined as:

$$\bar{\mu}_{eff} = \mu + \mu_{SGS} \tag{2.10}$$

where μ_{SGS} is the sub-grid-scale (SGS) eddy viscosity representing the residual motions. The simplest SGS eddy viscosity model is the Smagorinsky model:

$$\mu_{SGS} = \rho\left(C_s \Delta\right)^2 \sqrt{2\bar{S}_{ij}\bar{S}_{ij}} \tag{2.11}$$

where $\bar{S}_{ij} \equiv \left(\partial \bar{u}_i / \partial x_j + \partial \bar{u}_j / \partial x_i\right)$ denotes the filtered strain tensor and $\left(C_s \Delta\right)$, being the product of the Smagorinsky coefficient C_s and the filter width Δ, represents the Smagorinsky length scale. In the present model, the filter width, Δ, is defined by the grid size:

$$\Delta = \left(\Delta x_1 \Delta x_2 \Delta x_3\right)^{1/3} \tag{2.12}$$

where Δx_i $(i = 1, 2, 3)$ are the grid lengths. Under the isotropic turbulence condition, the Smagorinsky coefficient is $C_s \sim 0.2$ (Pope, 2001). However, C_s is usually not a constant and the value varies from 0.1 to 0.2 for different flows and numerical schemes. In the present simulations C_s is specified as 0.15, which is suggested by Lin and Li (2003) for free surface flows.

2.2.1. *Boundary conditions of LES*

In LES, the filtered velocity field is assumed to satisfy the no-slip condition on the solid boundary. The near-wall damping function derived by Cabot and Moin

(2000) is used to approximate the eddy viscosity in the first cell adjacent to the wall:

$$\frac{\nu_t}{\nu} = \kappa y_w^+ \left(1 - e^{-y_w^+/A}\right)^2 \tag{2.13}$$

where $y_w^+ = yu_* / \nu$ is the distance to the wall in wall units, $\kappa = 0.41$ and $A = 19$.

2.3. *VOF Numerical Algorithms*

The RANS equations and the filtered continuity and momentum equations in LES are solved by using a two-step projection method with the finite-volume discretization (Bussmann et al., 2002, Wu, 2004, and Liu et al., 2005). In order to track the air-water and fluid-obstacle locations on a fixed grid system, the volume of fluid (VOF) method (Hirt and Nichols, 1981; Bussmann et al., 2002) and the moving obstacle algorithm (Wu, 2004, Liu et al., 2005) are adopted. In the VOF method, a volume of fluid function, f, representing the volume fraction of water within a computational cell is introduced. The value of f equals to one, if the cell is full; zero if empty, and $0 < f < 1$ if the cell is partially occupied by water. Since f is advanced by fluid flows, it can be described by:

$$\frac{\partial f}{\partial t} + \frac{\partial \bar{u}_i f}{\partial x_i} = 0 \tag{2.14}$$

The momentum equation is also modified to be:

$$\frac{\partial (f\bar{u}_i)}{\partial t} + \frac{\partial (f\bar{u}_i\bar{u}_j)}{\partial x_j} = -\frac{1}{\rho}\frac{\partial f\bar{p}}{\partial x_i} + fg_i + \frac{1}{\rho}\frac{\partial}{\partial x_j}\left[\bar{\mu}_{eff}\left(\frac{\partial f\bar{u}_i}{\partial x_j} + \frac{\partial f\bar{u}_j}{\partial x_i}\right)\right] \tag{2.15}$$

The piecewise linear interface calculation (PLIC) (Rider and Kothe, 1998; Kothe et al., 1999) is adopted to reconstruct the air-water interface. Here we have used LES to briefly explain the VOF method, while same approach is applied to the RANS model.

3. The Computational Domain and Numerical Setup

The computational domain is a rectangular tank, which is 1.6m long, 0.61m wide, and 0.75m high. The tank is initially divided into two parts by a gate, which is located at 0.4 m from an end wall (see Figure 1). A water reservoir is created behind the gate with the height of 0.3m so that the total volume of the reservoir is 0.4m x 0.61m x 0.3m. A square cylinder with the cross-section of 0.12m x 0.12m is placed at 0.5m downstream of the gate and 0.24m from the near side wall of the tank. The numerical setup intends to duplicate the experiments conducted by Drs. Yeh and Petroff. In their experiments because of the difficulty of completely draining off the water on the downstream side of the gate and the leakage through the edges of the gate, a thin layer of water (approximately 1 cm deep) always appeared on the bottom of the tank before the gated was opened. As the gate was removed rapidly, the water in the reservoir collapsed at the gate and rushed into the tank. The water surged towards and past the square cylinder and collided with the downstream end wall. The time-histories of the total horizontal force acting on the cylinder and the time-histories of the fluid velocity at a fixed location were collected in the experiments.

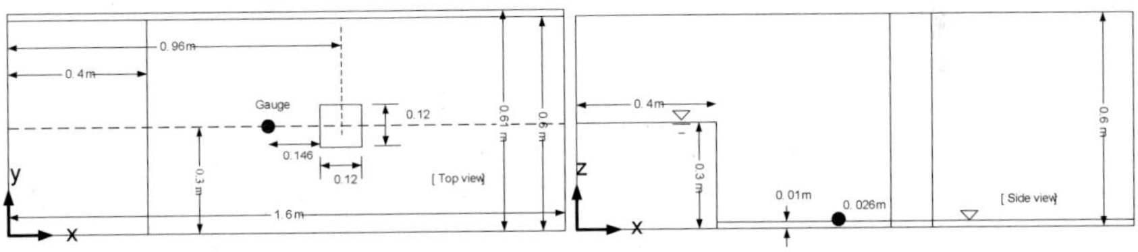

Figure 1. The setup for dam-break bore interacting with a square cylinder; Left: Top View and Right: side view.

The only difference between the computational domain and the actual experimental setup is that the ceiling of the computational domain is lowered down to 0.6 m to save CPU time. Because the width of the square cylinder is in the same order of magnitude of the channel width and the initial bore height, the vortices are mainly generated at the corners of the square cylinder and the complex flow is symmetric with respect to the centerline of the tank. This allows us to conduct numerical simulations in a half width to further save computation resources. The grid convergence tests have been performed with reduced grid sizes. The results with different grid sizes were compared for the wave forces acting on the cylinder. The differences become marginal when the uniform grid system with $\Delta x = \Delta y = \Delta z = 0.01$m or smaller is used. In order to make comparison of the velocity field, a numerical gauge is placed 0.146m downstream face of the cylinder, and 0.026m above the bottom. The nearest-neighbor interpolation method is adopted to obtain the velocity values at the gauge point. The details of the experimental setup and the computational domain are shown in Figure 1.

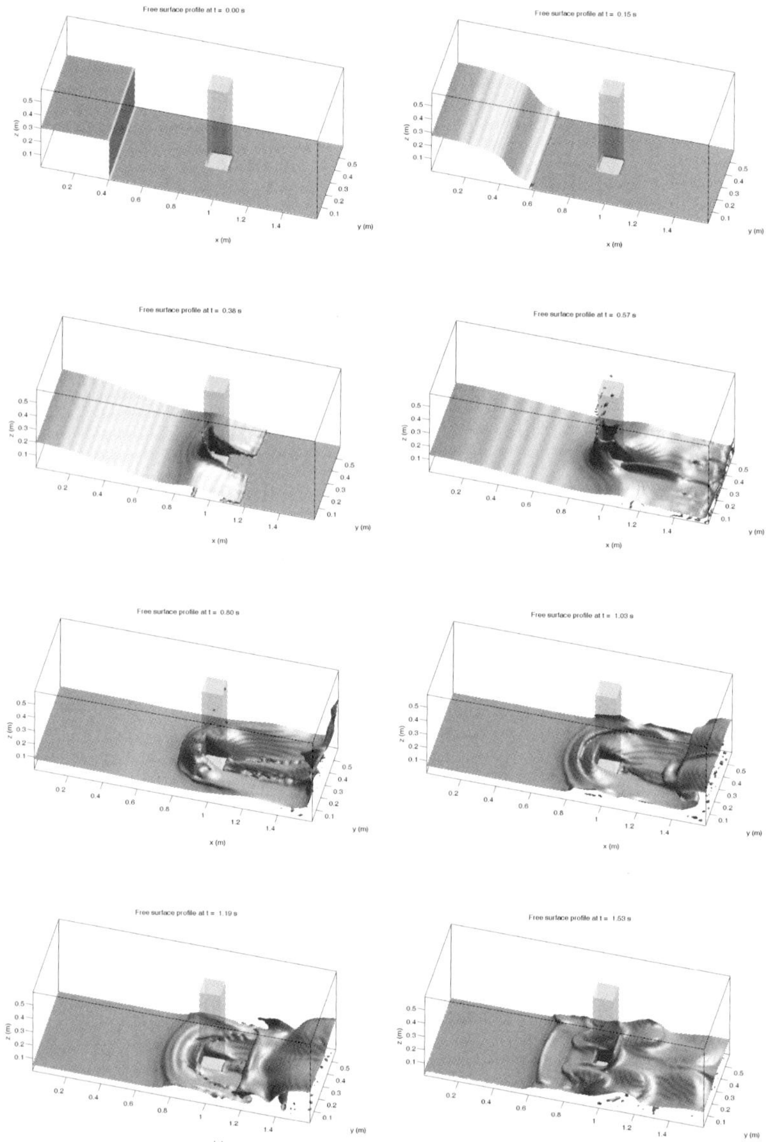

Figure 2. Snapshots of the free surface profiles obtained by the RANS model. The free surface location is represented by blue color, which is defined as the iso-surface where VOF of water $f = 0.5$. The gray color denotes the position of the square cylinder.

4. Results and Validations

4.1. *Simulation Results – A General Description*

Figure 2 shows the numerical results for a sequence of snapshots of free surface profiles from the RANS model. At t = 0.15s after the gate was lifted, a "mushroom head" is formed in the front of the surge. Up to this stage, the flow field is nearly 2D except for the weak boundary layer generated along the side walls. The mushroom head plunges into the floor and forms a bore moving towards the square cylinder; a part of "air bubble" is trapped in the water column. At t = 0.38s, the bore impinges on the cylinder and water piles up the front face of the cylinder. While vigorous splashes occur in front of the cylinder at t = 0.57s, flow separations can be seen behind the cylinder. The impinging bore moves farther downstream and merges in the wake zone. A series of eddies are generated at the edges of the cylinder and can be identified at t = 0.80s. At the same time, the runup on the cylinder is falling into the water body below generating outgoing waves. The main bore is reflected back by the end wall at t = 1.03s. Because of the flow convergence behind the cylinder, a strong plunging jet is formed in the reflected bore that hits the tank bottom and bounces back up at t = 1.19s. The reflected bore attacks the cylinder again at t = 1.53s and creates a complex flow pattern.

The general characteristics of the numerical results for the free surface profiles obtained from the LES model are very similar to those from the RANS model and will not be shown here. The differences in other aspects of numerical results from these two models will be discussed in the following sections.

4.2. *Validations*

To check the performance of the RANS model and LES, numerical results for the horizontal force acting on the cylinder are compared with experimental data (Figure 3). The time-histories of the horizontal force component, predicted by numerical models, are calculated by integrating the pressure acting on the cylinder, neglecting the shear force. As shown in Figure 4, the shear force is only a very small fraction of (about two orders of magnitude smaller than) the total force. In order to examine the importance of turbulence in computing the net force on a square cylinder, we also obtain numerical solutions by solving the DNS equations directly without resolving the turbulence. All three models use the same mesh system.

In the force comparison (Figure 3), it is shown clearly that before the bore reaches the square cylinder (t < 0.31s), the differences between three numerical models (RANS, LES, and DNS) and experimental data are barely visible, indicating that only a very small amount of turbulence has been generated. However, after the bore impacts on the square cylinder, the flow condition becomes turbulent and three-dimensional. Turbulence dissipation is under-estimated by the DNS since the grid size is too large to resolve the small-scale turbulence. The difference, in terms of the phase speed of reflected bore, among these three numerical models is visible at t = 1.25s. Because of the under-estimation of turbulence, the bore phase speed is over-estimated by the DNS. As a result, the reflected bore from the end wall reaches the cylinder earlier in the DNS than that in the other two solutions with turbulence considered. By the same reason, the net force obtained by the DNS contains stronger oscillations than those in the other two models.

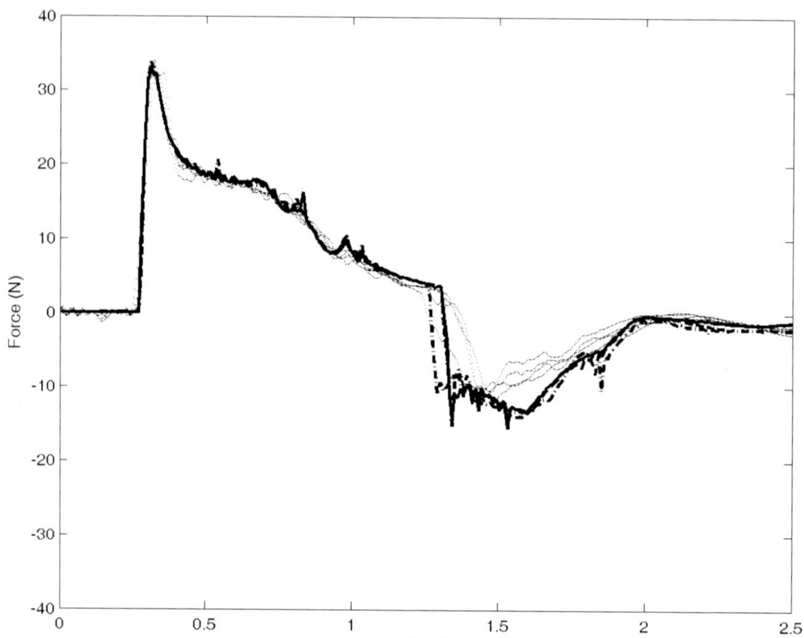

Figure 3. The time-histories of horizontal wave force on a square cylinder. The dots are four sets of experimental data. The solid line denotes the numerical solutions from the RANS model. The dashed line represents the numerical solutions from LES. The dashed-dot line corresponds to the numerical results from DNS.

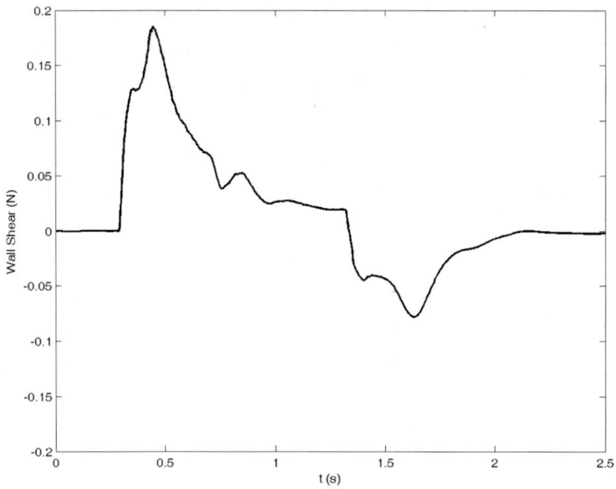

Figure 4. The time-history of horizontal shear force acting on a square cylinder. This solution is obtained from the RANS model.

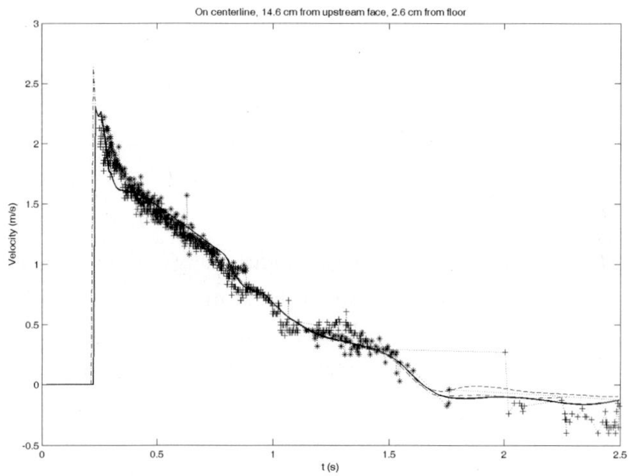

Figure 5. The time-histories of the horizontal velocities. The gauge is located on the centerline; 14.6 cm from upstream face of the square cylinder, 2.6cm from the floor. The dots are two sets of experimental data. The solid, dashed, dashed-dot lines denote the numerical solutions from the RANS model, DNS, and LES respectively.

Overall, all the numerical results from turbulence models are very close to the laboratory measurements, especially in terms of the maximum positive and negative forces acting on the square cylinder. It is interesting to note that the solutions predicted by the RANS model and LES turbulence models agree with each other, although these two turbulence models adopt different assumptions: the RANS model solves the ensemble averaged flow field, while the LES solve the spatially filtered velocity field. The agreement between RANS and LES results suggests that the turbulence is not very strong, especially the large scale turbulence.

Figure 5 shows the time-history horizontal velocity comparison of numerical simulations with the experimental measurements at the gauge position as shown in Figure 1. Once again the numerical results obtained from the RANS model, LES, and DNS and the experimental data agree well with each other. Numerical results and experimental data show a larger deviation after t = 0.17 s, when the reflected bore reaches the gauge position.

5. Discussions

5.1. *Spatial Distribution of Turbulent Characteristics*

Figure 6 shows a sequence of snapshots of turbulence intensity denoted by *TI*, defined as:

$$TI \equiv \frac{u'}{U} \tag{5.1}$$

where $$u' \equiv \sqrt{\frac{1}{3}\left(u_1'^2 + u_2'^2 + u_3'^2\right)} = \sqrt{\frac{2}{3}k}; \quad U \equiv \sqrt{\langle u_1 \rangle^2 + \langle u_2 \rangle^2 + \langle u_3 \rangle^2}$$

In the above equation, turbulence kinetic energy k is obtained from the RANS model and U is obtained from the local mean velocity. In Figure 6, the rolling bore tends to generate weak turbulence in the bottom and side wall boundary layers. A stronger *TI* is generated after the bore impinges upon the cylinder at t = 0.56 s. In the wake zone two series of vortices appear in which the *TI* is strong (t = 0.80 s). The magnitude of *TI* is up to 0.5 (t = 0.80 s and 1.03 s), which is in the same order of the magnitude as that observed in similar laboratory experiments performed by Amason (2005). We note that at t = 0.80s the runup on the cylinder only generates a very low level turbulence intensity, which is consistent with the observation that the total wave forces up to this time is not sensitive to turbulence. As the bore is reflected back from the downstream end wall, plunging breaking occurs and strong turbulence is generated (t = 1.53s), which influence the wave force calculations.

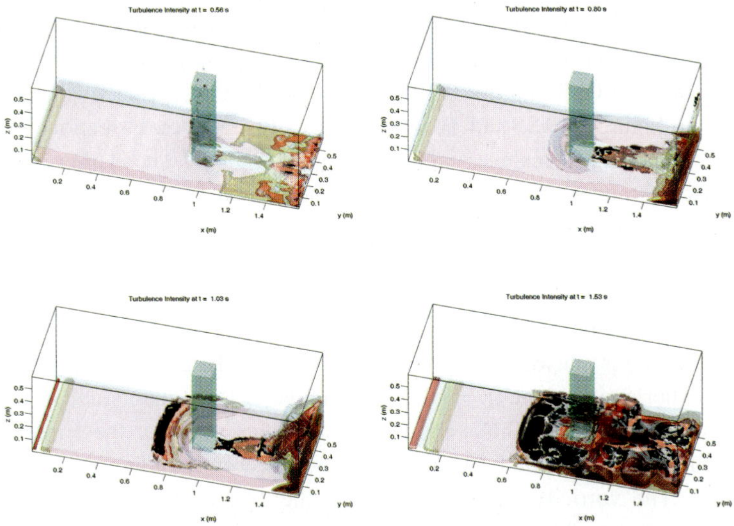

Figure 6. Snapshots of the turbulence intensity (*TI*) obtained by the RANS model. The iso-surfaces of turbulence intensity are denoted by various colors: Magenta: 0.05; Yellow: 0.1; Red: 0.2; Black: 0.5. The light blue indicates the free-surface elevation.

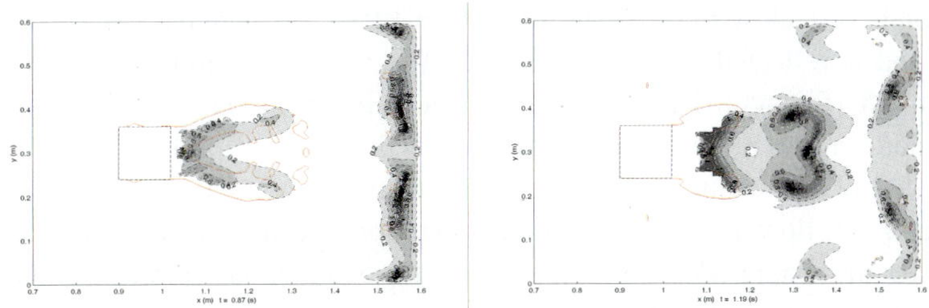

Figure 7-1. Snapshots of the turbulence intensity on z = 0.035m cross-section. Denser colors indicate stronger turbulence intensities. The values of colors can be referred to the contour lines.

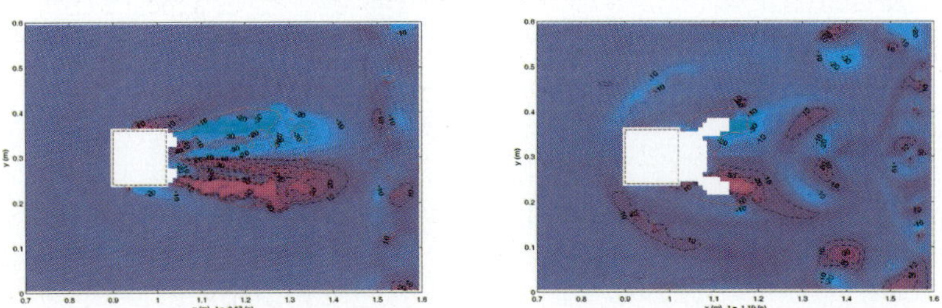

Figure 7-2. Snapshots of the *x*-component of vorticity on z = 0.035m cross-section. Color indicates the magnitude of vorticity. The values of colors can be referred to the contour lines.

Because the dam-break flow becomes fully three-dimensional after t = 0.38s, snapshots on various vertical and horizontal cross-sections should provide further insights on the flow complexity. Here we display results on three cross-sections at: z = 0.035m, y = 0.295m, and x = 1.255m, respectively. They are shown in Figures 7, 8, and 9, respectively. In each cross-section the *TI* is presented in a gray scale as well as in contour lines. The darker gray scale indicates a stronger *TI*. In Figure 7-1, the horizontal cross-section is chosen at z = 0.035m, which is roughly 1/3 of the bore height in the wake zone. As mentioned before, the flow separation of the incoming bore generates two sets of eddies in the wake region, which can be clearly observed at t = 0.87s. At t = 1.19s, a large region (x = 1.25 ~ 1.37m and y = 0.2 ~ 0.4m) is inundated with high *TI*, which is caused by the reflected bore from the end wall. In general, the *TI* on the horizontal plane (z = 0.035 m) is in the order of magnitude of 0.2 ~ 0.4 in the wake zone at t = 0.68s and 0.87s, and 0.4 ~ 0.8 after the bore has been reflected from the end wall. Figure 7-2 shows the corresponding x-component of vorticity. The vorticity patterns show that the *TI* and the vorticity are highly correlated. At t = 0.87s, two strips of strong vorticity, about $\pm$30 s^{-1}, can be seen in the wake zone indicating that a strong shear is generated by the interaction between the fast-moving bore and the square.

Figure 8 series shows the vertical cross-section at y = 0.295m. This cross-section is very close to the centerline. At t = 1.03 s, the bore has reached the end wall and is reflected back into the tank. The maximum turbulence intensity is about 0.8 (Figure 8-1). A region of relatively strong *TI* (~ 0.4) behind the cylinder can also be observed. The turbulence is mainly caused by the shear from the edge of the cylinder. Figure 8-2 shows the velocity vectors at t = 1.03s. Because the flow field is fully three-dimensional, different colors are used to indicate the magnitude of the velocity in the y-direction (spanwise direction). The velocity vector pattern suggests that a water jet is plunging into the current in the opposite direction and is bouncing up. The plunging jet does not reach the bottom of the tank. This feature can also be seen in the vorticity plot (Figure 8-3). A strong vorticity, about -30s^{-1}, is caused by the plunging wave.

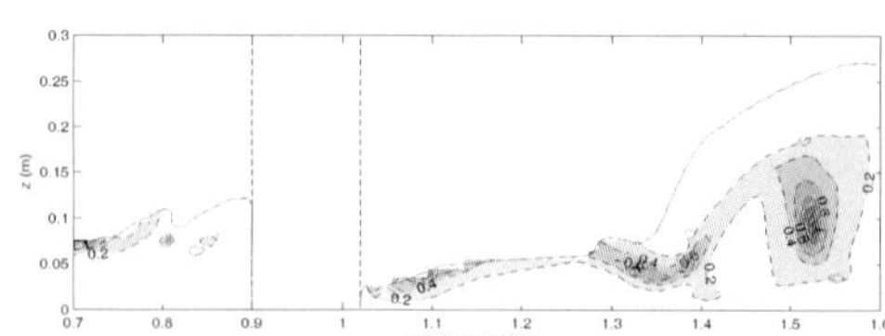

Figure 8-1. A snapshot of the turbulence intensity on y = 0.295m cross-section at t = 1.03s. The solid lines indicate the water surface where VOF of water f = 0.5. Denser colors indicate stronger turbulence intensities. The values of colors can be referred to the contour lines.

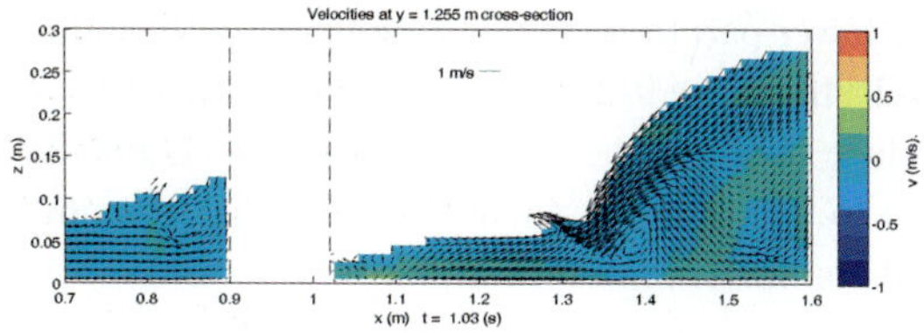

Figure 8-2. A snapshot of velocities on y = 0.295m cross-section at t = 1.03s. Vectors denote the velocity magnitude and direction on the vertical plane. Color indicates the value of the spanwise velocity.

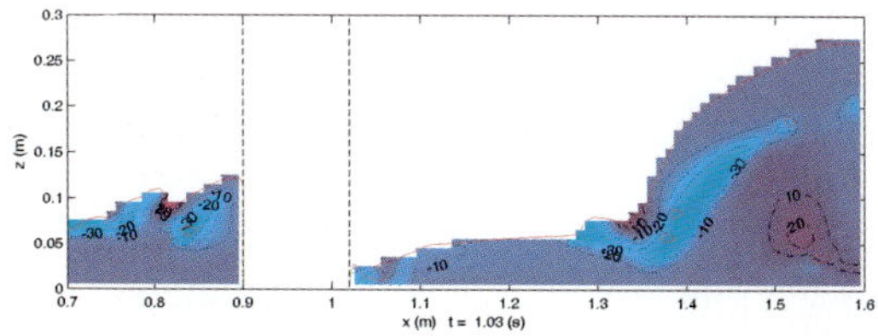

Figure 8-3. A snapshot of the y-component vorticity on y = 0.295m cross-section at t = 1.03s. Color indicates the magnitude of vorticity. The values of colors can be referred to the contour lines.

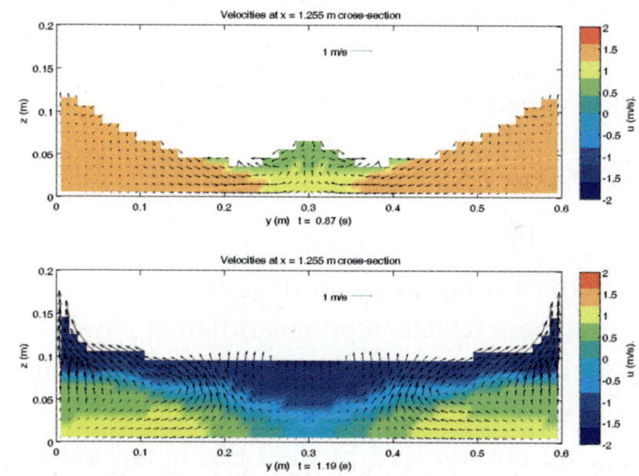

Figure 9-1. Snapshots of velocity fields on x = 1.255 m cross-section at two different times, t = 0.87s and 1.09s, respectively. Vectors denote the spatial distribution of velocities on the vertical plane. Color indicates the value of the streamwise velocity.

Figure 9 series shows the y-z (vertical) cross-section at x = 1.255m. This cross-section is located at a distance twice of the cylinder width behind the cylinder. Figure 9-1 shows the corresponding velocity fields at these instants. The velocity vectors at t = 0.87s demonstrate that the flow velocities in the wake zone are converging near the bottom and diverging in the upper part. At t = 1.19s, the contour color shows that the velocity in the x-direction changes from 1 m/s in the lower part to -1.5 m/s in the upper part and thus great *TI* can be seen in Figure 9-2. In Figure 9-2, at t = 0.8s, two jet-like eddies can be observed near the free surface in which the turbulence intensity is relatively strong (about 0.2) as compared to that in the ambient flow. At t = 1.19s, the bore is reflected from the end wall and the *TI* is as high as 0.8.

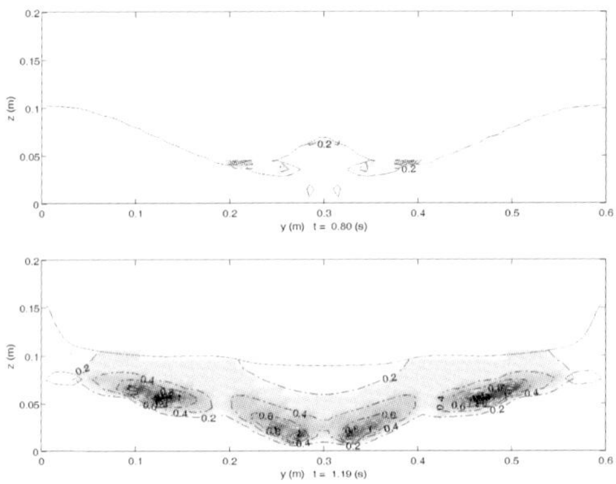

Figure 9-2. Snapshots of the turbulence intensity on the vertical y-z cross-section at x = 1.255m. The upper panel denotes t = 0.8s and the lower panel t = 1.19s. Darker colors indicate stronger turbulence intensities. The values of colors can be referred to the contour lines. The solid lines indicate the water surface where VOF of water f = 0.5.

5.2. *RANS Model versus LES*

As mentioned before, the RANS model is based on the ensemble averaging while the LES is derived using a spatial filtering method. Therefore, numerical results obtained from these models represent different physics and should not be the same. However, these two models should produce very similar results as long as the turbulence is weak and/or most of the turbulence energy is contained in small scales, which are not resolved by the LES. In this section, we shall make an attempt to qualitatively compare the numerical results from both models. Figure 10-1 shows the free-surface configuration at t = 1.03s. Results obtained from both models are very similar. However, the LES results clearly reveal more details than the RANS results. This is not surprising since the RANS results are ensemble averaged, while the LES results still contain the large eddies (fluctuations) larger than the grid resolution. Figure 10-2 shows the velocity field on the horizontal plane z = 0.035m at t = 1.03s. Although the flow patterns look alike in the results from both models, the locations of eddies are quite different and the flow structure from LES is slightly more complex. Figure 10-3 shows the

z-component of vorticity at t = 1.03s on z = 0.035 m cross-section. LES predicts slightly stronger vorticity than the RANS model. The spatial variation of the vorticity is also larger in the LES results.

RANS model LES

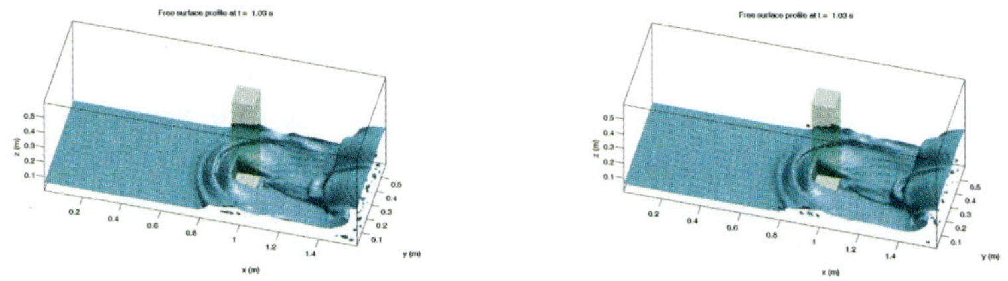

Figure 10-1. Free-surface profiles obtained by RANS model and LES. Snapshots are taken at t = 1.03s. The free surface location is represented by blue color, which is defined as the iso-surface where VOF of water f = 0.5. The green color denotes the position of the square cylinder.

RANS model LES

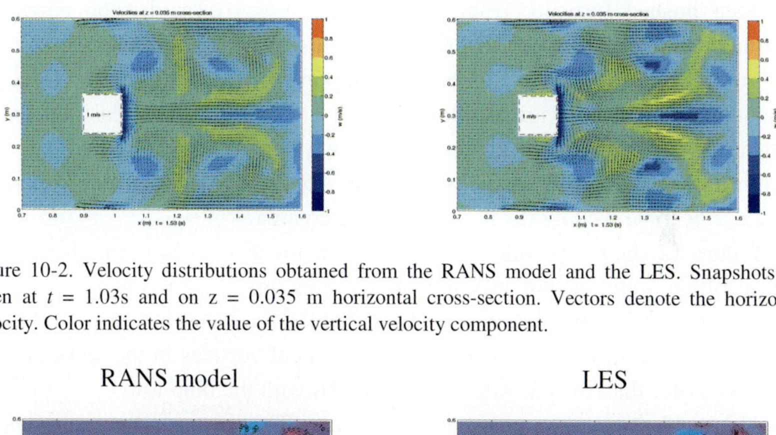

Figure 10-2. Velocity distributions obtained from the RANS model and the LES. Snapshots are taken at t = 1.03s and on z = 0.035 m horizontal cross-section. Vectors denote the horizontal velocity. Color indicates the value of the vertical velocity component.

RANS model LES

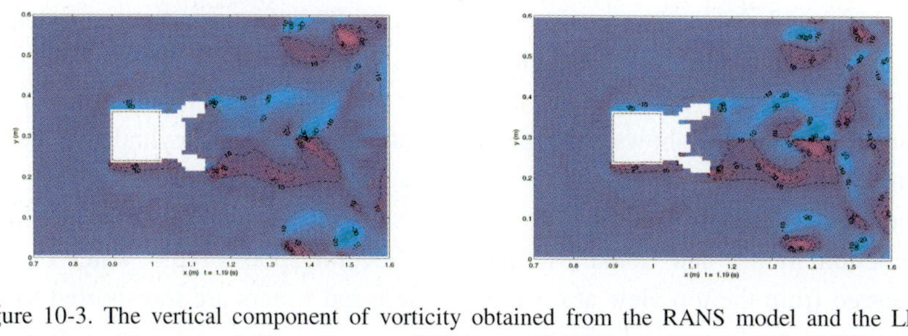

Figure 10-3. The vertical component of vorticity obtained from the RANS model and the LES. Snapshots are taken at t = 1.03s and on z = 0.035 m horizontal cross-section. Color indicates the vorticity magnitude. The values of colors can be referred to the contour lines.

5.3. *Effects of the Wet Floor*

Gomez-Gesteira and Dalrymple (2004) found that the shape of the bore front is sensitive to the floor condition downstream of the gate. When the floor is dry, the bore front looks like a tongue, which is very different from the mushroom head shape when a thin layer of water is present. Here, we further examine the interaction between the bore and thin layer of water in front of it.

In order to separately track the movement of fluid particles in the impoundment and those in the thin layer in front of the gate, we perform a new set of numerical experiments with LES. The numerical domain is exactly the same as that in the laboratory experiment including the full channel width (w = 0.61 m). Two VOF equations are used to track the interfaces between two fluid bodies, enabling us to visualize the fluid movement.

Figure 11 shows the overview of the simulation results. The 1cm thin layer of water in front of the gate is colored in red and referred as "red water". The impoundment is colored in blue and is referred as "blue water". The "mushroom head" can be observed at t = 0.15s. Clearly the water in the thin layer (red water) has been pushed forward, from x = 0.4m to x = 0.51m, by the collapsing impoundment, and the upper bore front stumbles over the piled-up red water. Up to t = 0.46s, there appears to be little mixing between the water from the impoundment and the original water in front of the gate. The splash-up on the cylinder is almost entirely made up of the blue water. At t = 0.82s, the blue water is on top of the red water in the wake zone. After t = 1.37s, the blue water and red water become well mixed in the downstream area.

Figure 12 shows the side view of the domain. It is even more clear to see that the blue water is not mixed with the red one before impinging on the cylinder at t = 0.29s. It is interesting to see that the red water is transported to the free-surface area through a series of strong vertical vortices in the wake zone at t = 0.88s. After that, the red water is well mixed with the blue water.

Figure 13 shows the top view of the results. This figure helps us to identify the symmetry of the flow field since it is a full domain simulation. We shall address again that the cylinder is actually not placed in the middle of the channel. The result shows that the flow pattern around the cylinder is nearly symmetric up to t = 0.91s. After that, the wave reflected from the end wall generates an asymmetric flow pattern. This might explain the accurate force prediction in the half domain simulation. Several spots of red water can be observed from the top view at t = 0.91s. These red waters are transported to the free surface by the vertical velocity.

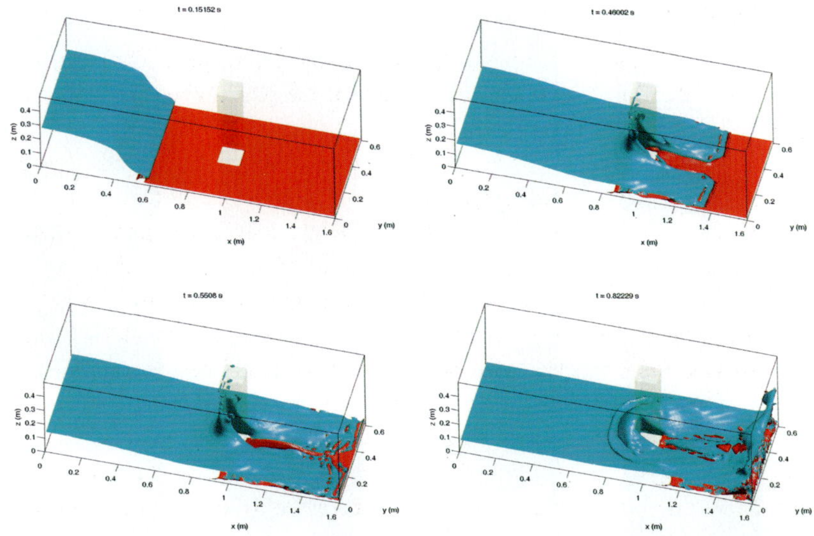

Figure 11. The overview of the two-water-body simulation. Red is the iso-surface of 1 cm thin water layer. Blue is the iso-surface of the impoundment.

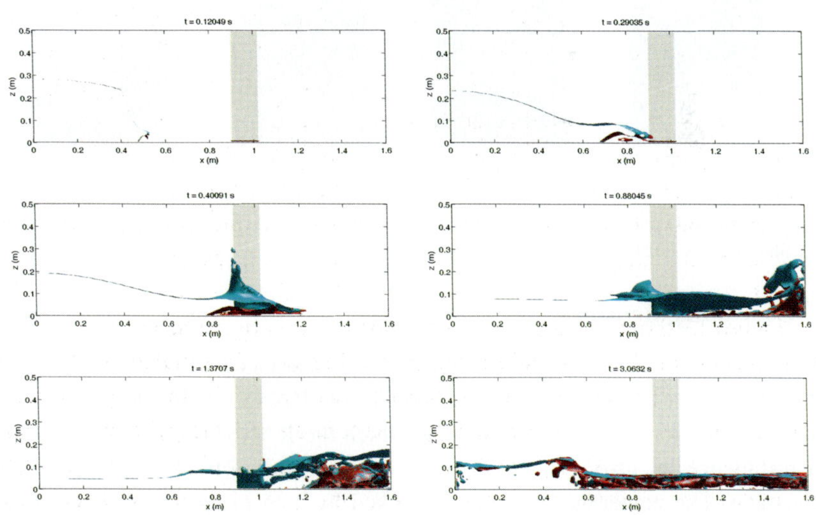

Figure 12. The side view of the two-water-body simulation. Red is the iso-surface of 1 cm thin water layer. Blue is the iso-surface of the impoundment.

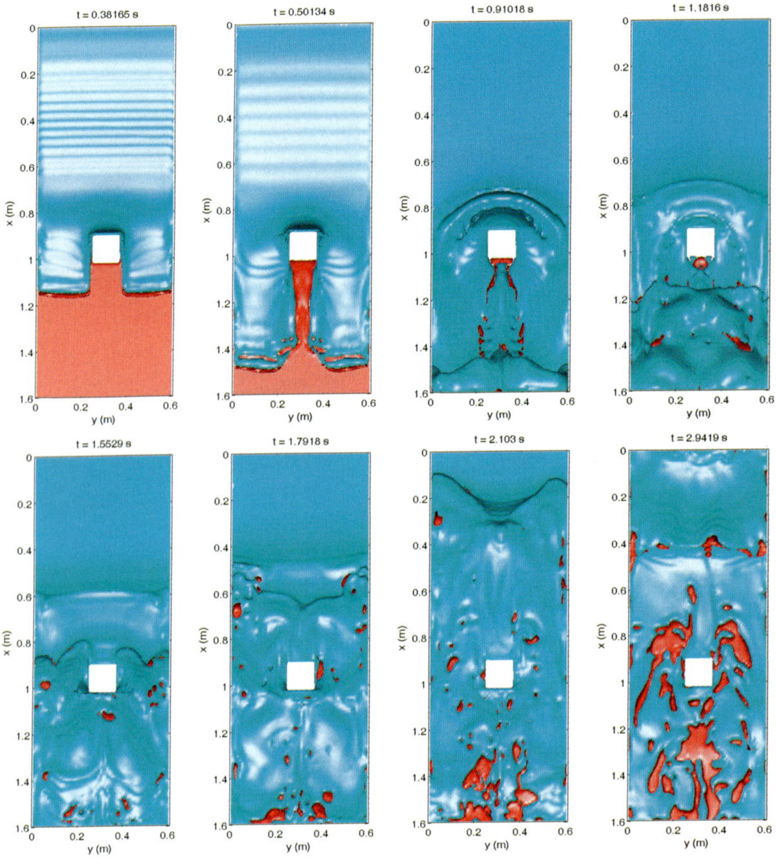

Figure 13. The top view of the two-water-body simulation. Red is the iso-surface of 1 cm thin water bed. Blue is the iso-surface of the impoundment.

6. Concluding Remarks

In this study, we have explored the physical phenomena of the bore-structure interaction. Three different numerical models, the RANS model, LES, and DNS, are performed for the 3D simulations in which the RANS model is used to find the turbulence intensity. The VOF method is adopted to track the free-surface movement. The numerical results are checked by comparing the predicted total horizontal forces on a rectangular cylinder and the associated velocity with the laboratory measurements. The results show that all three numerical models predict the force as well as velocity very well before the bore reaching the end wall. After the bore is reflected back from the downstream end wall, the RANS model and LES have better predictions than DNS. The shear stress is found to be two orders of magnitude smaller than the normal stress acting on the square cylinder.

The RANS model is also used to track the spatial distribution of turbulence. Strong turbulence intensity is observed in eddies and reflected waves. The magnitude of turbulence intensity is found to be in the range about 0.2 ~ 0.4 in the wake zone and 0.4 ~ 0.8 in the reflected plunging waves.

The results from the RANS model and LES are compared to each other. LES result reveals a finer structure on the free-surface and velocity field than the RANS model results. This observation is in agreement with the difference between the assumptions of ensemble averaging used in the RANS model and the spatial filtering used in LES. LES tents to predict stronger vorticity than the RANS model. However, both results predict similar force magnitude and the time when the reflected bore front reaching the cylinder from the end wall.

The thin water layer in front of the gate plays an important role in the initial development of the bore front. With a thin layer of water is present, the initial bore front has a "mushroom head" which is very different from the "tongue shape" when the tank floor is initially dry. The thin layer of water also prevents the direct contact of the bore front with the bottom of the tank. However, the water from the water bed can be transferred to the surface area through a series of vortex generated by the edge of the square cylinder.

Acknowledgment

The research presented here has been supported by research grants from National Science Foundation to Cornell University (CBET-0427115, 0635794; CMMI-0622758, 0710571).

References

1. Abul-Azm, A. G. 1994. Diffraction through wide submerged breakwaters under oblique waves. Oc. Engng., 21, 683-706.
2. Arnason, H. 2005. Interactions between an Incident Bore and a Free-Standing Coastal Structure. PhD dissertation, University of Washington.
3. Bussmann, M., Kothe, D. B. and Sicilian, J. M. 2002. Modeling high density ratio incompressible interfacial flows. Proc. FEDSM'02, 2002 ASME Fluids Engng Div. Summer Meeting. FEDSM 2002-31125, pp. 1-7.
4. Cabot, W. and Moin, P. 2000. Approximate wall boundary conditions in the large-eddy simulation of high Reynolds number flow. Flow Turb. Combust., 63, 269-291.

5. Chang, K.-A., Hsu, T.-J. and Liu, P. L.-F. 2001. Vortex generation and evolution in water waves propagating over a submerged rectangular obstacle Part I. Solitary waves. Coast. Engng., 44, 13-36.
6. Chen, G. et al. 1999. Two-dimensional Navier-Stokes simulation of breaking waves. Physics of Fluids, 11 (1): Jan, 121-133.
7. Dalrymple, R. A. and Rogers, B. D. 2006. Numerical modeling of water waves with the SPH method. Coastal Engineering, 53 (2-3): Sp. Iss. SI Feb, 141-147.
8. Ertekin, R. C. and Becker, J. M. 1996. Nonlinear diffraction of waves by a submerged shelf in shallow water. In: Chakrabarti, S. et al. (Eds), Offshore Mech. and Arctic Engng. 1996, vol. I Pt. B, ASME, New York, 31-39.
9. Gomez-Gesteira, M. and Dalrymple, R. A. 2004. Using a three-dimensional smoothed particle hydrodynamics method for wave impact on a tall structure. Journal of Waterway Port Coastal and Ocean Engineering-ASCE, 130 (2): Mar-Apr, 63-69.
10. Hirt, C. W. and Nichols, B. D. 1981. Volume of fluid (VOF) method for the dynamics of free boundaries. J. Comp. Phys., 39, 201-225.
11. Huang, C.-J. and Dong, C.-M. 1999. Wave deformation and vortex generation in water waves propagating over a submerged dike. Coast. Engng., 37, 123-148.
12. Irschik, K., Sparboom, U. and Oumeraci, H. 2003. Breaking wave characteristics for the loading of a slender pile. Proc. ICCE 2002. 1341-1352.
13. Lemos, C. M. 1992. Wave Breaking. Springer-Verlag.
14. Lin, P. and Liu, P. L.-F. 1998a. A numerical study of breaking waves in the surf zone. J. Fluid Mech., 359, 239-264.
15. Lin, P. and Liu, P. L.-F. 1998b. Turbulence transport, vortices dynamics, and solute mixing under plunging breaking waves in surf zone. J. Geophys. Res., 103(C8), 15677-15694.
16. Liu, P. L.-F., Lynett, P., Fernando, H., Jaffe, B. E., Fritz, H., Higman, B., Morton, R., Goff, J. and Synolakis, C. 2005. Observations by the International Tsunami Survey Team in Sri Lanka, Science, 308 (5728): Jun 10, 1595-1595.
17. Kothe, D. B., Williams, M. W., Lam, K. L., Korzewa, D. R., Tubesing, P. K. and Puckett, E. G. 1999. A Second-order accurate, linearity-preserving volume tracking algorithm, for free surface flows on 3-D unstructured meshes, Proceedings of the 3rd ASME/JSME Joint Fluids Engineering Conference. FEDSM99-7109, July 18-22.

18. Launder, B. E. and Spalding, D. B. 1974. The Numerical Computation of Turbulence Flows, Comp. Meth. Appl. Mech. Engng., Vol. 3, pp. 269-289.
19. Lin, P. and Li, C. W. 2003. Wave-current interaction with a vertical square cylinder, Ocean Engineering, 30, 855-876.
20. Liu, P. L.-F., Wu, T.-R., Raichlen, F., Synolakis, C. E. and Borrero, J. C. 2005. Runup and rundown generated by three-dimensional sliding masses, J. Fluid Mech., vol. 536, pp. 107-144.
21. Losada, I. J., Silva, R. and Losada, M. A. 1996. 3-D non-breaking regular wave interaction with submerged breakwaters. Coast. Engng., 28, 229-248.
22. Pope, S. B. 1975. A more general effective-viscosity hypothesis. J. Fluid Mech., 72, 331-340.
23. Pope, S. B. 2001. Turbulent flows. Cambridge University Press.
24. Rider, W. J. and Kothe, D. B. 1998. Reconstructing Volume Tracking, J. Comp. Phys., 141, 112-152.
25. Rodi, W. 1980. Turbulence models and their application in hydraulics – A state-of-the-art review. I.A.H.R. publication.
26. Sheridan, J., Lin, J.-C. and Rockwell, D. 1997. Flow past a cylinder close to a free surface. J. Fluid Mech., 330, 1-30.
27. Shih, T.-H., Zhu, J. and Lumley, J. L. 1996. Calculation of wall-bounded complex flows and free shear flows. International J. for Num. Meth. in Fluids, 23, 1133-1144.
28. Deardorff, J. W. 1974. Three-dimensional numerical study of the height and mean structure of a heated planetary boundary layer. Boundary-Layer Meterol., 7, 81-106.
29. Smagorinsky, J. 1963. General circulation experiments with the primitive equations: I. The basic equations. Mon. Weather Rev., 91, 99-164.
30. Tang, C. J. and Chang, J. H. 1998. Flow separation during solitary wave passing over submerged obstacle. J. Hydr. Engng., 124, 742-749.
31. Ting, F. C. K. and Kim, Y.-K. 1994. Vortex generation in water waves propagation over a submerged obstacle. Coast. Engng., 24, 23-49.
32. Wu, T.-R. 2004. A numerical study of three-dimensional breaking waves and turbulence effects, Ph.D. Thesis, Cornell University.
33. Zhuang, F. and Lee, J. J. 1996. A viscous rotational model for wave overtopping over marine structure. Proc. 25th Int. Conf. Coastal Eng., ASCE, pp. 2178-2191.